Annulenes, Benzo-, Hetero-, Homo-Derivatives, and their Valence Isomers

Volume I

Authors

Alexandru T. Balaban
Professor
Department of Organic Chemistry
The Polytechnic
Bucharest, Romania

Mircea Banciu
Associate Professor
Department of Organic Chemistry
The Polytechnic
Bucharest, Romania

Vasile Ciorba
First Economic Secretary
West Europe Direction
Foreign Trade Ministry
Bucharest, Romania

CRC Press, Inc.
Boca Raton, Florida

Library of Congress Cataloging-in-Publication Data

Balaban, Alexandru T.
  Annulenes, benzo-, hetero-, homo-derivatives and
their valence isomers.

  Includes bibliographies and indexes.
  1. Annulenes. I. Biancu, Mircea. II. Ciorba, Vasile.
III. Title.
QD341.A83B35 1987   547'.6   86-18393
ISBN-0-8493-6880-4 (v. 1)
ISBN-0-8493-6881-2 (v. 2)
ISBN-0-8493-6881-3 (v. 3)

Direct all inquiries to CRC Press, Inc., 2000 Corporate Blvd., N.W., Boca Raton, Florida, 33431.

© 1987 by CRC Press, Inc.

International Standard Book Number 0-8493-6880-4 (v. 1)
International Standard Book Number 0-8493-6881-2 (v. 2)
International Standard Book Number 0-8493-6881-0 (v. 3)

Library of Congress Card Number 86-18393
Printed in the United States

PREFACE

The idea of the present book dates many years back and originates in the authors' fondness and interest for aromaticity, valence isomerism, and for the chemistry of annulenes and their derivatives. Like many other chemists, the authors became fascinated by this area of chemistry. The ability to predict by means of primitive quantum chemistry (E. Hückel) or by more sophisticated methods (M. J. S. Dewar, W. E. Doering, R. Hoffmann, H. C. Longuett-Higgins, J. A. Pople, P. R. Schleyer, R. B. Woodward) novel interesting structures and reactions, coupled with the mathematical graph-theoretical basis for enumerating all possible valence isomers and heterocyclic aromatic systems (A. T. Balaban) made from the heuristic interplay between theory and experiment an exciting game to play and watch; along with some temporary failures and disappointments, many spectacular successes were registered by the chemists to whom this book is dedicated and by many others among whom we mention: J. Aihara, A. G. Anastassiou, V. Boekelheide, M. P. Cava, R. Criegee, D. Ginsburg, P. E. Eaton, R. C. Haddon, W. C. Herndon, B. A. Hess, Jr., M. Jones, Jr., L. Kaplan, T. J. Katz, S. Masamune, J. F. W. McOmie, R. H. Mitchell, M. Nakagawa, G. A. Olah, M. Randic, L. J. Schaad, H. A. Staab, E. E. van Tamelen, N. Trinajstić, E. Vedejs, and G. Wittig.

The authors had the privilege to work with Professor C. D. Nenitzescu and his wife Ecaterina Ciorănescu-Nenitzescu. It should be mentioned that C. D. Nenitzescu (1902 to 1970) and M. Avram (1920 to 1983) prepared the first valence isomer of an annulene, the Nenitzescu hydrocarbon $(CH)_{10}$,* and reawakened the interest for the chemistry of cyclo-butadiene simultaneously with, and independently from, R. Criegee, G. Schröder, and G. Maier. C. D. Nenitzescu, E. Ciorănescu-Nenitzescu, and co-workers prepared many new valence isomers of benzo- and dibenzoannulenes. We hope that the present review of a wide-ranging, difficult, and rapidly increasing area is a timely one. We are aware that the vast range of topics precludes an exhaustive treatment and we therefore apologize for the inherent omissions; we would appreciate any suggestions for improvement.

We thank the chemists who kindly provided us their reprints and preprints, often in impressive numbers: W. Adam, E. L. Allred, A. G. Anastassiou, A. G. Anderson, Jr., H. van Bekkum, V. Boekelheide, B. K. Carpenter, M. Christl, T. M. Cresp, P. E. Eaton, J. Elguero, R. Claramunt-Elguero, P. J. Garratt (who, along with T. M. Cresp, supplied not only their papers but also those of the late F. Sondheimer), T. L. Gilchrist, D. Ginsburg, R. Gleiter, M. J. Goldstein, H. Günther, K. Hafner, R. Hoffmann, C. W. Jefford, M. Jones, Jr., A. R. Katritzky, G. Maier, P. M. Maitlis, S. Masamune, A. de Meijere, R. H. Mitchell, I. Murata, M. Nakagawa, R. Neidlein, L. A. Paquette, H. Prinzbach, M. Rabinowitz, C. W. Rees, G. Schröder, L. T. Scott, M. Simonetta, W. Tochtermann, E. Vedejs, E. Vogel, and others.

For discussions in Bucharest and abroad we thank Professors P. R. Schleyer, K. Hafner, R. Neidlein, P. J. Garratt, and D. Ginsburg.

We thank CRC Press and especially Sandy Pearlman, Marsha Baker, and Paul Gottehrer for their cooperation and understanding. Finally the authors wish to thank their wives for putting up with having their homes in a continuous stress for several years during the preparation of the manuscript.

The literature coverage ends in 1985 with some data from 1986. The authors tried to summarize the logical thinking behind the stage on which the chemistry of annulenes, of their derivatives and of valence isomers is played. If one thinks that the area encompasses benzenoid and nonbenzenoid aromatics, including heterocyclics, it is easy to see that the

---

* The structural formula of Nenitzencu's hydrocarbon, $(CH)_{10}$, is a nonseparable cubic graph. Bitropyl, $(CH)_{14}$, whose formula is a separable cubic graph, had been prepared earlier by Doering and Knox.

authors' task was a difficult one. However, it is the authors' hope that they provided a clear, up-to-date picture, albeit inherently incomplete, of the many facets displayed by annulenes, their benzo-, homo-, and hetero-derivatives as well as by the corresponding valence isomers; what is more, we hope to convey to the reader the perspective of the still untrodden areas and side paths which may reveal new phenomena to be discovered in the future. If this book will succeed in transmitting to other chemists, especially younger ones, something from the beauty and excitement of this area, stimulating them to participate in its future development, then the authors will be rewarded by the thought that their arduous work has not been wasted.

DEDICATION

Dedicated to the memory of
E. Hückel, C. D. Nenitzescu, and **F. Sondheimer** without whose inspiration, enthusiasm, and diligence the field of annulenes and of their valence isomers might still be in its infancy.

Among those who contributed to the impressive development of this area we would like to pay a special tribute to Professors **Ecaterina Ciorănescu-Nenitzescu, M. J. S. Dewar, W. E. Doering, K. Hafner, G. Maier, L. A. Paquette, G. Schröder,** and **E. Vogel.**

THE AUTHORS

Alexandru T. Balaban is Professor of Organic Chemistry at the Polytechnic Institute, Bucharest, Romania. Born in Romania in 1931, he obtained the Diploma of Chemical Engineer at the Bucharest Polytechnic in 1953, and for the next 3 years he had a Ph.D. fellowship, followed by an Assistantship in Organic Chemistry at the Bucharest Polytechnic; his Ph.D. thesis, supervised by C. D. Nenitzescu (1902 to 1970), dealt with $AlCl_3$-catalyzed reactions and was defended in 1959. During the years 1956 to 1966 he collaborated with C. D. Nenitzescu on experimental organic chemistry and developed his own research group at the Laboratory of Labelled Organic Compounds, Institute for Atomic Physics, Bucharest. He became lecturer in Organic Chemistry and Professor of General Chemistry at the Bucharest Polytechnic. For 3 years (1967 to 1970) he was Senior Research Officer at the Chemistry Section of the International Atomic Energy Agency in Vienna, where his work involved international collaboration in the field of radiopharmaceuticals, organizing scientific meetings in Vienna, Mexico, Brazil, and Denmark, and editing three books on labeled compounds. Having been interested in theoretical chemistry and mathematics since his student years, he edited in 1976 a book entitled *Chemical Applications of Graph Theory* and published more than 100 papers on chemical graphs and topological indices in addition to over 250 papers on experimental organic chemistry (homogeneous catalysis, heterocyclic chemistry — pyrylium and pyridinium salts, oxazoles, furans, stable nitrogen-centered free radicals, organoboron systems, isotope effects, and labeled compounds). In addition to the books mentioned above, he is co-author of four other books: *Pyrylium Salts* (Academic Press, 1982), *Steric Fit in QSAR* (Springer, 1980), *Olefin Metathesis and Ring-Opening Polymerization of Cyclo-Olefins* (Wiley, 1985, Roumanian version, 1981), *Labelled Compounds and Radiopharmaceuticals Applied in Nuclear Medicine* (Wiley, 1986, Romanian version, 1979), and *Applications of Physical Methods in Organic Chemistry* (in Romanian, 1983). He also contributed book chapters in monographs on *Friedel-Crafts and Related Reactions* (edited by G. A. Olah), *Recent Trends in Heterocyclic Chemistry* (edited by R. Mitra), *Applications of Graph Theory* (edited by R. Wilson and L. W. Beinecke), *Physical Methods in Heterocyclic Chemistry* (edited by R. R. Gupta), etc.

He was elected as corresponding member of the Romanian Academy in 1963, and as Fellow of the New York Academy of Sciences in 1969. In 1962 he was awarded the chemistry prize of the Roumanian Academy and in 1966 the Romanian order of scientific merit.

He is currently one of the editors of *Revue Roumaine de Chimie, Mathematical Chemistry,* and *Journal of Labelled Compounds and Radiopharmaceuticals*; he also serves on editorial boards of *Scientometrics, Journal of Radioanalytical and Nuclear Chemistry, Advances in Heterocyclic Chemistry,* and *Organic Preparations and Procedures International.*

He has been invited as plenary lecturer at international congresses in the U.S. (1982, 1985), Czechoslovakia (1969), Belgium (1972), Yugoslavia (1972, 1979, 1981), France (1985), Portugal (1984), German Democratic Republic (1965, 1976, 1979), and West Germany (1969). He has made lecture tours in Britain (1963, 1966, 1970, 1979), West Germany (1964, 1968, 1973, 1979, 1983), the U.S. (1966, 1971, 1973, 1982), France (1966, 1980), and the U.S.S.R. (1960, 1967). He has published jointly with chemists and mathematicians from the U.S., Britain, France, West Germany, German Democratic Republic, the U.S.S.R., Bulgaria, and currently has scientific cooperation with the U.S., Bulgaria, France, and West Germany. He is married (Cornelia, Ph.D. dipl. chem. eng.) and has a son (Teodor-Silviu, dipl. chem. eng.) and a daughter (Irina-Alexandra).

In the present book he has written Chapters 1 to 4, 8, and 10.

Mircea D. Banciu is Associate Professor of Organic Chemistry at the Polytechnic Institute of Bucharest. Born in Romania in 1941, he obtained the Diploma of Chemical Engineer at the Bucharest Polytechnic in 1962, then became assistant at the chair of organic chemistry.

His Ph.D., thesis under the supervision of Professor Ecaterina Ciorănescu-Nenitzescu, was defended in 1969 and concerned solvolytic rearrangements of dibenzocycloalkanes. During the period 1972 to 1980 he was Assistant Professor and at present he teaches organic chemistry and applications of physical methods in organic chemistry. In 1969 he spent several months in Professor E. Vogel's laboratory at the University of Cologne (West Germany).

As a co-worker of C. D. Nenitzescu between 1962 and 1970 and Ecaterina Ciorănescu-Nenitzescu (since 1962), as well as an independent research worker, his main areas of investigation are (1) carbonium ion reactions in solvolyses and deaminations (the first observation in 1969 of a retro-π-route process); (2) valence isomerism (synthesis of new dibenzo[10]annulenes, interconversion of valence isomers); (3) chemistry of 3- and 4-membered ring compounds (thermolysis, solvolysis and photolysis); (4) spectrometric techniques and laboratory equipment.

He is author or co-author of 40 papers and three books in Romanian: *Physical Methods in Organic Chemistry* (1972), *Experimental Techniques in Organic Chemistry* (1977), and *Applications of Physical Methods in Organic Chemistry* (1983), the last one in collaboration with Professor A. T. Balaban.

He published several reviews on progress in the chemistry of annulenes, benzoannulenes, and their valence isomers.

He is a member of the Editorial Board of *Revue Roumaine de Chimie*. In 1969 he was awarded the scientific prize of the Romanian Ministry of Education, and in 1982 the Romanian Academy's chemistry prize for his investigations on the synthesis and reactions of dibenzocycloalkanes.

He is married (Anca, Ph.D. dipl. chem.) and has a daughter (Cristina).

In the present book he has written Chapters 5 to 7.

Vasile Ciorba, born in Romania in 1943, obtained the Diploma of Chemical Engineer at the Bucharest Polytechnic in 1966. His Ph.D. thesis, at the same Institute under the supervision of Professor A. T. Balaban, was finished in 1982 and concerned charge-transfer complexes between pyrylium cations and tetracyano-*para*-quinodimethane. During the period when he worked at the Bucharest Polytechnic for his Ph.D. he reviewed annulenes, homoannulenes, and their valence isomers. At present he works for the Romanian Ministry of Foreign Trade.

He is married (Cristina) and has two sons (Valentin and Cristian).

In the present book he has written Chapter 9.

TABLE OF CONTENTS

TABLE OF CONTENTS

Chapter 1

INTRODUCTION

When you have eliminated the impossible,
whatever remains, however improbable, is the truth.

A. Conan Doyle

In *The Sign of the Four* and elsewhere (e.g., *The Beryl Coronet*), Sherlock Holmes says that he always bases his reasoning on the above axiom, which is the motto of this book. Why does reading good detective stories captivate? Because the reader is not passive: he or she tries to match his/her wits against the sleuth's — and fails, mostly — be it because Chesterton's Father Brown perceives details unknown to the reader, or because Agatha Christie's Hercule Poirot is not fooled by the numerous candidates (who will profit if the victim is murdered) or by the omissions in the criminal doctor's diary.

Can ever (under)graduate chemistry textbooks provide as much fun and excitement as good thrillers? Perhaps, if one gets away from the passive attitude many students have towards learning. C. K. Ingold recalled that he became a chemist instead of becoming a physicist because physics was presented to students as a perfectly built monolith whereas chemistry was described as a scaffolding offering many challenges into the unknown.

One way of introducing the beauty of the research frontier is to let students create and discover chemistry by themselves. One of the authors of this book (A.T.B.) tried to do just that by the time he graduated: finding all possible monocyclic aromatic systems,[1] drawing all possible intermediates in molecular rearrangements,[2] etc. It soon became apparent that for such types of problems, a chemist must become acquainted with a branch of mathematics called "Graph Theory". After having browsed through the scanty literature available in the 1950s and early 1960s, he discovered with excitement that the problem of enumerating all possible valence isomers of annulenes[3] was not only a chemical, but also a mathematical unsolved problem; a formula for enumerating connected and disconnected general cubic graphs was published years later by mathematicians.[4,5] Even at present there is no simple way of finding the number and structures of connected planar cubic multigraphs which are relevant to the chemical counterpart, namely constitutions of annulene valence isomers; only by means of elaborate computer programs such as Lederberg's DENDRAL[6] is it now possible to obtain answers to some of these problems.

Graph theory is an excellent instrument for constitutional chemical problems because it born as a result of three types of questions, one of which is precisely the question of chemical isomerism (cf. Chapter 2). Graph theory is also necessary because one has to know in advance the field of combinatorial possibilities in order to limit the exploration area. In chemistry one has the feeling that the unknown lies closer than in other scientific areas such as physics or mathematics, but the diversity is too large; no good chemistry comes from experimenting "just by mixing" or by taking the "Beilstein approach" (i.e., all imaginable substituents), just as no good poetry comes out when an ape is punching a typewriter: there is too much "entropic noise". One needs a beacon to signal where the treasure is most likely to be hidden (this beacon is in some cases graph theory, and there are signs that advanced theoretical calculations will soon be the best beacon), and one also needs good diving gear to find the treasure (a good synthetic strategy, instruments for separating reaction mixtures, for characterizing compounds by physical methods, and for structure elucidation, etc.).

The present book is patterned according to Sherlock Holmes's method: (1) find out by means of graph theory all possible cubic graphs; (2) put the mathematical data in chemical context and prune off the impossible (disconnected or nonplanar graphs which are chemically irrelevant); and (3) now look for the improbable, most often due to steric (angle) strain, and try to match the list of possibilities with what is already known. What remains may contain a goal worthy of attention, but there is no substitute yet for a trained chemist's judgment to select the goal from the given possibilities, just as one needs the detective to pick out the culprit.

The fascinating problem of aromaticity is closely linked to the general topic of this book. The great theoretical advances of Hückel and Dewar served as the basis for the spectacular successes of the last 40 years. It is the authors' wish to present the corresponding area of chemistry as an exercise in inductive and deductive logic rather than as an area which developed spectacularly in a haphazard way assisted by serendipitous discoveries. It is this exercise in logic which makes chemists feel that they are collaborating with their predecessors and even their competitors. Two of the present authors (A.T.B. and M.B.) are authors of a book on structure elucidation by means of UV-VIS, IR, ^{1}H-NMR, ^{13}C-NMR, and mass spectra, where again graph theory assists in narrowing the search field.[7]

It is the authors' hope that these three volumes will be a short-cut to the vast literature devoted to this field, and that they will serve both for teaching and for research purposes. The tables of all possible valence isomers may entice enterprising chemists to synthesize yet unknown structures and to use them afterwards for obtaining new valence isomers by thermal, photochemical, or metal ion-catalyzed rearrangements. More extensive tables than the present ones are available in the literature cited.

One might ask the question whether bullvalene would have been easily available at present, had Doering[8] not predicted it to undergo rapid degenerate rearrangement, thereby prompting Schröder to report his synthesis[9] before Doering's stepwise synthesis of bullvalene.[10] The present authors believe that since bullvalene is obtained from other $(CH)_{10}$-valence isomers, it would probably have been known by now, but it would not have been easy to identify by NMR methods, and semibullvalene even less, owing to the rapid automerization. Discovery by good luck is too scarce, and becomes less likely as years go by, because more and more "easy-come" possibilities become exhausted. On the other hand, some discoveries may not be recognized because one does not know to look for them. The classical example is Merling's failure[10] to recognize tropylium bromide in 1891; it was purposefully prepared again by Doering and Knox in 1954.[11] Two less well-known examples involving one of the authors (A.T.B.) are the failure of many chemists who had investigated the Friedel-Crafts acylation of alkenes to look for pyrylium salts in the aqueous phase;[12-14] even when a crystalline pyrylium chlorozincate crystallized, it was mistaken for a β-diketonate complex.[15] On the other hand, after having found in the aqueous phase not only olefin diacylation products, i.e., pyrylium salts,[12-14] but also triacylation products,[16] A.T.B. failed to find tetra-acylation products; these were discovered by Roussel and co-workers[17] when they decomposed the Friedel-Crafts reaction mixture with ammonia instead of water. So the lesson is that, as a rule, "discoveries are made by a prepared spirit".

REFERENCES

1. **Balaban, A. T.,** *Stud. Cercet. Chim. Acad. R.P. Romania,* 7, 257, 1959; **Balaban, A. T. and Simon, Z.,** *Tetrahedron,* 18, 315, 1962; *Rev. Roum. Chim.,* 10, 1059, 1965.
2. **Balaban, A. T., Fărcasiu, D., and Bănică, R.,** *Rev. Roum. Chim.,* 11, 1205, 1966.
3. **Balaban, A. T.,** *Rev. Roum. Chim.,* 11, 1097, 1966.
4. **Read, R. C.,** *J. London Math. Soc.,* 34, 417, 1959; 35, 344, 1960.

5. **Palmer, E. M. and Harary, F.**, *Graphical Enumeration*, Academic Press, New York, 1973, 174.
6. **Lederberg, J., Sutherland, G. L., Buchanan, B. G., Feigenbaum, E. A., Robertson, A. V., Duffield, A. M., and Djerassi, C.**, *J. Am. Chem. Soc.*, 91, 2977, 1968.
7. **Balaban, A. T., Banciu, M., and Pogany, I.**, *Aplicatii ale Metodelor Fizice in Chimia Organică*, Editura Stiintifică, Bucharest, 1983.
8. **Doering, W. von E. and Roth, W. R.**, *Angew. Chem.*, 75, 27, 1963.
9. **Schröder, G.**, *Chem. Ber.*, 97, 3140, 1964; **Schröder, G., Oth, J. F. M., and Merenyi, R.**, *Angew. Chem. Int. Ed. Engl.*, 4, 752, 1965; **Schröder, G. and Oth, J. F. M.**, *Angew. Chem. Int. Ed. Engl.*, 6, 414, 1967.
10. **Doering, W. von E. et al.**, *Tetrahedron*, 23, 3943 (see footnote 30), 1967.
11. **Merling, G.**, *Ber. Dtsch. Chem. Ges.*, 24, 3108, 1891.
12. **Doering, W. von E. and Knox, L. H.**, *J. Am. Chem. Soc.*, 76, 3203, 1954.
13. **Balaban, A. T. and Nenitzescu, C. D.**, *Liebigs Ann. Chem.*, 625, 74, 1959; *J. Chem. Soc.*, 3553, 1961; **Praill, P. F. G. and Whitear, A. L.**, *J. Chem. Soc.*, 3573, 1961; *Proc. Chem. Soc.*, 312, 1959.
14. **Balaban, A. T., Schroth, W., and Fischer, W.**, *Adv. Heterocyclic Chem.*, 10, 241, 1969; **Balaban, A. T., Dinculescu, A., Dorofeenko, G. N., Fischer, G. W., Koblik, A. V., Mezheritskii, V. V., and Schroth, W.**, *Pyrylium Salts* (Adv. Heterocyclic Chem., Suppl. Vol. 2), Academic Press, New York, 1982.
15. **Byrns, A. C. and Doumani, T. F.**, *Ind. Eng. Chem.*, 35, 349, 1943.
16. **Balaban, A. T., Frangopol, P. T., Katritzky, A. R., and Nenitzescu, C. D.**, *J. Chem. Soc.*, 3889, 1962.
17. **Erre, C., Pedra, A., Arnaud, M., and Roussel, C.**, *Tetrahed. Lett.*, 25, 515, 1984.

Chapter 2

AROMATICITY OF MONO- AND POLYCYCLIC SYSTEMS

I. THE HISTORY OF THE AROMATIC SEXTET FROM KEKULÉ TO HÜCKEL

A. Kekulé's Benzene Formula

The flash-like intuition of genius of the 32-year-old August Kekulé in 1861 when he realized that carbon chains may be cyclic, like a snake biting its tail, has been recounted many times since Kekulé himself told this story in 1890[1] at the "Benzolfeier" of the German Chemical Society dedicated to the 25th anniversary of the publication date (1865) of the Kekulé formula.[2] It took Kekulé 4 years to publish this idea,[2] because in 1862 he married in Ghent a 19-year-old girl who died in 1863 in childbirth, leaving him with a son who needed all his care.

Why is Kekulé's formula one of the cornerstones of organic chemistry?

A glance in old chemistry textbooks is instructive before we attempt to answer this question. Gmelin's three-volume treatise of theoretical chemistry (the third volume was published in 1819 and contained all organic chemistry[3]) is fascinating: it is a collection of recipes, analyses, and natural occurrence sources. There is no system, no chemical symbol, and there is no formula, but only words, as in botany or archeology. Dalton, Lavoisier, Gay-Lussac, Dumas, Berzelius, Gerhardt, Laurent, Faraday, Wöhler, and Liebig made out of this repository of data a coherent science based on constitutional formulas. After the first chemical congress organized by Kekulé in Karlsruhe in 1860 when Avogadro's theory was restated by his countryman, Cannizzaro, the whole province of what we now call aliphatic chemistry was practically understood owing to Couper, Butlerov, Frankland, Williamson, and Kekulé. However, deep mystery still shrouded the larger area of what we now call aromatic and heteroaromatic chemistry. Kekulé's formula is famous not only because historically it is the first cyclic formula, or because it meant a lot to practical-minded chemists who developed on its basis around the turn of the century the powerful German industry of dyestuffs and pharmaceuticals, but mainly because it opened up the vista of deeper understanding of the largest class of chemical compounds we know: from the seven odd million substances, more than 90% have organic radicals in them; of these, more than 75% are aromatic. Kekulé's formula is the key to the first lock behind which aromaticity hid its secrets.

The next obstacle came soon when the numbers of isomers did not agree with a Kekulé formula with localized double bonds: *ortho*-di-A-substituted (**1**) and *meta*-di-A,B-substituted benzene derivatives (**2**) appeared always as a single substance, while a localized Kekulé formula predicted in these cases two isomers.* Kekulé surmounted this difficulty[1] by postulating the valence oscillation hypothesis, which was later reinterpreted electronically as the delocalization of the six π-electrons, and represented by the double-headed arrow.

It must be added that, in addition to having contributed decisively to the foundation of structure theory and to having proposed the formula for benzene, Kekulé originated the

---

* It will be seen that such isomers do appear in the anti-Hückel series (e.g., cyclobutadiene, cyclooctatetraene).

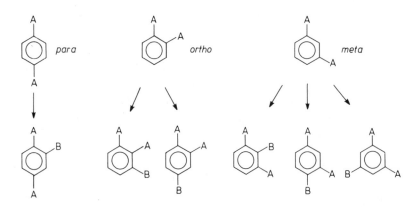

FIGURE 1. Körner's approach for determining substitution patterns for disubstituted benzenes $C_6H_4A_2$ by trisubstitution.

division of organic compounds into "aromatic" and "aliphatic" in his treatise of organic chemistry, as will be detailed in the next section.

Actually, the difficulty which faced chemists until Kekulé proposed his formula was that "aromatic compounds" analyzed like unsaturated substances, yet they differed in their chemical behavior from alkenes or alkynes, behaving rather like the more hydrogen-rich alkanes. Later, this dichotomy between aromatic and aliphatic compounds was conserved in terms of electronic bond structure: energies of aliphatic compounds resulted additively assuming constant increments for two-electron chemical bonds, whereas the aromatic compounds had lower energies than expected on the basis of such calculations.

The enthusiasm which followed Kekulé's revelation of the benzene structure is best reflected in Japp's obituary delivered[4] before the Chemical Society in London in 1897: "Kekulé's work represents a brilliant example for the force of ideas: a formula consisting of a few symbols connected by lines written on a piece of paper provided work and excitement for a whole generation of chemists working in scientific research, and a basis for the most complex industry which the world has seen."

None less than August Wilhelm von Hofmann said that he would give all his experimental discoveries for one of Kekulé's great ideas.[5]

In the wake of this enthusiasm, Graebe and Erlenmayer proposed a formula for naphthalene, patterned after Kekulé's benzene formula, and the structures of hundreds of benzene derivatives fell into place, as in a puzzle whose central piece has been correctly laid. The method for assigning structures was by stepwise substitution, starting with known structures. Körner[6] developed an "absolute method" not needing known reference compounds for assigning *ortho, meta*, and *para* structures of disubstituted benzene isomers with two identical substituents A; it consists in introducing a third substituent B, and analyzing the resulted reaction mixture for the number of trisubstituted isomers (irrespective of whether some of these isomers might be formed in minute amounts); if one trisubstituted isomer is obtained, the compound must be *para*-disubstituted; if two trisubstituted isomers are obtained the compound is *ortho*-disubstituted; if three trisubstituted isomers result, the compound is *meta*-disubstituted (Figure 1).

Kekulé had, however, to fight for his formula which, as it was found out much later, had been one of the many formulas considered in a book with small circulation published in 1861 by Loschmidt,[7] who, however, never sought to assert his priority; on one hand, Kekulé fought with Kolbe who had had some success as an empirical synthetic organic chemist, had developed an inflated ego expanding also into theory without solid background (he did not admit any constitutional formulas or stereochemical representations), and raised against Kekulé, Baeyer, van't Hoff, and others many ill-founded, exasperating objections;[8,9] on the

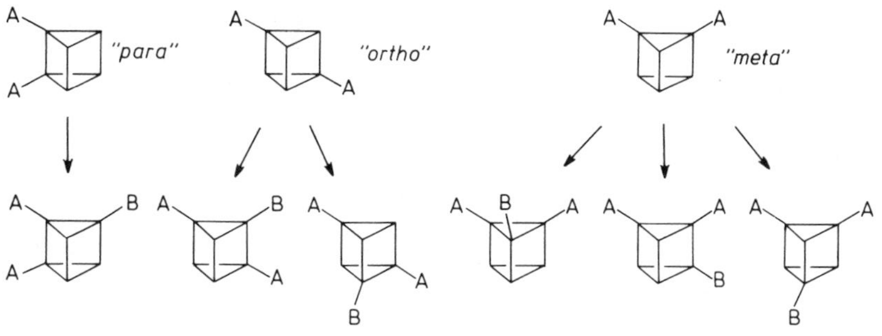

FIGURE 2.   Disubstituted benzprismanes $C_6H_4A_2$ and corresponding trisubstituted benzprismanes $C_6H_3A_2B$.

other hand, with scientifically based alternative formulas such as the Dewar[10] formula* (which, however, soon proved unable to explain the existence of just one isomeric monosubstituted derivative). One of the most interesting formulas was advocated by Ladenburg;[11] this was what we now call triprismane or benzprismane. At that time stereochemistry was in its infancy; Baeyer had not yet formulated his ring strain theory, so that all discussions were in terms of constitution. The prismane formula explained equally well the substitution pattern (Figure 2).

This dilemma was solved later by Baeyer,[12] who noted that (1) hydrogenation of *o, m, p*-disubstituted benzenes yields 1,2-, 1,3-, and 1,4-disubstituted cyclohexanes, respectively, a fact which could not be accommodated by Ladenburg's formula; (2) 2,5-dihydroxyterephthalic acid, obtained by Claisen condensation of succinic ester followed by oxidation, could have one of the two structures: **3** or **4**. Decarboxylation afforded hydroquinol (**5**) which had been definitely assigned a *para* structure and in agreement with the Körner method affords always only one isomer on introduction of one extra substituent. However, Ladenburg's formula ought to have led to an *ortho* structure (**6**). Finally, it was recognized after van't Hoff and le Bel's ideas had become widely accepted that (3) *ortho* prismane formulas should lead to optical activity which was never found, and that (4) according to Baeyer such formulas have a high angle strain.

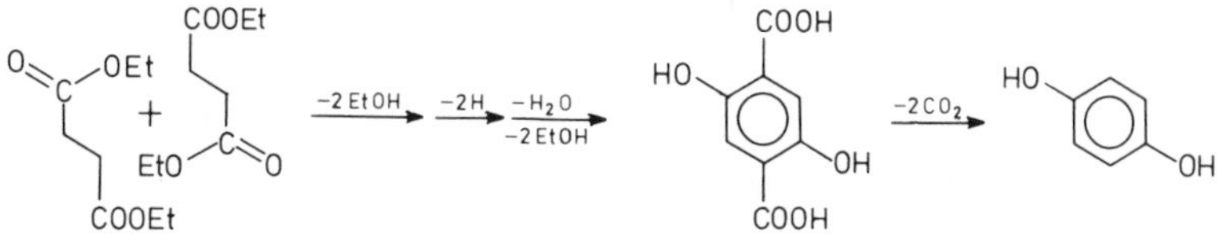

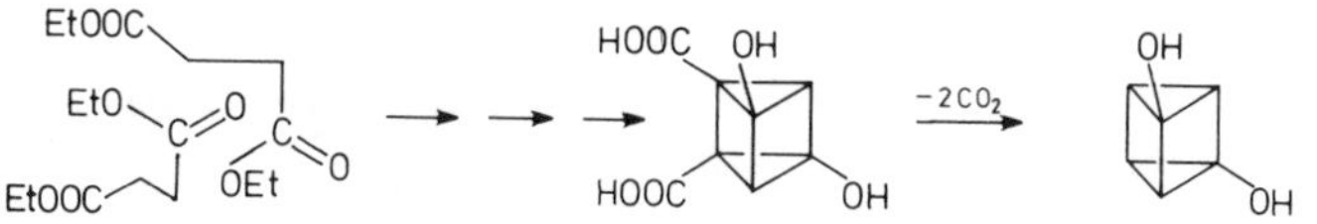

The discussion had been instructive because it had drawn attention to formulas which led to the same number of constitutional isomers, namely the delocalized benzene formula and Ladenburg's prismane formula, ignoring stereoisomerism. Such formulas are called co-isomeric.[13] The reason why they afford the same numbers of isomers is that they possess identical cycle indices in Polya's theorem. Other pairs of co-isomeric structures follow.

---

*   Actually, this formula was proposed by Wichelhaus and Städeler.[10]

In the $(CH)_8$ series there are three pairs of co-isomeric cubic multigraphs, **7** to **12**.

In the $(CH)_{10}$ series there are many co-isomeric structures from which only three pairs of constitutional isomers (**13** and **14**; pentaprismane, **15** and cyclodecapentaene, **16**; diademane, **17** and triquinacene, **18**), and a quintet (**19** to **23**) are presented. More details on this topic will be given in Chapter 3, Section VI.

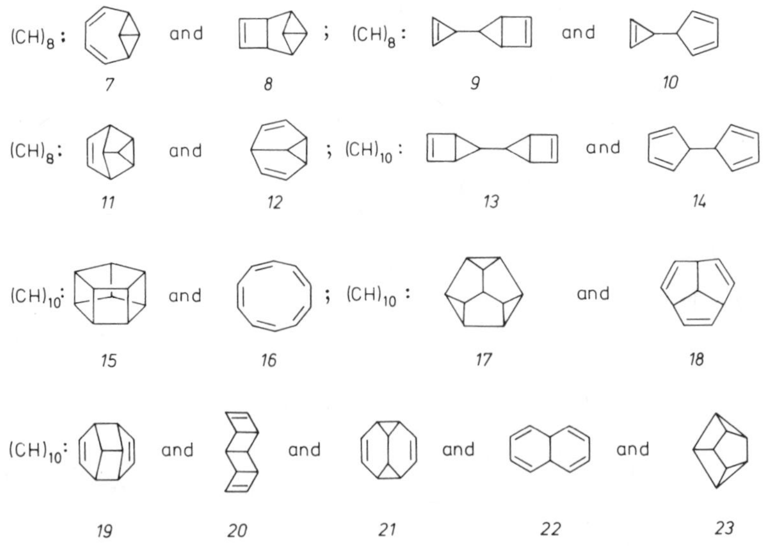

B. The Aromatic Sextet

Kekulé's *Treatise of Organic Chemistry*[14] sanctified the classification of organic compounds into aliphatic and aromatic ones, with the notion of aromatic character as a structural feature predictable *a priori* (even for a substance which is not yet prepared) rather than a particular one, connected with the measurement of any single property of an already existing compound.

Very soon afterwards, however, Erlenmayer in 1866 observed that aromatic compounds react preferentially by substitution rather than by addition, and applied this reactivity criterion as definition for aromaticity (see criterion 4 in Section III), requiring thereby the preparation of a compound and the testing of its reactivity before it can be classified as aromatic or nonaromatic. It will be seen that attempts to reconcile Kekulé and Erlenmayer's views failed, and that the solution came through theoretical work by Hückel, Coulson, and Dewar, and through generalizations which at present largely ignore the narrow view espoused by Erlenmayer, which is incompatible with the reactivity of most aromatic ions.

Thus the name "aromatic", which originally meant what it says, namely odor, became a rather vague concept connected with belonging to a certain class of compounds and exhibiting some particular type of chemical reactivity. As time passed, this notion of aromaticity became more and more fuzzy and some voices recently advocated[15,16] that the notion and words "aromatic" and "aromaticity" had better be relegated to the chemical burial grounds with phlogiston, alkahest, and philosopher's stone. Other chemists, on the contrary, tried to enlarge the sphere to include "tridimensional aromaticity".[17] There are also chemists who prefer to delineate precisely what they mean by the word "aromatic" and continue using epistemologically this fruitful and model-generating notion which has permeated organic and inorganic chemistry during the last 100 years.[18-25]

We shall briefly summarize how organic chemists gradually came to realize that the "magic number" *six* had something to do with aromaticity.

Collie and Tickle[26] observed that 4-pyrones (**24**) were able to undergo protonation or iodomethylation affording crystalline salts. Two structures could be considered for the cations of these salts: one involving the carbonyl oxygen (**25**) and the other involving the ethereal oxygen (**26**). Collie and Tickle[26] correctly assumed that the former alternative was true because thus the structure of the cation was benzene-like. Closer investigation of pyrones by Arndt and co-workers[27] led to the idea that their structure was "mesomeric" and that it could be formulated by a superposition or combination of two limiting structures, which we represent with the resonance symbol of a two-headed arrow.

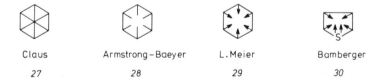

Thus pyrones and their cations (hydroxy- or alkoxy-pyrylium salts) have a great historic importance since subsequent developments due mainly to Ingold[28] and to Pauling[29] put the "mesomerism" theory on a sound physical basis with Pauling's resonance theory. In its simple valence-bond version,[30] this theory fails to account for the magic number six, but recent versions do,[31] using graph-theoretical improvements (Herndon's structure-resonance theory).[22]

Not only six-membered heterocycles such as pyrones or pyridine, but also five-membered heterocycles show aromatic-type reactivity; some of them, such as thiophene, resemble benzene so astonishingly that it took Meyer[32] an unsuccessful lecture experiment with benzene obtained from benzoic acid and intense subsequent investigations to realize that commercial benzene isolated from coal tar contained an unsuspected contaminant, which was surprisingly similar both chemically and physically, namely thiophene. It was Bamberger[33] who, by adapting earlier ideas put forward by Armstrong, Baeyer, and Meier with their "centric formulas" **28** and **29** of benzene (derived, in turn, from the Claus "diagonal formula" **27** of benzene), first came to state that six "centric bonds" as in thiophene **30** are a structural requirement for aromatic character.

Claus   Armstrong–Baeyer   L.Meier   Bamberger
27   28   29   30

Thiele's false but stimulating hypothesis[34] of "partial valencies" prompted Willstätter to attempt the synthesis of cyclobutadiene (unsuccessfully) and of cyclo-octatetraene. He succeeded in obtaining the latter substance which presented olefinic, rather than aromatic, character,[35,36] thereby disproving Thiele's assumptions and showing that three double bonds, rather than any even number, in cyclic conjugation were a prerequisite for aromaticity.

These experimental findings required a theoretical rationalization. Organic chemists developed ad hoc explanations, advocating steric strain in the four- and eight-membered rings, and absence of steric strain in benzene with sp^2-hybridized orbitals possessing bond angles of exactly 120° as required by the geometry of the planar regular hexagonal ring. However, it was felt that this was not a satisfactory explanation since it did not explain the "magic number" six which was dubbed "aromatic sextet" by Robinson, who proposed the circle

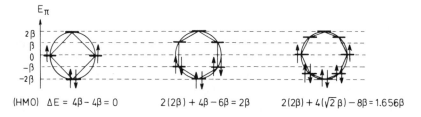

FIGURE 3.   Hückel delocalization energies for [4], [6], and [8]annulenes, and the Frost-Musulin geometrical approach.[40]

inside the benzene ring to symbolize this sextet.[37] However, it was from physics that the solution came, since quantum mechanics had just been formulated, and Hückel had adapted it to calculating properties of π-electron delocalized systems.[38] He advocated that σ-electrons were localized and could therefore be separated from π-electrons. Actually the neglect of σ-electrons is one of the major weaknesses of HMO theory since today we know that σ-electrons have an important role in aromaticity. This point should be remembered in the following discussion, although we shall discuss mostly the π-electrons. The role of σ-electrons is lucidly exposed in the book by Lewis and Peters.[18] Neglecting all electronic interactions except those between adjacent atoms, Hückel calculated the energy levels of molecular orbitals (MOs) in terms of β units, equal to the energy of a π-electron in ethene. In a fully conjugated ring ([*n*]annulene) whose *n* carbon atoms contribute each with one π-electron, MOs are obtained by an approximate solution of the Schrödinger equation leading to a determinant which can be obtained from the adjacency matrix of the graph by introducing x values on the main diagonal and equating the resulting system of secular equations to zero. In concise form the energy levels are[23,39]

$$x_j = 2 \cos(2\pi j/n), \quad \text{where } j = 0,1,2, \ldots, n - 1$$

leading to a total π-electron energy of a conjugated neutral polyene equal to:

$$E_\pi = \begin{cases} 4 \cot g(\pi/n) & \text{for } n = 4m, \text{or } n \equiv 0 \pmod 4 \\ 4 \csc(\pi/n) & \text{for } n = 4m + 2 \equiv 2 \pmod 4 \end{cases}$$

In the simple Hückel MO (HMO) picture, the stabilization energy of cyclopolyenes, ΔE, results by subtracting from this E_π value the energy of isolated ethenic double bonds, leading to the situation presented in Figure 3.

E_π (CH)$_4$ (CH)$_6$ Planar(CH)$_8$

$\Delta E = 4\beta - 4\beta = 0$ $\Delta E = 2(2\beta) + 4\beta - 6\beta = 2\beta$ $\Delta E = 2(2\beta) + 4(1.414\beta) - 8\beta = 1.656\beta$

Frost and Musulin[40] developed a geometrical approach for obtaining the same result by inscribing a regular polygon in a circle of radius 2β with the center in the origin of the energy axis and a vertex at the lowest energy level. The heights of all vertices indicate the MO energies.

If one subtracts from the total π-electron energy (calculated according to the number of electrons on each energy level) the energy of isolated π-electrons (β for one electron in an ethylene π-orbital), one obtains the Hückel delocalization energy, DE (vertical resonance energy).

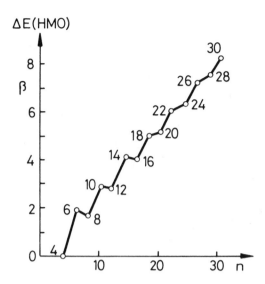

FIGURE 4. Hückel delocalization energies for [*n*]annulenes.

The resulting diagram of DE vs. ring size *n* is a zigzag line (see Figure 4) indicating higher stabilization for systems having $4n + 2$ cyclically conjugated π-electrons. (Hückel's rule for aromaticity, with the aromatic sextet as the particular case for $n = 1$.)

This DE overemphasizes the energies, however, so that it appears wrong that antiaromatics have less stabilization, when they actually have *negative* stabilization.

A better model is Dewar's Resonance Energy,[41] ΔE_D, or DRE, calculated as E_π minus the DE of an acyclic conjugated polyene possessing *n* carbon atoms.[19,23,40,42] This latter energy is

$$E'_\pi = 2 \operatorname{cosec}(\pi/n)$$

hence $\Delta E_D = E_\pi - E'_\pi$ resulting in positive stabilization energies for [$4n + 2$]-annulenes and negative (destabilization) energies for [$4n$]-annulenes.

The influence of ring size on the DRE is presented in Figure 5. Evidently, the situation is now as it should be expected to be: aromatic [$4n + 2$]-π-electron systems are stabilized (have a positive DRE value) while antiaromatic [$4n$]-π-electron systems are destabilized (i.e., they have negative DRE value). It is not yet certain for which large *n* value the decreasing difference between such systems will become negligible.

Figure 6 presents the various ways in which one may define the DE of benzene, that makes impossible a simple additive calculation as in aliphatic chemistry. The Hückel DE is too high (actually it is equal to the energy of four, not three, π-electron pairs of ethylene), and the DRE is a far better measure of the delocalization, predicting correctly which molecules are aromatic and which are antiaromatic. In addition, the so-called "experimental" resonance energy may be measured by calorimetry either from combustion data or from hydrogenation. In both cases the result depends on the standards one adopts for C–C, C=C, and C–H bond energies. According to Klages[43a] (data are summarized in Wheland's[18] or Lewis and Peters's books[30]) the best values for combustion refer to *cis*-disubstituted ethylenes in a six-membered ring, and yield an experimental value of 36 kcal/mol which corresponds to Hückel's standard, three isolated ethene molecules. From heats of hydrogenation, taking as standard three times the hydrogenation enthalpy of cyclohexene, one obtains the same value for the experimental DE of benzene, 36 kcal/mol. However, had we chosen in both cases acyclic ethene as standard, much larger DEs (about 50 kcal/mol) would have resulted.

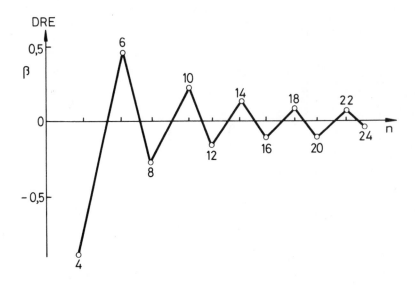

FIGURE 5.   Dewar resonance energies (DREs) for [*n*]annulenes.

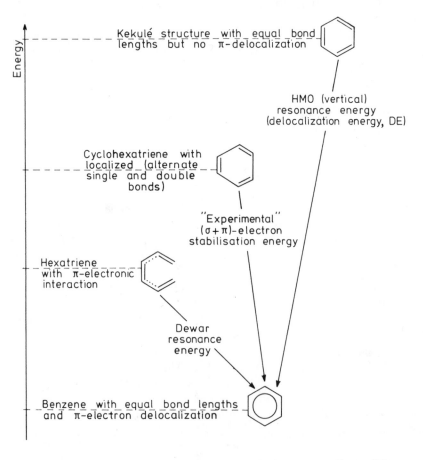

FIGURE 6.   Various types of delocalization energies for benzene according to different definitions and standard molecules.

A new paradigm was recently advocated first by Epiotis, and then confirmed by via *abinitio* calculations by Shaik and Hiberty:[160] benzene is π-destabilized and it has a hexagonal D_{6h} geometry not because, but in spite of, its π-electrons; it is the σ-bonding which is responsible for this geometry by resistance to adistrotion which would lead to a trigonal D_{3h} geometry. On the contrary, cyclobutadiene adopts a rectangular geometry by paying in σ-bond stretching the cost for "antiaromaticity avoidance". Hypothetical molecules such as hexazine (N_6) as well as H_6 and Li_6 are calculated to be unstable towards dissociation into three disconnected N_2, H_2, or Li_2 molecules, because they lack the strong σ-bonds present in benzene.

To conclude this historical review, it should be mentioned that Reference 25 has two historical tables: pre-Hückel, encompassing chronologically the period from 1808 when benzoic acid was discovered until World War II and post-Hückel, presenting alphabetically some names connected with remarkable discoveries in the field of aromatic compounds; in this latter period, Dewar was instrumental in reviving interest in novel aromatic compounds.

II. KEKULÉ STRUCTURE COUNT

A. Corrected (Algebraic) Structure Count

In benzenoid systems, there is a satisfactory correlation between the number of Kekulé structures which can be written for a given system and its stability, e.g., phenanthrene with five Kekulé structures has a higher stability and resonance energy than anthracene with four Kekulé structures. The number of Kekulé structures for polycyclic benzenoid hydrocarbons can be calculated simply by means of an algorithm published by Gordon and Davison.[120] For nonbranched *cata*-condensed systems, Balaban and Tomescu[130] published simple algebraic formulas giving the number of Kekulé structures. A whole chapter is devoted to this topic in Trinajstić's book.[23]

Nonbenzenoid systems necessitate, however, a new idea in order to accommodate the Hückel rule in the valence bond theory. This idea is to assign a parity ($+$ or $-$) to Kekulé structures of nonbenzenoid hydrocarbons;[22,42,44-46] then the algebraic (or corrected) structure count (ASC), K is

$$ASC = K = K_{(+)} - K_{(-)} \geq 0$$

By convention, the positive parity is assigned to the more numerous set of Kekulé structures having the same parity, so as to obtain $K \geq 0$. A simple method for assigning parities is the superposition technique: by considering only the π-bond in double bonds of two Kekulé structures k_i and k_j, the superposition $k_i + k_j$ is the graph according to the rules:

$$k_i + k_j = k_j + k_i; \qquad k_i + k_i = k_i$$

The result of superpositions is a graph $k_i + k_j$ consisting of cycles and/or bonds. If the graph $k_i + k_j$ contains an even number (this includes zero) of $4m$-membered rings, then the two structures k_i and k_j share the same parity; if the graph $k_i + k_j$ contains an odd number of $4m$-membered rings, then the two structures k_i and k_j have opposite parities.

An example is benzocyclobutene, **31**, with limiting Kekulé structures k_1, k_2, and k_3. By superposition we obtain the three superposition graphs **32** to **34**. From them, **33** ($k_1 + k_3$) has an even number (zero) of $4m$-membered rings and the two other ones have one $4m$-membered ring each. It can be seen that k_1 and k_3 have the same parity ($+$) but k_2 has opposite parity ($-$); hence

$$K = K_{(+)} - K_{(-)} = 2 - 1 = 1$$

By the same procedure it can be shown that any mono- or polycyclic benzenoid system has the same parity for all Kekulé structures; therefore, benzene has K = 2; however, cyclobutadiene and cyclooctatetraene have K = 0. This is the valence-bond equivalent of the Hückel rule.

| 31A (k_1) | 31B (k_2) | 31C (k_3) |

| $k_1 + k_2$ | $k_1 + k_3$ | $k_2 + k_3$ |

III. AROMATICITY CRITERIA

The problem of aromaticity is one of the most controversial in chemistry today. On one hand, all textbooks teach it; international meetings,[47-51] books,[18,20,52-87] sections in *Chemical Abstracts, Annual Reports,* and *Specialist Periodical Reports* of the Chemical Society (London),[58,59] are devoted to this topic; as indicated above, from the seven million chemical substances characterized till now, more than 75% are aromatic. Some of the most spectacular discoveries in organic chemistry during 1945 to 1975 were made in the quest for novel aromatic systems.

On the other hand, recently some well-known chemists have questioned the usefulness of this concept which does not have a unique clear-cut definition. The dispute is largely philosophical and semantical, but it serves to reveal weak points in our reasoning.

One should remember that several authors proposed, at various periods, different criteria for aromaticity. Some of these defined aromatic compounds in a qualitative sense (structural definitions such as the presence of continuous cyclic conjugation involving $4n + 2$ delocalized π-electrons), whereas other ones tried to assess quantitatively the extent of aromaticity. It is here that most failures occur, since scales do not agree.

Labarre expressed very aptly the problem as follows: "Chemists and physicists are at present in the middle of a cavern which Plato would not have disavowed; they observe on the walls of the cavern certain shadows resulting from the lighting of an unknown subject (aromaticity) by the different sources of light represented by their various chemical or physical techniques of observation: an agreeable odour, an aptitude to nitration and sulphonation, a ring current, a magneto-optical excess, a diamagnetic anisotropy, a resonance energy, a UV bathochromic effect, and even for some, a mathematical term. The question is: Do these shadows all belong to the same invisible reality? No one is able, at present, to answer such a question."

Elsewhere Labarre stated: "The term aromaticity has lost all simple meaning, and, in fact, one finds oneself incapable of defining it. A further difficulty arises from the fact that it is practically impossible to measure the aromaticity of a molecule, and that even the relative scales proposed by means of the different chemical or physical phenomena mentioned above do not generally agree. We believe that the term aromaticity should be eliminated from the scientific vocabulary in order to avoid the ambiguity that consists in postulating that 'whoever says "aromatic" says "benzenic" and vice versa'. Moreover the considerable number of prefixes joined to the term "aromaticity" (non-, anti-, quasi-, pseudo-, homo-, etc.) indicate sufficiently that this term is outdated. The solution to this problem is not as Coulson has remarked (private communication) to look for a new word to replace aromaticity. It is preferable to propose several (as meaningful as possible) new words, so that each of them refers to one of the multiple chemical or physical properties that were intended to be expressed by the term "aromaticity". We have considered introducing the concept of potential strobilism to describe the phenomenon measured by the Faraday effect and by NMR in the case of cyclic molecules that are the seat of a pi-electronic delocalization, that is, of a ring current."

Binsch,[15] in a paper with a provocative title "Aromaticity — an Exercise in Chemical Futility", stated: "It is indeed suspicious how often magic rules had and have to serve as an alibi for creating an aura of intellectual respectability for chemical research which is on the verge of turning stale . . . For us in pure chemistry the embarrassing question arises with alarming frequency whether we are not simply contributing another tombstone to an already enormous data cemetery."

It was Heibronner who opposed most strongly the notion of "aromaticity". In the discussion of the first review paper at the Jerusalem Symposium "Aromaticity, Pseudo-Aromaticity, Anti-Aromaticity" he said: "Somebody has to be the devil's advocate. To begin with, I think we should all realize that we are united here in a symposium on a non-existent subject. It must also be stated quite clearly at the beginning that aromaticity is not an observable property, i.e., it is not a quantity that can be measured and it is not even a concept which, in my own experience, has proved very useful. Indeed my experience as a teacher is that the amount of confusion caused by the term 'aromaticity' in the student's mind is not compensated for by a gain in the understanding of the chemistry and physics of the molecules so classified . . . Quantum chemistry has finally succeeded in providing us with a pretty good picture of the molecules we deal with, and it yields reasonably good predictions concerning the behaviour of strange molecules even before they have been synthesized. Nobody can claim that the vague concept of 'aromaticity' must be introduced at any stage, to make these theories work. 'Aromaticity', if to be used at all, should be a purely structural concept. One thing is quite certain: 'aromaticity', let alone the 'degree of aromaticity' should not depend on the chemical and physical properties of a given compound. . . . Depending on the physical method which you are using, you will come to a different grading, to a different classification of 'aromaticities' for the same set of compounds. Even for a given type of physical measurement, according to which a compound is going to be classified, it becomes quite arbitrary where you want to draw the line that discriminates between 'aromaticity' and 'non-aromaticity'. As an unescapable consequence, you can have many compounds that according to one method will be classified as 'aromatic' and according to another as 'nonaromatic'. It might be a good idea to introduce the term 'schizoaromatic' for such molecules.

That 'aromaticity' cannot be measured, it not being a physical quantity, is perhaps best shown by the well-known but elusive 'resonance energy' . . . Resonance energies are defined quantities and not observed ones. And this is the reason for the widely diverging values quoted by different authors for the same compound. . . .

"I am primarily a chemist, and my view of theoretical chemistry is terribly pragmatic. I want to know what my molecules will do when subjected to certain experiments . . . Let us assume that I want to make such a prediction. There is absolutely no instance, at any time, where I have to bring the aromaticity. In other words, from the point of view of the chemist who wants to know the behaviour of his molecules, the question of whether a molecule is aromatic, pseudo-aromatic, anti-aromatic, or whatever aromatic, is completely irrelevant. We want to know the properties of molecules, and we don't want to know whether a molecule falls into a certain abstract category that we have invented. Looked at from this point of view, an argument concerning 'aromaticity' must lead to the same sort of discussion that theologians had in the Middle Ages when they wanted to determine how many angels could sit on the head of a pin."

Lloyd and Marshall[62] stated: "The term 'aromatic' was interpreted at different times in terms of molecular structure, of reactivity and of electronic structure, and, in consequence, there has been much confusion over its precise meaning and definition. We suggest that because of this confusion, it would be better if the use of the term 'aromatic' were discontinued, save perhaps with its general and original connotation of 'perfumed', and that it should pass with other technical terms which have outlived their precision and usefulness into the realm of the historian of chemistry." Then Lloyd went on to propose classifications as "benzenoid" for structural features; cyclic conjugated systems should be termed either "Hückel" or "anti-Hückel"; for chemical reactivity which leads to retention of type in unsaturated systems, the term "meneidic" or "regenerative" was suggested.

Two types of opposite tendencies, namely to generalize the notion of aromaticity, should be mentioned. A few words should be said about three-dimensional aromaticity. Many reactions of carboranes suggest that they have a closed shell of molecular orbitals and indeed theoretical treatment confirms this idea. In this and in other similar cases, several authors[17] advocated that it would be appropriate to use the term "three-dimensional aromaticity" for indicating (1) the exceptional stability, (2) the tendency to retain the type, and (3) the presence of doubly occupied bonding MOs, of unoccupied antibonding MOs, and the absence of nonbonding MOs.

Theoretical calculations[161] and experiments performed by Smalley and coworkers indicate an exceptional stability for clusters of 60 carbon atoms.[162] A spherical structure with 5- and 6-membered rings similar to a football (truncated icosahedron, named buckminsterfullerene) was proposed for this cluster; the number of Kekulé structures is 12,500; one foreign atom may be included, e.g. $C_{60}La$ detected mass-spectrometrically like C_{60}.[163]

On the other hand, Rassat,[63] following a suggestion by Woodward, defined aromaticity as the possibility of formal exchange between two localized Kekulé structures and concluded that ethylene and [4n + 2]-annulenes are aromatic in the ground state, while cyclobutadiene is aromatic in the excited state.

We have dealt at length with the critiques addressed to the notion of aromaticity in order to show how serious the problem is.

Several reviews on criteria for aromaticity are available: Chapter 2 of Lewis and Peter's book,[18] a review by Cook et al. on the aromaticity of heterocycles,[64] and Bergmann and Agranat's introductory remarks at the Jerusalem Symposium.[65]

The following criteria for aromaticity were proposed until now:

1.   Structural formula containing three conjugated double bonds in six-membered rings.[2]
2.   Presence of six "centric bonds" (Bamberger, 1891),[33] refined later into the idea of aromatic sextet.[37]
3.   Presence of [4m + 2]-π-electrons in a cyclic conjugated planar ring.[38]
4.   Chemical reactivity favoring electrophilic substitution over addition. As expressed by Lloyd,[53] this shift from common structural features to similar chemical behavior and the possibility of alternative interpretations has subsequently bedeviled organic chemistry.
5.   Ability to sustain an induced diamagnetic ring current.[67,68]
6.   Diamagnetic anisotropy,[69,70] diamagnetic susceptibility exaltation,[71,72] and Faraday effect (magneto-optical excess).[73]
7.   Uniform proton-proton vicinal coupling constants.[74,75]
8.   Presence of a substantial resonance energy relative to an acyclic standard.[40] Further developments include topological resonance energy.[76-78]
9.   Uniformization of the length of peripheral bonds;[79-81] further refinements include bond orders instead of bond lengths;[82,83] absence of first- or second-order bond fixation.[84]
10. Other criteria: Craig's criterion[71] involved the symmetry of the ground state wave function; Kruszewsky and Krygowski[85] as well as Dixon[86] advocated a calculated propensity (specific DE) towards substitution vs. addition; Anet[87] measured solvent

shifts induced in the ^{1}H-NMR spectra of dipolar molecules, namely acetonitrile vs. cyclohexane as standard.

From these criteria it is apparent that initially (criteria 1 to 3) aromaticity was defined qualitatively in terms of structure: a compound was or was not aromatic, so that the question was: "aromatic ?" and the answer was "yes/no". The advantage was that one was able to predict *a priori* the character of unknown molecules. Starting with Erlenmeyer (criterion 4), chemists tried to ask: "how much ?" on the basis of chemical behavior. The drawback is that only experiments with molecules that have already been prepared are able to give a reply to such questions; sometimes different experiments give contradictory replies. In addition, with quantitative data, one has to draw an arbitrary line of demarcation between aromatic and nonaromatic properties. Finally, in many instances reactivities are difficult to compare (for instance, electrophilic substitutions in cations, neutral compounds, or anions), and all reactions are complicated by transition state energetics, solvation, and many other factors.

The magnetic criteria (6 to 8), especially if they are treated in a qualitative rather than quantitative fashion, offer the highest promise, and indeed will be highlighted in this book. They are not exempt from criticism[68,70,88-90] but if care is exercised they remain the most indicative of all experimental data.

Theoretical calculations (criteria 8 and 9) like structural criteria have an advantage over all preceding criteria, namely that they can be applied to compounds which have not yet been prepared, or are too elusive to withstand the experimental determinations.

IV. RESONANCE ENERGY OF AROMATICS

It will be assumed that the reader is familiar with Hückel's molecular orbital (HMO) theory.[38,39,91] The adjacency matrix of the hydrogen-depleted graph corresponding to the aromatic hydrocarbon having all carbon atoms with sp^2-hybridization is converted into the secular determinant by introducing x values on the main diagonal; this determinant may be developed into the characteristic polynomial by simple algebraic operations. On equating the determinant or the polynomial to zero, one finds the roots x_i which are the eigenvalues of the secular equation:

$$x_i = (E_i - \alpha)/\beta$$

where the energies E_i of MOs may be computed in terms of resonance integrals, β, and Coulomb integrals, α. For benzenoid hydrocarbons, $\alpha_c = 0$ and $\beta_{cc} = 20$ kcal/mol. The available π-electrons are allowed to occupy the lower available MOs according to the Aufbau principle, and the total energy is computed.

The problem is to calculate the resonance energy relative to an acceptable standard. In Hückel's method, the standard was formed by isolated ethene molecules wherein each π-electron had a bond energy [β]. Dewar[19,40] argued convincingly that this was a poor standard since it gave positive resonance energies for a host of unstable molecules: "the HMO method has been known almost from its inception to involve approximations so gross as to be almost outrageous; however, its early successes led chemists to believe that it worked well in practice, regardless of its theoretical imperfections. We can now see that these early successes were due essentially to chance: the first applications of the HMO method were to benzenoid aromatic hydrocarbons, where the method might be expected to give reasonable results . . . The HMO method is so unreliable as to be useless in treating compounds containing heteroatoms, or even hydrocarbons that are non-aromatic or contain odd-numbered rings. Much time has been wasted on the synthesis of otherwise uninteresting compounds solely because

Table 1
RESONANCE ENERGIES FOR PLANAR
[n]ANNULENES (IN β UNITS UNLESS OTHERWISE
STATED)

n	HMO	TRE	TREPE	DRE(HMO) (kcal/mol)	DRE(SCF) (kcal/mol)
4	0.000	− 1.228	− 0.307	− 1.07	− 17.8
6	2.000	0.270	0.045	0.39	20.1
8	1.657	− 0.584	− 0.073	− 0.48	− 2.5
10	2.944	0.160	0.016	0.26	7.8
12	2.928	− 0.384	− 0.032	− 0.29	1.8
14	3.976	0.112	0.008	0.23	3.5
16	4.109	− 0.304	− 0.019	− 0.18	2.8
18	5.035	0.090	0.005	0.22	3.0
20	5.255	− 0.240	− 0.012	− 0.11	2.8
22	6.107	0.066	0.003	0.22	2.8
24	8.383	− 0.192	− 0.008	—	—

the HMO method had erroneously predicted that they would be aromatic."[92] Thus the Hückel DE per π-electron is 0.33β for benzene, 0.31β for the unstable pentalene, 0.30β for the unstable heptalene, etc.

Dewar[19,40] combined the Pople method and the perturbational molecular orbital (PMO) method of Coulson and Longuet-Higgins and took as standard a system having the same atoms in a single acyclic conjugated polyene. The reason for choosing such a standard was that in conjugated systems, C=C and C–C bond energies differ in polyenes from those of "pure double" and even more of "pure single" bonds, owing to differences in hybridization: a single $C(sp^3) - C(sp^3)$ bond as in alkanes has a different energy from a $C(sp^2) - C(sp^3)$ bond as in polyenes. He parametrized the SCF π-MO theory so as to reproduce the ground-state properties and obtained excellent results. This Dewar Resonance Energy (DRE) is positive for lower [4n + 2]-annulenes, negative for lower [4n]-annulenes, and becomes increasingly smaller as n increases so that beyond [30]annulene it was predicted that the Hückel rule no longer mattered. The onset of alternating single/double bonds in larger annulenes was confirmed by calculations carried out by Longuet-Higgins and Salem.[93] Breslow et al.[94] used a similar approach.

Simpler ways of defining resonance energies in close agreement with DRE values were proposed by Hess and Schaad,[24,78] Trinajstić and co-workers, and Aihara.[77] These authors did not employ elaborate SCF calculations but had recourse to the Hückel theory using, however, a different standard. Hess and Schaad[24,78] used an eight-bond parameter scheme, while Trinajstić's Zagreb group calculated the polyene-like reference structure by a two-bond parameter scheme, obtaining thereby the resonance energy per π-electron (REPE).[23,76]

Nonparametric resonance energies are even simpler to obtain, but they necessitate a deeper knowledge of graph theory. The characteristic polynomial of a graph, whose roots (eigenvalues) give the MO energies, was known to result from contributions of various subgraphs. According to the Sachs theorem, one is able to remove from this polynomial all contributions due to cycles in the graph. The resulting "reference polynomial" is of the same degree as that of the initial graph, and its roots are to be treated as MO energies of an imaginary acyclic polyene. This idea was developed independently by Aihara[77] and by Trinajstić, Gutman, and co-workers,[23,76] leading to the topological reasonance energy (TRE) or the same energy divided by the number of π-electrons (TREPE). Table 1 presents numerical data for annulenes.

Not only MO methods, but also valence bond methods, were developed recently. Two treatments were proposed, and were shown to be essentially similar. Herndon and co-

Table 2
PARAMETERS R_n AND Q_n (IN eV) FOR
RANDIĆ'S CONJUGATED CIRCUITS
MODEL[21,95] VS. THE NUMBER β OF DOUBLE
BONDS IN THE CONJUGATED CIRCUIT

n	[4n + 2]-membered circuit		[4n]-membered circuit	
	β	R_n	β	Q_n
1	3	0.869	2	−1.60
2	5	0.246	4	−0.45
3	7	0.100	6	−0.15
4	9	0.041	8	−0.006
5	⩾11	0	⩾10	0

workers[22] wrote all Kekulé structures of a conjugated cyclic hydrocarbon and considered that resonance energies are composed equally from combinations of pairs of structures; however, the contribution of various pairs of Kekulé structures depends on common and differing bonds. A few examples follow.

$$RE = \frac{2}{2} H_{AB} \qquad RE = \frac{2}{3} (H_{CD} + H_{CE} + H_{DE})$$

It is easy to see that Kekulé structures **C** and **D** of naphthalene have two common bonds, and the only differing bonds are the other three, exactly as in the two Kekulé structures of benzene. Herndon therefore equates H_{CD} with H_{AB}, and therefore: $H_{AB} = H_{CD} = H_{CE} = 0.841$ eV; $H_{DE} = 0.336$ eV.

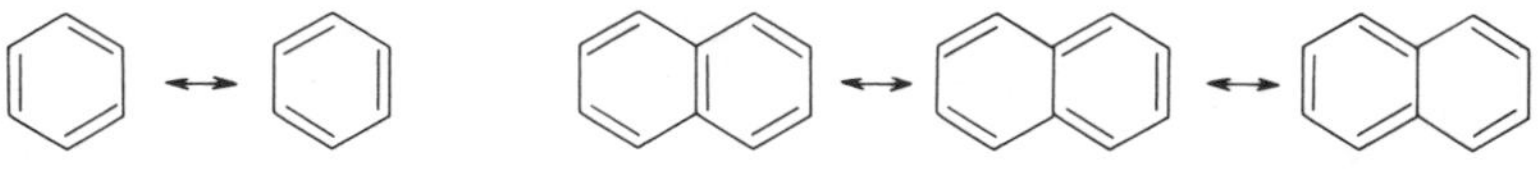

On the other hand, the resonance integrals between the two Kekulé structures of cyclobutadiene (−0.650 eV) and of cyclooctatetraene (−0.260 eV) are negative in Herndon's parametrization.

Separately and independently, and based on quite different starting assumptions, Randić,[21] on one hand, and Gomes and Mallion, on the other,[21] developed another approach (the Conjugated Circuits Model). The former author enumerated all conjugated circuits in all Kekulé structures (a conjugated circuit is a closed path consisting alternatively of single and double bonds, e.g., for naphthalene **C** has two six-membered circuits, **D** and **E** have each one six- and one ten-membered conjugated circuit). For each [4n + 2]-membered circuit one adds positive R_n terms and for each [4n]-membered circuit one adds negative Q_n terms (Randić's values R_n and Q_n are given in Table 2), obtaining the resonance energy.

An example is provided by bicyclo[4.2.0]decapentaene, **35A** to **35C**.

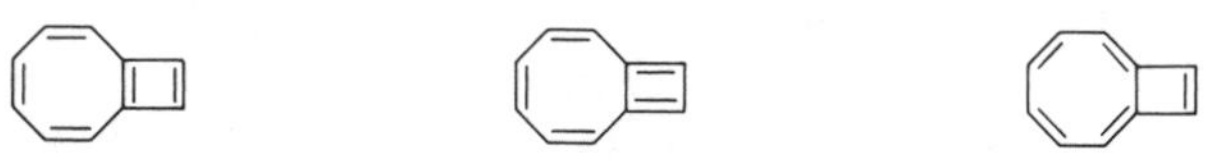

$$3RE = (Q_1 + Q_2) \qquad + \qquad (R_2 + Q_1) \qquad + \qquad (R_2 + Q_2)$$

This molecule is found to be antiaromatic, with a resonance energy of -1.2 eV. Even in the crude HMO treatment, the zero-atom bridge reduces the total energy with respect to cyclodecapentaene, whereas if the bridge is placed to afford azulene or naphthalene, the total energy increases. The only difference of principle between the Herndon and Randić methods is that Randić discards the largest circuit in any linearly dependent set, i.e., whenever a large circuit results by sums or differences of smaller circuits. Of course, the parametrizations are different but the results are very close.

Schaad and Hess[24] compared all these new methods and found close parallels between REPE and Herndon's valence bond method. Gutman[96] showed that the TRE and the parametrized REPE method are also related:

$$TREPE \cong REPE + (0.69/N) \ln K$$

where N is the number of π-electrons and K is the number of Kekulé structures.

V. AROMATICITY OF POLYCYCLIC BRIDGED SYSTEMS

An interesting problem arises concerning the aromaticity of polycyclic systems having an annulenic periphery maintained in a certain geometry by various types of bridges.

It should be stressed that saturated $-(CH_2)_n-$ bridges with $n > 0$ change only the geometry and steric constraints of the system; any π-electronic changes are solely due to changes in hybridization caused by angle constraints. However, zero-atom bridges (as in **35** to **37**) or bridges which have π-electrons or unshared electron pairs, able to interact with the annulenic π-electrons, may interfere with the electronic delocalization of the latter. Actually the Hückel rule is strictly valid for monocyclic systems. The fact that naphthalene, **37**, and azulene, **36**, possess aromatic character was interpreted as indicating that one might extend the Hückel rule also to systems possessing "zero-atom bridges". Similarly, the aromaticity of pyrene, **38**, was taken to indicate that the central double bond did not interact with the [14]annulenic periphery. On the other hand, the aromaticity of perylene, **39**, was explained[97] as indicating two noninteracting naphthalene subunits, and indeed the two bonds linking these two naphthalenes are very long. In the above formulas the "zero-atom bridges" were indicated by dotted lines. However, this interpretation is a very crude approximation; one of the most serious limitations is the antiaromaticity of bicyclo[6.2.0]decapentaene, **35**. If the zero-atom bridge theory would hold, this compound with a [10]annulene periphery should be slightly less aromatic than azulene because of higher ring strain. Actually, calculations carried out by Allinger and Yuh[98] using the molecular mechanics MMP2 program (which takes into account both the electronic energy of the π system and the strain energy in the σ system) show that this compound is antiaromatic having a negative resonance energy. A derivative with this structure analyzed by X-rays[99] proved to be planar and to have alternating bond lengths very similar to those required by structure **35C** (the transannular bond has a length of 1.535 Å; all other single bonds are in the range 1.41 to 1.45 Å; all double bonds are in the range 1.34 to 1.40 Å). Table 3 presents some data calculated by Allinger and Yuh.[98]

Unlike azulene, **36**, for which MMP2 calculations lead to a single minimum, for compound **35** a second local energy minimum was found, corresponding to structures **35A** $\leftrightarrow$ **35B**, which is higher in energy than **35C**. It corresponds to a strongly antiaromatic system with calculated resonance energy -39.84 (planar resonance energy -16.42 kcal/mol), and a smaller strain energy of 46.23 kcal/mol. This latter structure might fluctuate to the energetically lower-lying structure **35C**, but there is no resonance between these two local energy minima (**35C** on one hand, and the less stable **35A** $\leftrightarrow$ **35B**). The resonance energy and "planar resonance energy" were calculated according to Dewar[19] and Breslow,[94] with slightly differing parametrization; the latter energy is calculated for a flattened conformation by

Table 3
RESONANCE AND STRAIN ENERGIES FOR A FEW ANNULENES

Compound	Resonance energy (planar) (kcal/mol)	Strain energy (kcal/mol)
Cyclobutadiene	−20.06	59.90
Cyclooctatetraene (tub)	−21.38 (−2.48)	12.18
[10]Annulene (c.c.t.c.t.)	−17.29 (8.64)	28.34
1,6-Methano[10]annulene	−7.55 (12.83)	28.02
Naphthalene (**37**)	33.70	−2.00
Azulene (**36**)	6.74	9.67
Bicyclo[6.2.0]decapentaene (**35C**)	−4.02	54.97

omitting the direction cosine terms in the expressions used to calculate the resonance integral (and, of course, taking the difference between this energy and that of the corresponding acyclic polyene). One may conclude from Allinger's results that: [4]annulene is unstable owing both to the antiaromaticity and to the strain energy; while the nonbridged [10]annulene in the energetically most favorable conformation (*cis-cis-trans-cis-trans*, corresponding to the naphthalene perimeter) has unfavorable resonance energy, a 1,6-methano-bridge keeps the steric strain constant but is more favorable electronically (both the planar and nonplanar resonance energies are higher algebraically); the 1,6-transannular interaction in naphthalene, **37**, leads to a high resonance energy, the 1,5-interaction in azulene, **36**, to a low resonance energy (associated with a charge separation making the five-membered ring negatively, and the seven-membered ring positively charged, whereas the 1,4-interaction in the last compound, **35C**, is destabilizing, leading to antiaromaticity.

Even the less precise methods (conjugated circuits model, etc.) lead to similar results concerning the antiaromaticity of **35C**; therefore, all inferred aromaticities in conjugated polycyclic systems having "zero-atom bridges" must be viewed with caution, especially when the rings they form are 4*n*-membered. On the other hand, saturated (methano, polymethylene) bridges do not affect the π-system insofar as they do not interfere with the planarity of the π-system.

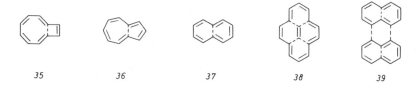

35   36   37   38   39

Systems **35** to **37** were all *cata*-condensed. Another case in which the zero-atom bridge model fails is the following. Whenever a 4*n* + 2 fully conjugated periphery has more than two zero-atom bridges forming either [4*n*]- or [4*n* + 1]- and [4*n* + 3]-membered rings, the resulting *cata*-condensed systems may have filled antibonding orbitals, vacant bonding orbitals, or nonbonding orbitals. All such systems do not have a closed electronic shell. Many examples were analyzed. A few selected ones are indicated below.[100]

1. Alternant nonbranched systems, **40** and **41**, with nonbonding orbitals

$40 : 88^{0}8^{0}8$ $41 : 86^{0}8$

2.   Nonalternant nonbranched systems, **42** to **44**, with vacant bonding orbitals[100,101]

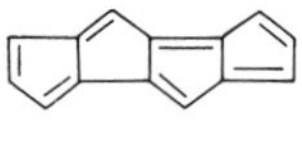

$42:55^15^25$ $43:55^16^05^25$ $44:59^39^45$

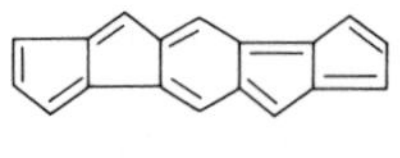

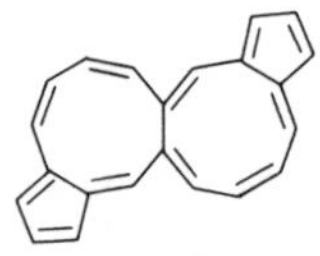

3.   Nonalternant nonbranched systems, **45** and **46**, with occupied antibonding orbitals

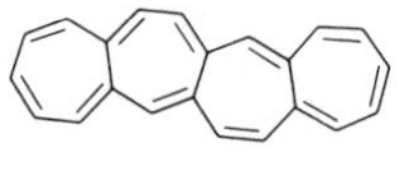

77^17^27
46

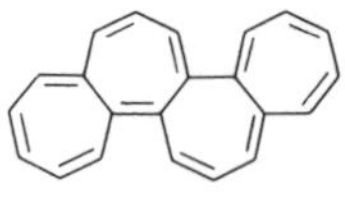

77^37^47
45

The above two cases (2 and 3) were due to an excess of eigenvalues of one sign over those of the other sign.

4.   Nonalternant nonbranched systems, **47** and **48**, with nonbonding levels and/or excess of eigenvalues of one sign

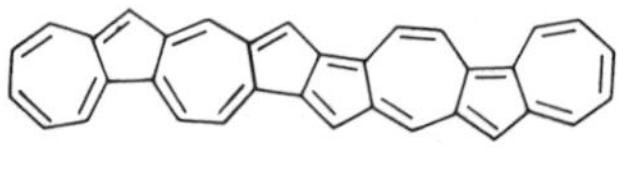

$47:75^17^25^15^27^15^27$
Occupied antibonding MO

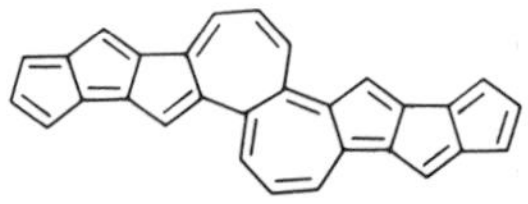

$48:55^15^27^37^45^15^25$
Nonbonding MO

5.   Branched systems, **49**, with nonbonding MOs

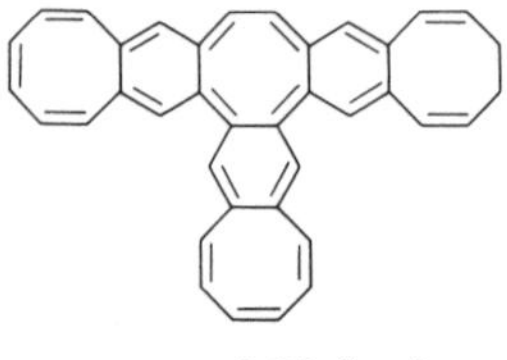

$49:86^08^0(^36^08)6^08$

All the above systems were found by applying Dewar's PMO method[102] and checked by HMO calculations. In all these cases, despite the presence of a [4n + 2]-atom perimeter joined by zero-atom bridges, no closed electronic shell exists. However, this situation is strongly dependent on the bonding topology because closely related systems to those shown above but with different annelation do have a closed electronic shell configuration. The annelation is specified according to the dualist graph conventions: the numbers indicate the ring size, and the exponents show the annelation angle (exponent zero stands for 180°; the higher the exponent, the smaller the angle). Details on dualist graphs will be presented in Chapter 3, Section VIII.

Randić's conjugated circuits model accounts for the fact that such systems are not aromatic, though they are bridged $[4n + 2]$-annulenes; for instance, the eight resonance structures of **40** lead to a resonance energy $(20Q_2 + 8R_3)/3 = -1.03$ eV; the three resonance structures each of **42** and **45** lead to resonance energies $(2R_3 + Q_2)/3 = -0.08$ eV and $2Q_3/3 = -0.1$ eV, respectively. The four resonance structures of **41** yield a resonance energy $(2R_1 + 2R_4 + 4O_2 + 4Q_3)/4 = -0.58$ eV.

Similarly, by considering the algebraic structure count, one can arrive at similar conclusions: the algebraic signs of structures of **35A** and **35B** are opposite; the four resonance structures of **41** and the eight resonance structures of **40** are partitioned in equal sets with different algebraic signs; from the three resonance structures each of **42, 44, 45,** and **46**, two have one sign and one has the other sign.

The general feature of structures presented in the last paragraph, having zero-atom bridges on a $[4n + 2]$-periphery, is the fact that their algebraic structure count is 0 or 1.

VI. CONDENSED BENZENOID POLYCYCLIC HYDROCARBONS

Several interesting problems are associated with the large class of polycyclic aromatic hydrocarbons (PAHs) formed from condensed benzenoid rings:

1. Structure, isomerism, coding, and nomenclature
2. Radicals, diradicals, and related systems devoid of Kekulé structures
3. Number and parity of Kekulé structures
4. Local aromaticity
5. Stability and resonance energy
6. Relationships between structure and properties, including carcinogenicity
7. Conclusions

In order to simplify the drawings, unless otherwise indicated, in all formulas carbon atoms are assumed to have sp^2-hybridization and six-membered rings are assumed to be benzenoid.

A. Structure, Isomerism, Coding, and Nomenclature

It will be described in Chapter 3, Section VIII how a PAH (called for simplicity a polyhex) may be classified as *cata*-condensed PAH (catafusene), *peri*-condensed (perifusene), or *corona*-condensed (coronafusene), by means of the dualist graph structure proposed by Balaban and Harary:[103] acyclic, with three-membered rings, or with larger rings, respectively. For catafusenes, a coding and nomenclature system was proposed on this basis,[103,104] presenting some simplifications relative to the IUPAC or *Chemical Abstract* nomenclature.

The dualist-graph definition makes all catafusenes with the same number of condensed benzenoid rings isomeric among themselves and allows the enumeration of all these isomers. In such enumerations one has the choice of viewing polyhexes either as portions of the honeycomb tessellation of the plane (i.e., of the graphite lattice), or one may allow for the existence of two or more superimposed atoms on this lattice, e.g., **50** to **53**; the latter case is closer to reality because helicenes are known and stable (e.g., heptahelicene, **50**). Though it is unlikely that the two bond-localized forms (**51** and **52**) or the delocalized forms (**51** $\rightleftarrows$ **52**) of [10]annulene will assume this geometry, it is interesting to speculate if chiral benzenoid hydrocarbons such as **53** might be stable.

50   *51*

52   53

For peri- and coronafusenes one needs, however, a different approach. The enumeration of isomers requires complicated computer programs.[105,106] The last reference includes extensive tables of polyhexes but does not allow for superimposed atoms on the graphite lattice.

Dias[107] investigated polyhexes in terms of their molecular formulas and numbers of benzenoid rings and produced tables arranged accordingly.

For the coding of cata- or perifusenes, two systems were derived: the first, by Bonchev and Balaban,[108] is based on the dualist graph of the polyhex and on the generalization of the topological center concept. A vertex belonging to the center of the dualist graph is taken as the focal point, and the coding follows from a few simple conventions. The basic idea is that around the focal point, the vertices of the dualist graph may occupy only fixed positions on the graphite (honeycomb) lattice; these positions are given numbers in clockwise fashion in each successive shell around the focal point; in the code, successive shells are separated by a slash. Thus the code for dibenzo[*a,h*]pyrene, **54**, is 1,2,3,5/3. It is easily seen from the drawing which is the focal point and how the potential vertices of dualist graphs in successive shells are numbered.

The second coding was recently proposed by Cioslowski and Turek,[109] and places the dualist graph, **55**, so that it overlaps with the plane oblique net whose cells are parallelograms. This positioning is required to have the minimal number of (1) rows and (2) columns. The minimal circumscribed parallelogram is symbolized by a numerical matrix with vertex occupancy specified by 1, and 0 otherwise. The rows, read sequentially, form binary numbers which are then converted into decimal numbers. The same polyhex is, according to this idea, [4.15.2]-hexacene.

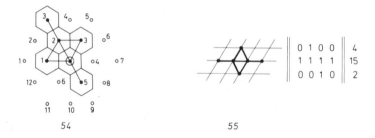

54   55

Other nomenclature proposals[110] in this field seem to be less easy to implement.

B. Radicals, Diradicals, and Related Systems Devoid of Kekulé Structures

A complication which arises only in perifusenes is the possibility of obtaining structures with unpaired electrons (free radicals). This is an expected feature for structures with an odd number of carbon atoms such as perinaphthyl, **56**.

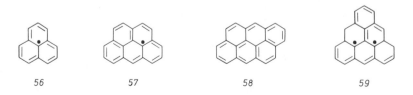

56   57   58   59

However, diradicals are less easy to rationalize since they are isomeric to normal non-radicalic systems. On adding a trimethine chain to the free radical **57**, one can obtain either a normal perifusene, **58**, or the diradical **59**, for which no Kekulé structure can be written without leaving two unpaired electrons on nonadjacent carbons. Similarly, the free radical **60** may be augmented by $(CH)_3$ to the hydrocarbon zethrene, **61**, which has six Kekulé structures, all of which have two localized double bonds (indicated on **61**) or the diradical **62**, which has two odd electrons. In graph-theoretical terms, the σ-skeletons of **58** or **61** admit a decomposition into 1-factors but **59** or **60** do not.

It is easy to see that the triangles in the dualist graphs of **61** and **62** point up or down. It was inferred[111] that equal numbers of such triangles were the prerequisite for a perifusene to have Kekulé structures, whereas an excess of up- or down-pointing triangles over down- or up-pointing ones indicates (2k)-radicals. However, this is not a sufficient condition.

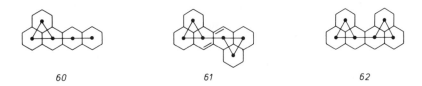

          60   *61*   *62*

Since all PAHs are alternant hydrocarbons, corresponding to bipartite graphs, an alternative rule for finding whether a PAH has, or does not have, Kekulé structures is to count the number of starred and unstarred vertices: if the two numbers are equal then the PAH has Kekulé structures; the difference between these numbers gives the number of unpaired spins.

C. Number and Parity of Kekulé Structures

In Pauling's valence-bond method and its newer versions (structure-resonance theory developed by Herndon, and Randić's conjugated circuits model) the number of Kekulé structures which may be written for a given PAH is an important parameter: this number is related to the resonance energy; the bond orders between carbon atoms can be related to the bond order between the same pair of carbons in each of the Kekulé structures, and the conjugated circuits are found from each Kekulé structure.

The parity of Kekulé structures, introduced by Dewar and Longuet-Higgins[112] and redefined by Gutman and Trinajstić,[113,114] was discussed in Section II. A. For benzenoid condensed PAHs all Kekulé structures have the same parity. A simple relationship exists between the numbers K of Kekulé structures in such systems and the absolute value of the last coefficient a_o of the characteristic polynomial (i.e., the constant or the coefficient of x^o): $|a_o| = K^2$. It is thus fairly easy to compute the number of Kekulé structures for PAHs, especially if one uses efficient speedy methods based on the Sachs theorem[115] or on matrix multiplication for obtaining the coefficients of the characteristic polynomial from the adjacency matrix.[116] By means of the parity concept it is possible to calculate the algebraic structure count (ASC) as the difference between the numbers of Kekulé structures with different parities. Thus ASC = 0 for the [10]annulene with two zero-atom bridges forming one six- and two four-membered rings **63**; this compound illustrates Randić's postulate that [4*m*]-membered rings destabilize more strongly chemical structures than the same number of [4*m* + 2]-membered rings stabilize these structures.

              63A (+)   *63B* (+)   *63C* (−)   *63D* (−)

Another similar case of a $[4n + 2]$-periphery with two zero-atom bridges leading to a nonaromatic system is the known and stable dicyclooctatetraeno[*a,d*]benzene, or tricyclo[10.6.0]octadecanonaene, **41**.

41A   41B   41C   41D

From the four limiting structures which may be written for this compound, two (**41C**, **41D**) have a [18]annulene periphery, yet the actual chemical structure is best described as a resonance between the other two formulas having three double bonds in the benzenoid ring and nonplanar marginal rings (**41A** ↔ **41B**). The compound was prepared by Sondheimer and co-workers[117] in 10 to 12% yield via the unstable 1,3,5-cyclooctatrien-7-yne (**66**) formed from bromocyclooctatetraene and *t*-BuOK. The structure of **41** was demonstrated by catalytic addition of 6 mol of hydrogen affording the benzenoid derivative **67**. If the molecule was planar, then one might expect it to show evidence for a nonbonding MO. Since no such evidence exists, the structure is very probably nonplanar.

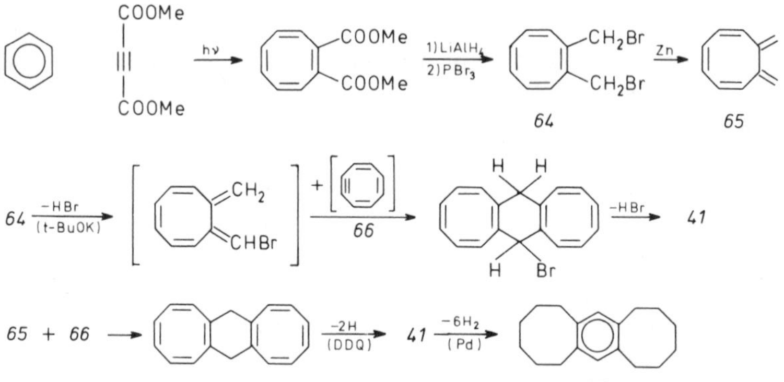

In the ^{1}H-NMR spectrum of the solid **41**, m.p. 181°C, the protons of the eight-membered rings appear in the olefinic region; the absence of an aromatic 18π-electron system is the consequence of [4*m*]-membered rings formed by the zero-atom bridges in the *cata*-condensed system **41**.

An extreme case is that of butalene, **68**, with a zero-atom bridge in benzene forming two four-membered rings. According to theoretical calculations, Aihara[118] obtained a large negative DRE ($-0.604|\beta|$) in agreement with the high instability of this substance, despite its substantial DE of $+1.657|\beta|$: it cannot be isolated from the frozen matrix, as shown by Breslow.[119] It also contradicts Bredt's rule.

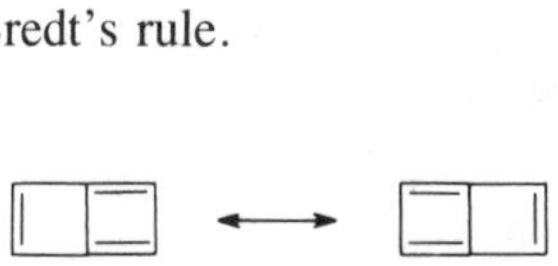

68A   68B

For the enumeration of Kekulé structures many methods were devised.[120-130] They are well reviewed.[131] In the characteristic polynomial, the coefficient a_o of x^o is related to the number K of Kekulé structures: $K = |a_o|^{1/2}$, as shown above.

For nonbenzenoid structures which contain [4*m*]-membered rings, the algebraic structure count (ASC) replaces K: $ASC = |a_o|^{1/2}$.

The number of Kekulé structures may also be calculated by means of the acyclic polynomial, the Hosoya polynomial, the sextet polynomial, or the Wheland polynomial.

A very simple algorithm of finding the numbers of Kekulé structures for catafusenes was devised by Gordon and Davison.[120] For nonbranched systems one starts at an endpoint of the dualist graph, which receives the digit 2. The subsequent ring receives a label 2 + 1 = 3; if the next ring is condensed linearly, it is labeled 3 + 1 = 4, but if the next ring is condensed angularly, it receives as label the sum of the preceding two ring labels. For branched systems the algorithm works also.

Combinatorial[122] or algebraic expressions[130] were published for the numbers of Kekulé structures. Interestingly, the numbers of Kekulé structures for the series of zigzag catafusenes (and of all isoarithmic polyhexes whose three-digit codes differ only by 1/2 interchanges) form the Fibonacci series: 2 for benzene, 3 for naphthalene, 5 = 2 + 3 for phenanthrene, 8 = 3 + 5 for chrysene, 13 = 8 + 5 for picene, etc.[120,122,130] For details on isoarithmic polyhexes and three-digit codes, see Chapter 3, Section VIII.

D. Local Aromaticity

Clar[55,132] observed that annelation of PAHs has different effects on electronic absorption bands and assigned these bands to electronic transitions with polarizations connected with the topology of the PAH. From these initial concepts, he developed an empirical notation wherein some benzenoid rings were believed to be less aromatic than other rings. Thus the marginal rings in phenanthrene, triphenylene, and dibenzo[a,h]perylene are considered to be fully aromatic, while the other rings are less aromatic. The greater aromaticity of phenanthrene than of anthracene (by 7 to 12 kcal/mol) is attributed to the fact that phenanthrene has two fully aromatic sextets while anthracene has only one which may occupy any of the three rings; the arrow symbolizes the possibility of one sextet to be written anywhere along the rings it crosses.

Polansky and Derflinger[133] were the first to put Clar's ideas on a quantitative quantum-chemical basis, by assigning "local aromaticity indices" to various rings in PAHs; if the aromaticity index of benzene is 1.000, then according to their quantum-chemical data each ring of naphthalene has an index 0.912; the central rings of anthracene and phenanthrene have indices 0.840 and 0.813, respectively, while the marginal ones have indices 0.893 and 0.928, respectively. These and other data are in agreement with the intuitive idea that a marginal ring in phenanthrene is more benzene-like than in anthracene while the central rings in both these hydrocarbons are much less benzene-like than the marginal ones since they readily undergo additions. Alternative ring indices based on other (MO, VB, or empirical) approaches followed soon thereafter by Dewar,[134] Kruszewski,[135] Herndon,[45] Gutman,[136] and Aihara.[137] Two simple approaches were developed by Randić: the first[128,138] considers the benzene character of a ring as equal to the fraction of Kekulé structures where the respective ring is formally benzenoid, i.e., appears with three double bonds. Thus dibenzo[a,c]anthracene, **69**, has 13 Kekulé structures; the two equivalent marginal rings are formally benzenoid in 12 Kekulé structures, the other marginal ring in 8, the ring next to it in 8, and the ring condensed with three other rings (the "empty" ring in Clar's formalism) in 2 Kekulé structures; the resulting figures (**69A**) agree fairly well with Clar's picture (**69B**) and with experimental data for ring currents (**69C**).[68,139] Randić's second approach[128,138] is based on his model of conjugated circuits; by counting conjugated circuits with 6, 10, 14, and 18 vertices and ascribing to each of these circuits decreasing positive contributions to the resonance energy (Section IV) one may calculate the contribution of each ring to the total resonance energy in percent (**69D**).

An even simpler relationship was observed by Zander[140] between Polansky's benzenoid character orders and a very simple invariant of the dualist graph, namely the distance sums (sums of topological distances from a vertex of the dualist graph to each other vertex).

E. Relationships Between Structure and Properties Including Carcinogenicity

Among the many relationships between structure and properties of PAHs, the first one which attracted attention was the regular bathochromic shift of one of the absorption bands in the electronic absorption spectra of acenes. On this basis, Clar[55,132] built up his well-known band classification; a pronounced shift is observed in the acene series for Clar's *para*-band (Platt's ^1L_a band), giving a linear plot of $\lambda_{max}^{1/2}$ vs. the number N of condensed benzenoid rings in the acene. Better linear correlations for selected classes of PAH with characteristic annelation patterns were obtained by plotting the excitation energy for the *para*-band vs. $(N - 2)/N^2$.[141]

Similar correlations were found for the redox potentials and other data which depend on the frontier orbitals HOMO and LUMO, such as an experimental parameter (triplet zero-field splitting) or a theoretical parameter (Dewar reactivity number, which explains the regioselectivity in additions or substitutions involving PAHs).

The calculated resonance energy E_π of *cata*-condensed PAHs was found[142] to be linearly related to the total number, N, of benzenoid rings and to the number, a, of zeros in the three-digit code; such relationships explored by Sahini[143] (using as parameter a, the number of "imperfect" benzenoid rings) were interpreted by Balaban[142] in terms of graph-theoretical invariants of the dualist graph:

$$E_\pi = 1.75N - 0.15a + 0.25 \text{ [in beta units]}$$

The number a_{max} of zeros in the three-digit code (i.e., the number of benzenoid rings) of the longest linearly condensed portion of catafusenes was found[144] to be related to the rate of Diels-Alder addition between these catafusenes and maleic anhydride. In a biparametric correlation, the experimental rate constants determined by Biermann and Schmidt (using more than 100 catafusenes from Clar's collection)[145] was expressed as a function of w and a_{max} (for w *vide infra*); earlier correlations with quantum mechanical data gave smaller correlation coefficients, although they afforded a better insight into the nature of the transition state for cycloadditions.[146]

The above cycloadditions involve two *meso*-anthracenic positions of one benzenoid ring. It is known that *peri*-condensed systems possessing bay regions *(vide infra)* react with maleic anhydride with dehydrogenation forming a new benzenoid ring at the bay-region site; in these cycloadditions *o,o'*-biphenylic sites of two different benzenoid rings are involved,[147] and satisfactory rate correlations with "butadienic order" calculated with Polansky's PARS or Dewar's PMO methods were found.[141,148]

By counting the number, w, of 120° angles, N of points, and s of lines in the dualist graph, Zander et al. proposed the following topological index for PAHs (cata- and perifusenes):

$$z = (w + 1)/(5N^2 - Ns + N)$$

This topological index shows a good correlation with the Dewar and Hückel resonance energies, though other topological indices such as the Wiener index (sum of all topological distances in the distance matrix) for the constitutional or dualist graph of the PAH did not show any such correlations.[149]

The most important biochemical property of PAHs is the carcinogenicity of those compounds which have a bay region, i.e., a re-entrant position marked by an arrow in benzo[*a*]pyrene (**70**) and benz[*a*]anthracene (**71**), two representatives of the carcinogenic PAHs (especially if the latter compound is substituted, e.g., with methyl groups, at positions remote from the bay region). An important condition is that the bay region must be bordered by one unsubstituted and nonannelated benzenoid ring, labeled B in formula **71**.

It was proved by several investigators (Boyland, Borgen, Sims, Jerina et al., Harvey) that the ultimate carcinogen is a product of biological "detoxification" (which in this case turns out to be harmful) via oxidation of the benzenoid ring B which yields a diol epoxide **72** (several stereoisomers are possible). Owing to its intercalating ability in the DNA double helix and to its sensitivity to ring-opening on attachment by nucleophilic amine groups (H − Nu) of the purinic and pyrimidinic nucleotides, the PAH becomes permanently attached to the DNA strand and the DNA repair mechanism is no longer able to excise this defect in the DNA, leading to malignancies.[50-55]

Since the presence of bay regions is a structural feature, many correlations between structure and carcinogenicity or mutagenicity were proposed, starting with the K-region theory of the Pullmanns[56] and the Daudels.[57] Actually, there are links between these obsolete theories and the bay-region feature.[58] The problem is made difficult by the fact that at present there are no numerical (quantitative) scales for measuring these biological effects; earlier attempts to provide such numerical data are today considered to lack sound backing. There exist good correlations between the calculated stability of carbenium ions as in **73** and the carcinogenicity. Other correlations employed the structure-resonance theory. Computer programs were developed to provide the three-digit code of PAH and to indicate the number of bay regions.[59,60]

F. Conclusions

Can condensed PAHs be viewed as annulenes with zero-atom bridges?

cata-Condensed benzenoid hydrocarbons composed only of six-membered rings with continuous conjugation have been often considered as Hückel annulenes with zero-atom bridges; while this is formally always true (both for linearly condensed systems such as the phenes or the acenes and for branched systems such as triphenylene) or even for *corona*-fused systems such as Kekulene, one must remember that the Hückel rule holds true only for monocyclic systems below a certain ring size. The common bond to two *cata*-condensed benzenoid rings is such a strong perturbation that even in the HMO approach from the five bonding orbitals only one from each degenerate pairs of MOs in [10]annulene maintains its energy in naphthalene. More sophisticated methods indicate major quantitative changes too, but the qualitative picture remains that of an aromatic system.

If a Hückel annulene has zero-atom bridges which lead to the formation of $[4m + 1]$- and $[4m + 3]$-membered rings in a *cata*-condensed system, the perturbation is larger and may lead to qualitative differences; the resulting nonalternant system may not have a closed electronic shell.

If, on the other hand, the Hückel annulene has zero-atom bridges which lead to the formation of $[4m]$-membered rings, then the resulting *cata*-condensed system is no longer aromatic.

Extension of the Hückel rule to *peri*-condensed systems with several zero-atom bridges connecting a $[4m + 2]$-membered periphery to an isolated central system may also lead to false conclusions; if this qualitative idea works in the case of pyrene or a few other systems, it fails in the case of Kekulene which does not have two Hückel peripheries but local benzenoid rings connected by single and double bonds. In the case of perylene there are two connected naphthalene moieties whose π-electron systems do not interact.

One can conclude, therefore, that the Hückel rule is rigorously valid for monocyclic systems; for *cata*-condensed benzenoid systems one can think in terms of zero-atom bridges without getting a wrong qualitative picture; for zero-atom bridges which lead to the formation of $[4m + 1]$- and $[4m + 3]$-membered rings in *cata*-condensed systems, or to benzenoid rings in *peri*-condensed systems one may obtain a wrong qualitative picture; finally, when zero-atom bridges lead to the formation of $[4m]$-membered rings in *cata*- or *peri*-condensed systems, then very probably the picture is wrong both qualitatively and quantitatively.

The structures and valence isomers of antiaromatic compounds were reviewed by Glukhovtsev et al.[160]

REFERENCES

1. **Benzolfest,** *Ber. Dtsch. Chem. Ges.,* 23, 1265, 1890.
2. **Kekulé, A.,** *Bull. Soc. Chim. France,* (2)3, 98, 1865; *Liebigs Ann. Chem.,* 137, 169, 1866.
3. **Gmelin, L.,** *Handbuch der Theoretischen Chemie,* Vols. 1 and 2, F. Varrentrapp, Frankfurt-am-Main, 1817; Vol. 3, 1819, contains organic chemistry.
4. **Japp, F. R.,** Obituary address for A. Kekulé, *Chem. Soc. London,* 1897.
5. **Hofmann, A. W., von, cited after Volhard, J.,** *Ber. Dtsch. Chem. Ges.,* 35(III), 165, 1902.
6. **Körner, W.,** *Bull. Acad. R. Belg.,* (2)24, 166, 1867.
7. **Loschmidt, J.,** *Chemische Studien,* Vienna, 1861; see also **Anschütz, R.,** *Ostwalds Klassiker Exacten Wiss.,* 1913, 190; *Ber. Dtsch. Chem. Ges.,* 45, 539, 1912; Mentzer, Bull. Soc. Chim. France, p.2671, 1964.
8. **Kolbe, H.,** Das chemische Laboratorium der Universität Leipzig, Vieweg, Braunschweig, 1872, 163; Kurzes Lehrbuch der organischen Chemie, 1881, 366.

9. **Schorlemmer, C.,** *Der Ursprung und die Entwicklung der Organischen Chemie,* Vieweg, Braunschweig, 1889 (reprinted, Akad.-Verlag, Leipzig, 1879); **Lachmann, A.,** *The Spirit of Organic Chemistry,* Macmillan, New York, 1899, chap. 3; **Tilden, W. A.,** *A Short History of the Progress of Scientific Chemistry,* Longmans, London, 1899; **Ihde, A. J.,** *The Development of Modern Chemistry,* Harper & Row, New York, 1964; **Findlay, C.,** *A Hundred Years of Chemistry,* Duckworth, London, 1948; **Farber, E.,** *The Evolution of Chemistry,* Ronald Press, New York, 1952.

10. **Dewar, J.,** *Proc. R. Soc. Edinburgh,* p, 84, 1866—67; **Wichelhaus, H.,** *Ber. Dtsch. Chem. Ges.,* p.197, 1869; **Städeler, G.,** *J. Prakt. Chem.,* 103, 106, 1868.

11. **Ladenburg, A.,** *Ber. Dtsch. Chem. Ges.,* 2, 140, 272, 1869; 5, 322, 1872.

12. **Baeyer, A., von,** *Ber. Dtsch. Chem. Ges.,* 19, 1797, 1888; 23, 1272, 1890; *Liebigs Ann. Chem.,* 245, 103, 1887; 251, 257, 1889; 256, 1, 1890; 258, 145, 1890; 266, 169, 1892; 276, 259, 1893.

13. **Balaban, A. T.,** *Rev. Roum. Chim.,* 19, 1323, 1974.

14. **Kekulé, A., von,** *Lehrbuch der Organischen Chemie,* Vol. 2, Enke, Eilangen, 1866, 493.

15. **Binsch, G.,** *Naturwissenschaften,* 60, 369, 1973.

16. **Marchand, A. P.,** *Chem. Ind.,* p.161, 1965.

17. **Jemmis, E. D. and Schleyer, P. von R.,** *J. Am. Chem. Soc.,* 104, 4781, 1982; **Muetterties, E. L. and Knoth, W. A.,** *Polyhedral Boranes,* Marcel Dekker, New York, 1968; **Lipscomb, W. N.,** *Acc. Chem. Res.,* 6, 257, 1973; **Balaban, A. T. and Rouvray, D. H.,** *Tetrahedron,* 36, 1851, 1980; **King, R. B. and Rouvray, D. H.,** *J. Am. Chem. Soc.,* 99, 7834, 1977; **Aihara, J.,** *J. Am. Chem. Soc.,* 100, 3339, 1978.

18. **Lewis, D. and Peters, D., Eds.,** *Facts and Theories of Aromaticity,* Crane-Russak, Co., London, 1975.

19. **Dewar, M. J. S., Ed.,** *The Molecular Orbital Theory of Organic Chemistry,* McGraw-Hill, New York, 1969.

20. **Garrat, P. J.,** *Aromaticity,* McGraw-Hill, London, 1971; **Barton, D. and Ollis, W. D., Eds.,** *Comprehensive Organic Chemistry,* Vol. 1, Pergamon, Oxford, 1979, 361; **Garratt, P. J. and Sargent, M. V.,** *Advances in Organic Chemistry, Methods and Results,* Vol. 6, Interscience, New York, 1969, 1.

21. **Randić, M.,** *J. Am. Chem. Soc.,* 99, 444, 1977; *Chem. Phys. Lett.,* 38, 68, 1976; **Gomes, J. A. N. F. and Mallion, R. B.,** *Rev. Port. Quim.,* 21, 82, 1979; **Gomes, J. A. N. F.,** *Croat. Chem. Acta,* 53, 561, 1980; *Theor. Chim. Acta,* 59, 333, 1981.

22. **Herndon, W. C.,** *J. Am. Chem. Soc.,* 95, 2404, 1973; *Isr. J. Chem.,* 20, 270, 1980; **Herndon, W. C. and Ellzey, M. L., Jr.,** *J. Am. Chem. Soc.,* 96, 6631, 1976.

23. **Trinajstić, N., Ed.,** *Chemical Graph Theory,* Vol. 1, CRC Press, Boca Raton, Fla., 1983, chap. 6; Vol. 2, Chap. 1 to 3.

24. **Hess, B. A., Jr. and Schaad, L. J.,** *Pure Appl. Chem.,* 52, 1471, 1980; **Schaad, L. J. and Hess, B. A., Jr.,** *Pure Appl. Chem.,* 54, 1097, 1982.

25. **Balaban, A. T.,** *Pure Appl. Chem.,* 52, 1409, 1980.

26. **Collie, J. N. and Tickle, T.,** *J. Chem. Soc.,* 75, 710, 1859.

27. **Arndt, F., Scholtz, E., and Nachtwey, P.,** *Ber. Dtsch. Chem. Ges.,* 56, 1903, 1924.

28. **Ingold, C. K., Ed.,** *Structure and Mechanism in Organic Chemistry,* Bell, London, 1953.

29. **Pauling, L., Ed.,** *The Nature of the Chemical Bond,* Cornell University Press, Ithaca, 1940.

30. **Wheland, G. W., Ed.,** *Resonance in Organic Chemistry,* John Wiley & Sons, New York, 1955.

31. **Klein, D. J.,** *Pure Appl. Chem.,* 55, 299, 1983.

32. **Meyer, V.,** 1882; cited after **Joule, J. A. and Smith, G. F.,** *Heterocyclic Chemistry,* Van Nostrand-Reinhold, London, 1972, 220.

33. **Bamberger, E.,** *Ber. Dtsch. Chem. Ges.,* 24, 1758, 1891; 26, 1946, 1893; *Liebigs Ann. Chem.,* 273, 373, 1893.

34. **Thiele, J.,** *Liebigs Ann. Chem.,* 306, 87, 1899.

35. **Willstätter, R. and Waser, E.,** *Ber. Dtsch. Chem. Ges.,* 44, 3423, 1911; **Willstätter, R. and Heidelberger, M.,** *Ber. Dtsch. Chem. Ges.,* 46, 517, 1913; see also **Baker, W.,** *J. Chem. Soc.,* p.258, 1945.

36. **Schröder, G., Ed.,** *Cyclooctatetraen,* Verlag-Chemie, Weinheim, 1965.

37. **Armit, J. W. and Robinson, R.,** *J. Chem. Soc.,* 127, 1604, 1925.

38. **Hückel, E.,** *Z. Physik,* 70, 204, 131; 72, 310, 1938; *Z. Electrochem.,* 61, 866, 1937; *Grundzüge der Theorie unges ättigter und aromatischer Verbindungen,* Berlin, 1940, 71.

39. **Heilbronner, E. and Bock, H.,** in *Das HMO Modell und Seine Anwendung,* Vol. 1, Verlag-Chemie, Weinheim, 1968, 116; **Günthard, H. H. and Primas, H.,** *Helv. Chim. Acta,* 39, 1645, 1956.

40. **Frost, A. A. and Musulin, B.,** *J. Chem. Phys.,* 21, 572, 1973.

41. **Dewar, M. J. S.,** *Tetrahedron,* Suppl. 8, 75, 1966; **Dewar, M. J. S. and de Llano, C.,** *J. Am. Chem. Soc.,* 91, 789, 1969; see also **Baird, N. C.,** *J. Chem. Educ.,* 48, 509, 1971.

42. **Graovac, A., Gutman, J., and Trinajstić, N.,** *Topological Approach to the Chemistry of Conjugated Molecules;* Lecture Notes in Chemistry, Vol. 4, Springer-Verlag, Berlin, 1977.

43a. **Klages, A.,** Cited after References 18 and 30.

43b. **Gordon, M. and Davison, W. H. T.,** *J. Chem. Phys.,* 20, 428, 1952.
44. **Wilcox, C. F., Jr.,** *Tetrahed. Lett.,* p.795, 1968; *J. Am. Chem. Soc.,* 91, 2732, 1969.
45. **Herndon, W. C.,** *Tetrahedron,* 29, 3, 1973; *Tetrahed. Lett.,* p.671, 1974; *J. Chem. Educ.,* 51, 10, 1974.
46. **Gutman, I. and Trinajstić, N.,** *Croat. Chim. Acta,* 45, 423, 1973; 45, 539, 1973; 47, 95, 1975; **Crethovic, D., Gutman, I., and Trinajstić, N.,** *J. Chem. Phys.,* 61, 2700, 1974.
47. Aromaticity, in *Int. Symp. Sheffield,* Spec. Publ., Chemical Society, London, 1967, 21.
48. **Bergmann, E. D. and Pullman, B., Eds.,** Aromaticity, pseudo-aromaticity, anti-aromaticity, in *Int. Symp. Jerusalem,* Israel Academy of Science, Humanities, 1971.
49. Theoretical organic chemistry, in *Kekulé Symp.,* Chemical Society, London, 1959.
50. **Graovac, A. and Trinajstić, N., Eds.,** *Int. Symp. Aromaticity,* Dubrovnik, 1979; *Pure Appl. Chem.,* 52 (6), 1980; **Trinajstić, N., Graovac, A., and Babić, D.,** *Croat. Chim. Acta,* 56 (2 and 3), 1983.
51. **Agranat, I., Ed.,** *4th Int. Symp. Chem. of Novel Aromatic Compounds,* Jerusalem, 1981; *Pure Appl. Chem.,* 54 (5), 1982.
52. **Ginsburg, D., Ed.,** *Non-Benzenoid Aromatic Compounds,* Interscience, New York, 1959.
53. **Lloyd, D. M. G., Ed.,** *Carbocyclic Non-Benzenoid Aromatic Compounds,* Elsevier, Amsterdam, 1966.
54. **Badger, G. M., Ed.,** *Aromatic Character and Aromaticity,* Cambridge University Press, 1969.
55. **Clar, E. Ed.,** *The Aromatic Sextet,* John Wiley & Sons, London, 1972.
56. **Snyder, J. P., Ed.,** *Nonbenzenoid Aromatics,* Academic Press, New York, 1969; Vol. 2, 1971.
57. **Nozoe, T., Breslow, R., Hafner, K., Ito, S., and Murata, I. I., Eds.,** *Topics in Nonbenzenoid Aromatic Chemistry,* Vol. 1, Hirokawa, Tokyo, 1972; Kekulé Centennial, *Am. Chem. Soc. Adv. Chem. Ser.,* 61, 1966.
58. **Bird, C. W. and Cheeseman, G. W., Eds.,** *Aromatic and Heteroaromatic Chemistry,* Vols. 1 to 5, Specialist Periodical Reports, Chemical Society, London, 1973 to 1977.
59. Annual Reports on the Progress of Chemistry (B), to 80, 1983, The Chemical Society, London; chapters on Aromatic Compounds and Heterocyclic Compounds.
60. **Labarre, J. F.,** in *Int. Symp. Jerusalem,* Israel Academy of Science, Humanities, 1971, 55, 56; see also **Labarre, J. F. and Crasnier, F.,** *Top. Curr. Chem.,* 24, 33, 1971.
61. **Heilbronner, E.,** in *Int. Symp. Jerusalem,* Israel Academy of Science, Humanities, 1971, 21.
62. **Lloyd, D. and Marshall, D. R.,** in *Int. Symp. Jerusalem,* Israel Academy of Science, Humanities, 1971, 87.
63. **Rassat, A.,** *Tetrahed. Lett.,* p.4081, 1975.
64. **Cook, M. J., Katritzky, A. R., and Linda, P.,** *Adv. Heterocyclic Chem.,* 17, 255, 1974.
65. **Bergmann, E. and Agranat, J.,** in *Int. Symp. Jerusalem,* Israel Academy of Science, Humanities, 1971, 9.
66. **Erlenmeyer, E.,** *Liebigs Ann. Chem.,* 137, 327, 1866.
67. **Elvidge, J. A. and Jackman, L. M.,** *J. Chem. Soc.,* p.859, 1961.
68. **Abraham, R. J., Sheppard, R. C., Thomas, W. A., and Turner, S.,** *Chem. Commun.,* p.43, 1965.
69. **Pople, J. A.,** *J. Chem. Phys.,* 41, 2559, 1964.
70. **Musher, J. I.,** *J. Chem. Phys.,* 43, 4081, 1965.
71. **Craig, D. P.,** in *Kekulé Symp.,* Chemical Society, London, 1959, 20; **Ginsburg, D., Ed.,** *Non-Benzenoid Aromatic Compounds,* Interscience, New York, 1959, 1.
72. **Dauben, H. J., Wilson, J. D., and Laity, J. L.,** *J. Am. Chem. Soc.,* 90, 811, 1968; 91, 1991, 1969; **Snyder, J. P., Ed.,** *Nonbenzenoid Aromatics,* Vol. 2, Academic Press, New York, 1969, 167.
73. **Labarre, J. F. and Crasnier, F.,** *J. Chim. Phys.,* 64, 1664, 1967; see also Reference 60.
74. **Laszlo, P. and Schleyer, P., von R.,** *J. Am. Chem. Soc.,* 85, 2017, 1963.
75. **Günther, H.,** *Tetrahed. Lett.,* p.2960, 1967.
76. **Gutman, I., Milun, M., and Trinajstić, N.,** *Math. Chem.,* 1, 171, 1975; *J. Am. Chem. Soc.,* 99, 1692, 1977; see also References 23 and 42.
77. **Aihara, J.,** *J. Am. Chem. Soc.,* 98, 2750, 1976; 99, 2048, 1977; **Mizoguchi, N.,** *J. Am. Chem. Soc.,* 107, 4419, 1985.
78. **Hess, B. A., Jr. and Schaad, L. J.,** *J. Am. Chem. Soc.,* 93, 305, 1971; **Schaad, L. J. and Hess, B. A., Jr.,** *J. Am. Chem. Soc.,* 94, 3068, 1972; see also Reference 24.
79. **Albert, A.,** in *Heterocyclic Chemistry,* The Athlone Press, London, 1959, 201.
80. **Julg, A.,** in *Int. Symp. Jerusalem,* Israel Academy Science, Humanities, 1971, 383; **Julg, A. and François, P.,** *Theor. Chim. Acta,* 8, 249, 1967; *Tetrahedron,* 21, 717, 1965.
81. **Leroy, G. and Jaspers, S.,** *J. Chem. Phys.,* 64, 470, 1967.
82. **Kemula, W. and Krygowski, T. M.,** *Tetrahed. Lett.,* p.5135, 1968.
83. **Jug, K.,** *Jerg. Chem.,* 48, 1344, 1983.
84. **Binsch, G. and Heilbronner, E.,** *Tetrahedron,* 24, 1215, 1968; **Binsch, G.,** in *Int. Symp. Jerusalem,* Israel Academy Science, Humanities, 1971, 25.
85. **Kruszewski, J. and Krygowski, T. M.,** *Tetrahed. Lett.,* p.319, 1970; **Krygowski, T. M.,** *Tetrahedron Lett.,* p.1311, 1970.

86. **Dixon, W. T.,** *J. Chem. Soc. B,* p.612, 1970.
87. **Anet, F. A. L. and Schenck, G. E.,** *J. Am. Chem. Soc.,* 93, 556, 1971.
88. **Jones, A. J.,** *Rev. Pure Appl. Chem.,* 18, 23, 1968.
89. **Mallion, R. B.,** *J. Mol. Spectrosc.,* 35, 491, 1970; *J. Chem. Phys.,* 75, 793, 1981; 76, 4063, 1982; *Pure Appl. Chem.,* 52, 1541, 1980; **Haigh, C. W. and Mallion, R. B.,** *Prog. NMR Spectrosc.,* 13, 257, 1979.
90. **Abraham, R. J. and Thomas, W. A.,** *J. Chem. Soc. B,* p.127, 1966.
91. **Streitwieser, A.,** *Molecular Orbital Theory for Organic Chemists,* John Wiley & Sons, New York, 1961.
92. **Dewar, M. J. S.,** in *Int. Symp. Sheffield,* Chemical Society, London, 1967, 178.
93. **Longuet-Higgins, H. C. and Salem, L.,** *Proc. R. Soc.,* A251, 172, 1959; A257, 445, 1960; see also **Coulson, C. A. and Dixon, W. T.,** *Tetrahedron,* 17, 215, 1962; **Dewar, M. J. S. and Gleicher, G. J.,** *J. Am. Chem. Soc.,* 87, 685, 1965; **Müllen, K.,** *Chem. Rev.,* 84, 603, 1984; **Vogler, H.,** *Tetrahedron,* 41, 5383, 1985; *Croat. Chem. Acta,* 57, 1177, 1984.
94. **Breslow, R., Brown, J., and Gajewski, J. J.,** *J. Am. Chem. Soc.,* 89, 4383, 1967.
95. **Randić, M.,** *Tetrahedron,* 33, 1905, 1977; *Mol. Phys.,* 34, 849, 1977; see also Reference 23, Vol. 2, chap. 3.
96. **Gutman, I. M.,** *Bull. Soc. Chim. Biograd.,* 43, 191, 1978.
97. **Sidman, J. W.,** *J. Am. Chem. Soc.,* 78, 4217, 1956.
98. **Allinger, N. L. and Yuh, Y. H.,** *Pure Appl. Chem.,* 55, 191, 1983.
99. **Kabuto, C. and Oda, M.,** *Tetrahed. Lett.,* p.103, 1980.
100. **Balaban, A. T.,** *Rev. Roum. Chim.,* 17, 1531, 1972.
101. **Bochvar, D. A., Stankevich, I. V., and Tutkevich, A. V.,** *Izv. Akad. Nauk SSSR, Otd. Khim. Nauk,* p.1185, 1969.
102. **Dewar, M. J. S. and Dougherty, R. C., Eds.,** *The PMO Theory of Organic Chemistry,* Plenum Press, New York, 1975; **Dewar, M. J. S.,** *Pure Appl. Chem.,* 52, 1431, 1980.
103. **Balaban, A. T. and Harary, F.,** *Tetrahedron,* 24, 2505, 1968.
104. **Balaban, A. T.,** *Tetrahedron,* 25, 2949, 1969.
105. **Klarner, D.,** *Fibonacci Q.,* 3, 9, 1965; **Lunnon, W. F.,** in *Graph Theory and Computing,* Read, R. C., Ed., Academic Press, New York, 1972, 87.
106. **Trinajstić, N., Jericević, Z., Knop, J. V., Müller, W. R., and Szymanski, K.,** *Pure Appl. Chem.,* 55, 379, 1983; **Knop, J. V., Szymanski, Z., Jericević, Z., and Trinajstić, N.,** *J. Comput. Chem.,* in press; **Szymanski, K.,** *Polyhexagons der Ordnungen 1-10 klassifiziert nach der Anzahl der inneren Knoten,* University Düsseldorf, 1982; **Knop, J. V., Müller, W. R., Szymanski, K., and Trinajstić, N.,** *Computer Generation of Certain Classes of Molecules,* Kemija u industriji, Zagreb, 1985. **Knop, J. V., Szymanski, K., Klasinc, L., and Trinajstić, N.,** *Comput. Chem.,* 8, 107, 1984.
107. **Dias, J. R.,** *J. Chem. Inf. Comput. Sci.,* 22, 15, 139, 1982; *Math. Chem.,* 14, 83, 1983.
108. **Bonchev, D. and Balaban, A. T.,** *J. Chem. Inf. Comput. Sci.,* 21, 2, 1981.
109. **Cioslowski, J. and Turek, A. M.,** *Tetrahedron,* 40, 2161, 1984.
110. **Elk, S. B.,** *Math. Chem.,* 8, 121, 1980; 13, 239, 1982.
111. **Balaban, A. T.,** *Rev. Roum. Chim.,* 26, 407, 1981; *Pure Appl. Chem.,* 54, 1075, 1982.
112. **Dewar, M. J. S. and Longuet-Higgins, H. C.,** *Proc. R. Soc. (London),* A214, 482, 1952; **Ham, N. S. and Ruedenberg, K.,** *J. Chem. Phys.,* 29, 1215, 1958.
113. **Gutman, I. and Trinajstić, N.,** *Croat. Chem. Acta,* 47, 35, 1975.
114. **Gutman, I., Randic, M., and Trinajstić, N.,** *Rev. Roum. Chim.,* 23, 383, 1978.
115. **Sachs, H.,** *Publ. Math. (Debrecen),* 11, 119, 1964; **Graovac, A., Gutman, I., Trinajstić, N., and Zivković, T.,** *Theor. Chim. Acta,* 26, 67, 1972; *Chemical Graph Theory,* Vol. 1, CRC Press, Boca Raton, Fla., 1983, chap. 5.
116. **Balasubramanian, K.,** *Theor. Chim. Acta,* 65, 49, 1984.
117. **Sondheimer, F. et al.,** in *Int. Symp. Sheffield,* Chemical Society, London, 1967, 75.
118. **Aihara, J.,** *Pure Appl. Chem.,* 54, 1115, 1982.
119. **Breslow, R., Napienski, J., and Clarke, T. C.,** *J. Am. Chem. Soc.,* 97, 6275, 1975.
120. **Gordon, M. and Davison, W. H. T.,** *J. Chem. Phys.,* 20, 428, 1952.
121. **Aihara, J.,** *Bull. Chem. Soc. Jpn.,* 50, 2010, 1977; 51, 2729, 1978.
122. **Yen, T. F.,** *Theor. Chim. Acta,* 20, 399, 1971.
123. **Gutman, I.,** *Croat. Chim. Acta,* 55, 371, 1982; *Math. Chem.,* 13, 173, 1982; 11, 127, 1981; **Gutman, I. and Cvetkovic, D.,** *Math. Chem.,* 46, 15, 1974.
124. **Gutman, I. and Hosoya, H.,** *Theor. Chim. Acta,* 48, 279, 1978.
125. **Herndon, W. C.,** *J. Chem. Educ.,* 51, 10, 1979.
126. **Dzonova-Jerman-Blazić, B. and Trinajstić, N.,** *Croat. Chim. Acta,* 55, 374, 1982; *Comput. Chem.,* 6, 121, 1982.
127. **Polansky, O. E. and Gutman, I.,** *Math. Chem.,* 8, 269, 1980.
128. **Randic, M.,** *Tetrahedron,* 31, 1477, 1975; *Pure Appl. Chem.,* 52, 1587, 1980.

129. **El-Basil, S.,** *Chem. Phys. Lett.,* 89, 145, 1982.
130. **Balaban, A. T. and Tomescu, I.,** *Math. Chem.,* 14, 155, 1983; *Croat. Chem. Acta,* 54, 391, 1985.
131. **Trinajstić, N.,** *Chemical Graph Theory,* Vol. 2, CRC Press, Boca Raton, 1983, chap. 2.
132. **Clar, E., Ed.,** *Polycyclic Hydrocarbons,* Academic Press, London, 1964.
133. **Polansky, O. S. and Derflinger,** *Int. J. Quantum Chem.,* 1, 379, 1967.
134. **Dewar, M. J.,** *Angew. Chem. Int. Ed. Engl.,* 10, 761, 1971.
135. **Kruszewski, J.,** *Acta Chim. Lodz,* 16, 77, 1971.
136. **Gutman, I. and Bosanac, S.,** *Tetrahedron,* 33, 1809, 1977.
137. **Aihara, J.,** *J. Am. Chem. Soc.,* 99, 2048, 1977.
138. **Randić, M.,** *J. Am. Chem. Soc.,* 99, 444, 1977; *Tetrahedron,* 30, 2067, 1974; 33, 1905, 1977; *Mol. Phys.,* 34, 849, 1977.
139. **Memory, J. D.,** *J. Chem. Phys.,* 38, 1341, 1963; **Haigh, C. W. and Mallion, R. B.,** *Mol. Phys.,* 18, 767, 1970.
140. **Zander, M.,** *Math. Chem.,* 14, 183, 1983.
141. **Zander, M.,** in *Handbook of Polynuclear Aromatic Hydrocarbons,* Bjørseth, A., Ed., Marcel Dekker, New York, 1983; *Z. Naturforsch.,* 33a, 1398, 1978.
142. **Balaban, A. T.,** *Rev. Roum. Chim.,* 15, 1243, 1970.
143. **Sahini, V. E.,** *J. Chim. Phys.,* p.177, 1962; *Rev. Chim. Acad. R. P. Roum.,* 7, 1265, 1962.
144. **Balaban, A. T., Biermann, D., and Schmidt, W.,** *Tetrahedron,* 40, 100, 1984.
145. **Biermann, D. and Schmidt, W.,** *J. Am. Chem. Soc.,* 102, 3163 and 3173, 1980.
146. **Biermann, D. and Schmidt, W.,** *Isr. J. Chem.,* 20, 312, 1980.
147. **Clar, E. and Zander, M.,** *J. Chem. Soc.,* p.4616, 1957; **Zander, M.,** *Angew. Chem.,* 72, 513, 1960.
148. **Zander, M.,** *Z. Naturforsch.,* 33a, 1395, 1978.
149. **Gundermann, K. D., Lohberger, C., and Zander, M.,** *Z. Naturforsch.,* 36a, 276, 1217, 1981; *Croat. Chem. Acta,* 56, 357, 1983.
150. **Phillips, D. H. and Sims, P.,** in *Chemical Carcinogens and DNA,* Vol. 2, Grover, P. L., Ed., CRC Press, Boca Raton, 1979, 29.
151. **Jerina, D. M.,** *Pure Appl. Chem.,* 50, 1033, 1978; **Jerina, D. M. and Daly, J. W.,** in *Drug Metabolism,* Parke, D. V. and Smith, R. L., Eds., Taylor & Francis, London, 1976; **Yagi, H., Hernandez, O., and Jerina, D. M.,** *J. Am. Chem. Soc.,* 97, 688, 1975.
152. **Gelboin, H. and Ts'o, P. O. P., Eds.,** *Polycyclic Hydrocarbons and Cancer,* Academic Press, New York, 1978.
153. **Dipple, A.,** in *Chemical Carcinogens,* A.C.S. Monograph 173, Searle, C. E., Ed., American Chemical Society, Washington, D.C., 1976, 245.
154. **Belard, F. A. and Harvey, R. G.,** *J. Chem. Soc. Chem. Commun.,* p.84, 1976.
155. **Tsang, W. S. and Griffin, G. W., Eds.,** *Metabolic Activation Polynuclear Aromatic Hydrocarbons,* Pergamon Press, Oxford, 1979.
156. **Pullman, A. and Pullman, B.,** *Adv. Cancer Res.,* 3, 117, 1955; *Physico-Chemical Mechanisms of Carcinogenesis,* Proc. 1st Jerusalem Symp., Bergmann, E. D. and Pullman, B., Eds., Academic Press, New York, 1969, 9, 325.
157. **Daudel, P. and Daudel, R.,** *Biol. Med.,* 39, 201, 1950.
158. **Pullman, B.,** *Int. J. Quantum Chem.,* 16, 669, 1979.
159. **Balasubramanian, K., Kaufman, J. J., Koski, W. S., and Balaban, A. T.,** *J. Comput. Chem.,* 1, 149, 1980.
160. **Epiotis, N. D.,** *Nouv. J. Chim.,* 8, 11, 1984; *Pure Appl. Chem.,* 55, 229, 1983; *Unified Valence Bond Theory of Electronic Structure. Applications, Lecture Notes in Chemistry,* 34, 358, 1983; **Shaik, S. S. and Hiberty, P. C.,** *J. Am. Chem. Soc.,* 107, 3089, 1985; **Hiberty, P. C., Shaik, S. S., Lefour, J. M., and Ohanessian, G.,** *J. Org. Chem.,* 50, 4659, 1985; **Shaik, S. S. and Bar, R.,** *Nouv. J. Chim.,* 8, 411, 1984.
161. **Bochvar, D. A. and Galpern, G. E.,** *Dok. Akad. Nauk SSSR,* 209, 610, 1973; **Davidson, R. A.,** *Theor. Chim. Acta,* 58, 193, 1981; **Haymet, A. D. J.,** *J. Am. Chem. Soc.,* 108, 319, 1986; **Klein, D. J., Schmalz, T. G., Seitz, W. A., and Hite, G. E.,** *J. Am. Chem. Soc.,* 108, 1301, 1986.
162. **Kroto, H. W., Heath, J. R., O'Brien, S. C., Curl, R. F., and Smalley, R. E.,** *Nature London,* 318, 162, 1985; **Zhang, Q. L. et al.,** *J. Phys. Chem.,* 90, 525, 1986.
163. **Heath, J. R.** et al., *J. Am. Chem. Soc.,* 107, 7779, 1985.
164. **Glukhovtsev, M. N., Simkin, B. I., and Minkin, V. I.,** *Vsp. Khim.,* 54, 86, 1985.

Chapter 3

GRAPH THEORY AS A KEY TOOL FOR THE DEFINITION AND ENUMERATION OF VALENCE ISOMERS

I. GRAPH-THEORETICAL PRELIMINARIES

Since the discussion in the present and subsequent chapters is based on graph theory, a few elementary graph-theoretical notions will be introduced, and a brief description of the interplay between graph theory and chemistry will be sketched. For more detailed accounts of graph-theoretical definitions, books[1-4] and reviews[5-7] are available.

Graph theory is a branch of mathematics closely related to topology and combinatorial analysis; it studies graphs that can be visualized as collections of points (which may also be called vertices) and lines joining these vertices (lines may also be called edges or bonds). Shapes, lengths, or angles of lines are irrelevant (as in topology, which may be regarded as geometry in a rubber space or on a rubber sheet). The simplest way to draw edges is as straight lines, but this is not mandatory.

Organic chemists learn many aspects of graph theory unconsciously when they are taught how to find constitutional isomers. Their approach is not rigorous, however: instead of saying: "butane and ethanol have each two isomers because I cannot draw more than two formulas in each case" (maybe somebody else can?) it would be better to say: " . . . because one can demonstrate that in each of these cases only two graphs can exist". Organic chemists feel as surprised by the ease with which they can master graph theory as Molière's main character in *Le Bourgeois Gentilhomme* when he realized that he was speaking prose.

Three important graph parameters are the number n of vertices (sometimes called the "order" of the graph), the number q of edges, and the number μ of fundamental rings (cycles); the latter number, also called "cyclomatic number", is $\mu = q - n + 1$.

The mathematical definition of a graph is "the application of a set on itself"; the elements of the set are the vertices, and the nonordered binary relations (pairs of vertices) specifying the application on itself are the edges. Should the pairs of vertices be ordered, we would obtain a "directed graph" (or, more briefly, a "digraph"). Though digraphs can be applied in chemistry, e.g., in chemical kinetics, they will not be further discussed here.

Two vertices joined by an edge are said to be "adjacent". Two edges in a simple graph meeting at the same vertex are also said to be adjacent. An edge is "incident" to the vertices it joins. If any two vertices are joined by at most one line, we have a "simple graph"; if multiple (double, triple) edges connect some pairs of vertices, we have a "multigraph". If one vertex may be joined to itself we have a "loop-graph". Graphs which are the same time loop-graphs and multigraphs are called "general graphs" or "pseudographs". We illustrate some of the above definitions with examples involving graphs of order six (with six vertices): **1, 5, 6,** and **7** are simple graphs; **2** is a multigraph, **3** is a loop graph, and **4** is a pseudograph (general graph).

In rare applications, double bonds or loops are counted as two- and one-membered rings, respectively, whereas in most applications the cyclomatic number μ is calculated without counting separately multiple edges or loops (one ignores the loops, and the multiple edges are counted as single edges in the formula for μ). We say therefore that benzene is a monocyclic compound with $\mu = 1$, and that Dewar-benzene is bicyclo[2.2.0]hexa-2,5-diene with $\mu = 2$. It is important to realize that in polycyclic compounds with $\mu \geq 2$ the number of all rings is actually larger than the number μ of fundamental rings; thus Dewar-benzene with $\mu = 2$ has two four-membered rings (fundamental rings) and a six-membered ring.

Two graphs are "isomorphic" if there exists a one-to-one correspondence between their point sets which preserves adjacency; therefore one may draw a graph in many different ways which will all correspond to isomorphic graphs.

A "path" of a graph is a sequence, beginning and ending with points, in which incident vertices and edges alternate, and in which all points are distinct. If the initial and final points of a path coincide, we have a "circuit" (cycle or ring). A loop may be considered sometimes as a one-membered ring, and a double bond as a two-membered ring.

A graph is "connected" if each possible pair of vertices can be joined by a path, and "disconnected" otherwise. A connected graph is "separable" if it becomes disconnected by deleting one edge (cut-edge) or one vertex (cut-point); spiro-compounds are examples for the latter case. A graph with no cycles is called "tree" if it is connected, and a "forest" if it is disconnected: **5** and **6** are two isomorphic trees (vertices are numbered so as to indicate how to preserve adjacencies), while **7** is a forest. Graphs **2** to **6** and **8** are connected ones, while **1** and **7** are disconnected graphs.

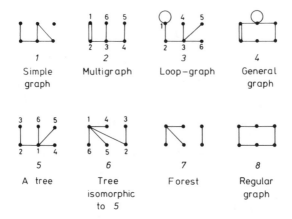

The "degree" of a point is the number of lines incident with it; a loop is counted twice in the degree, e.g., **3** has vertices 4, 5, and 6 of degree one ("endpoints"), vertex 2 of degree 2, vertex 1 of degree 3, and vertex 3 of degree 4. In a "regular graph", all points have the same degree, e.g., **8** is regular of degree 2 but all other graphs (**1** to **7**) are irregular, i.e., their vertices have different degrees. Thus in **2**, 5 and 6 are endpoints (degree 1), vertices 1 and 4 have degree 2, and vertices 2 and 3 have degree 3. Regular graphs of degree 3 are called "cubic graphs".

The "partition" of a graph is the sequence of degrees of its points, often listed in nondecreasing order. Thus the partitions of graphs **2**, **3**, and **8** are 1,1,2,2,3,3; 1,1,1,2,3,4; and 2,2,2,2,2,2, respectively. Since every line is incident with two vertices, it is counted twice in the sum of all degrees. We thus arrive at the first theorem of graph theory due to Euler: the sum of all degrees of graph vertices is twice the number of edges. One can verify in the above three cases that the sums in the partitions equal $2 \times 6 = 12$ for **2**, **3**, and **8**. Two important corrollaries of Euler's theorem are that the number of vertices of odd degree must be even, and that cubic graphs, e.g., **10** and **12** have even numbers n of vertices.

The "distance" (or topological distance) between two points is the number of lines in the shortest path joining them; in trees, two vertices are joined by a unique path. Two graphs are "homeomorphic" if both yield the same graph on ignoring all points of degree 2 (excepting endpoints of degree 2 in multigraphs with a marginal double bond), e.g., **9** is homeomorphic to **10**.

A "subgraph" of a given graph with n points and q lines has either an order $\leq n$ or a number of lines $\leq q$; a "spanning subgraph" has n points but fewer than q lines. Such a spanning subgraph is the "Hamiltonian circuit" in the case of cyclic graphs, i.e., a cycle containing all points of the given graph. In a "complete graph" K_n (see graphs **10** and **11**) all pairs of points are adjacent; it can be easily shown that complete graphs of order n are regular of degree $n - 1$. A "bipartite" or "alternant graph" ("bigraph") has just two subsets of vertices V_1 and V_2 so that every line joins vertices belonging to different subsets. Every tree is bipartite and so (according to König's theorem) are all graphs devoid of odd-membered rings. If each vertex of one subset (k vertices) is joined to each vertex of the other subset (j vertices, $j + k = n$) we have a complete bigraph $K_{j,k}$, e.g., **12**.

A "planar graph" can be drawn on a plane so that it has no crossing lines. Kuratowski has demonstrated that the necessary and sufficient condition for any graph to be planar is not to have a subgraph homeomorphic to the complete graph K_5 or to the complete bigraph $K_{3,3}$, **11** and **12**, respectively.

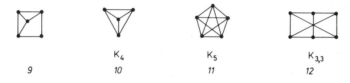

	K_4	K_5	$K_{3,3}$
9	*10*	*11*	*12*

II. INTERPLAY BETWEEN GRAPH THEORY AND CHEMISTRY

Immediately after Kekulé's foundation of structure theory based on the postulate of carbon having valency four, constitutional formulas became the tool through which chemists tried to rationalize the known phenomenon of isomerism. The valence bond introduced by Couper and Crum Brown led to easily understandable formulas. Butlerov showed that the number of predicted isomeric butanols agreed with chemical reality by preparing the still unknown *tert*-butanol.

It was easy to see that constitutional formulas of covalent compounds are graphs where points correspond to atoms, and lines to covalent bands. In particular, hydrocarbons such as isobutane **13** are graphs wherein points have either degree four (carbon atoms) or one (hydrogens); however, chemists introduced a significant simplification by using hydrogen-depleted graphs and specifying only the carbon atoms. In this case, which will be used exclusively henceforth, isobutane is depicted as **14**; one sees that for alkanes $C_nH_{2n + 2}$ hydrogen-depleted graphs have order n instead of $3n + 2$ as in graphs such as **13** containing hydrogens as endpoints. It is also easy to see that in hydrogen-depleted graphs of hydrocarbons the degrees of vertices are no longer either 1 or 4, but may take all integer values in this range. The number of hydrogen atoms attached to a given carbon atom is 4 minus the degree of the corresponding vertex.

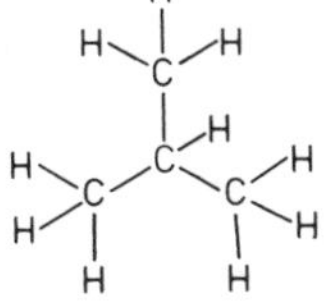

13    *14*

Mathematicians were challenged to try and find a formula or an algorithm for the number of constitutionally isomeric alkanes (and later, after Le Bel and Van't Hoff developed stereochemistry, of enantiomeric alkanes). In particular, Sylvester[8,9] (who coined the term "graph") and Cayley discussed this problem in the context of enumerating trees with given numbers of vertices. Cayley found that no algebraic formula was possible but developed an algorithm and a recursive formula, first for enumerating "rooted trees",[10] then with restric-

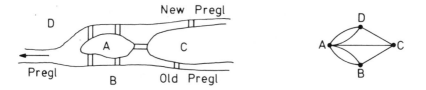

FIGURE 1.   The seven Königsberg bridges in 1736 and the corresponding graph.

tions of vertex degrees (≤ 4) for enumerating monosubstituted alkanes;[9,10] only about 17 years later did he succeed in finding an algorithm for enumerating constitutionally isomeric alkanes,[11] i.e., unrooted trees with degrees ≤ 4. Finally, after another 15 years,[12] he found an extremely simple formula for enumerating all t_n "labeled trees" with n vertices: $t_n = n^{n-2}$.

Cayley's method for enumerating hydrocarbons was cumbersome and impractical, so that it made little impact on chemists despite further attempts[13-15] to improve it during 1880 to 1920.

In the late 1920s two papers which were published by Redfield[16] and by Lunn and Senior[17] foreshadowed the fundamental development that was to come through Pólya.

During the 1930s two American chemists, Henze and Blair, used recursive formulas to enumerate various homologous series of chemical isomers.[18-20] Today, by means of computers, this method is quite fast but then it involved lengthy computations and was always endangered by errors.

The breakthrough came when Pólya achieved through his theorem[21] a unification of previous efforts by Burnside, Redfield, and Senior. The chemical and mathematical implications of this theorem were reviewed[1b,2,22] and will be discussed in more detail in a subsequent section of this chapter.

We end this historical overview by mentioning that Otter[23] developed a powerful mathematical formula for enumerating trees, that an alternative approach to chemical enumeration problems was developed by Ruch[24] (which is sometimes simpler to apply[25] than Pólya's theorem), and that Robinson et al.[26] solved one of the last remaining problems in tree enumeration (achiral trees).

III. CHEMISTRY AS A BREEDING GROUND FOR GRAPH THEORY

Historically, graph theory was born from the merging of three independent sources: (1) a mathematical one illustrated by the famous problem of the "Königsberg bridges". Königsberg (today's Kaliningrad) lies on the river Pregl (Pregolya) which is formed by the merging of the New and Old Pregl rivers (Figure 1); an island (Kneiphof) is formed; this island (A) and the three land portions B, C, and D delineated by the river were at that time joined by seven bridges. The problem was whether one might walk and cross once and only once each bridge. Euler published the demonstration that the problem had no solution; this paper marks the birthdate of graph theory, and its date (1736, when Euler was 29 years old) is in the title of a book with excerpts and comments on the most important papers in graph theory (a chapter thereof is entitled "Chemical Graphs").[9]

Euler showed that the problem is soluble only for graphs having no more than two vertices of odd degree (refer to Figure 7); these vertices are then the start and endpoint of the walk.

The second source is represented by Kirchhoff's investigations on the flow of electricity in a network of wires. The network is a graph. Kirchhoff proved that the number of fundamental cycles in any graph is the cyclomatic number μ.

The third source is the problem of chemical isomerism which attracted Sylvester (he introduced the term "graph theory") and especially Cayley.

Table 1
THE 11 ISOMERS C_4H_4 AND THE 4 CLASSES OF VALENCE ISOMERS

Partition	Valence isomers
3,3,3,3	
2,3,3,4	
2,2,4,4	
1,3,4,4	

A few excerpts from Sylvester's writings close this historical review on the links between chemistry and graph theory.

By the *new* atomic theory, I mean that sublime invention of Kekulé which stands to the *old* in a somewhat similar relation as the astronomy of Kepler to Ptolemy's or the system of Nature of Darwin to that of Linnaeus. Like the latter, it lies outside of the immediate sphere of energetics, basing its laws on pure relation of form and like the former as perfected by Newton, these laws admit of exact arithmetical definitions . . .

Every invariant and covariant thus becomes expressible by a *graph* precisely identical with a Kekuléan diagram or chemicograph . . .

Chemistry has the same quickening and suggestive influence upon the algebraist as a visit to the Royal Academy, or the old masters may be supposed to have on a Browning or a Tennyson. Indeed it seems to me that an exact homology exists between painting and poetry on the one hand and modern chemistry and modern algebra on the other. In poetry and algebra we have the pure idea elaborated and expressed through the vehicle of language, in painting and chemistry the idea enveloped in matter, depending in part on manual processes and the resources of art for its due manifestation.

Thus, organic chemistry can be considered as the third midwife for the birth of graph theory. Today it is the turn of graph-theoretical methods and results to assist in solving some newer problems raised by chemistry, such as the definition and enumeration of valence isomers.

IV. DEFINITION OF VALENCE ISOMERS

In chemistry, valence isomers are usually considered to be isomers differing only by rearranging σ and/or π bonds, without disturbing atom groupings.[27a,b] Therefore, an alternative name, bonding isomers, is occasionally used. Cope,[27a] who introduced the term "valence (or valence bond) isomerization", defined it as follows: "Only electronic displacements corresponding to interconversion of double and single bonds occur in this isomerization, with the corresponding changes in bond angles and distances."

This definition lacks precision and admits different interpretations, leaving doubt as to whether *o*-, *m*-, and *p*-disubstituted benzenes are to be considered valence isomers; are 2H- and 4H-pyran valence isomers? Is bicyclopropenyl a valence isomer of benzene?

The graph-theoretical definition of valence isomers is the following:[28-30] constitutional formulas (hydrogen-depleted graphs) with the same partition (of vertex degrees) correspond to valence isomers. On the basis of this definition the reply to all three questions raised above is yes. One can also see that **15** and **16** have, among the isomers of hexane, the same partition of vertex degrees: 1,1,1,2,2,3; hence they are valence isomers. All 11 graphs which correspond to formula C_4H_4, displayed in Table 1 are isomeric and can be further grouped into four classes of valence isomers.[30] Owing to steric strain, many of these are unstable species.

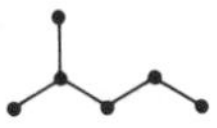

15    16

Based on the above graph-theoretical definition, we consider 1,2-dihydro-1,2-di(cyclopropenyl)benzene and the corresponding 1,4-derivative to be valence isomers $(CH)_{12}$; there is no 1,3-counterpart, for the same graph-theoretical reason which precludes the existence of a *meta*-benzoquinone: in bipartite (alternant) graphs, quinonic oxygens must be bonded to a starred and an unstarred carbon atom.

It is easy to see that alkanes C_nH_{2n+2} have $q = n - 1$ edges and no cycles. Their cyclomatic number $\mu = q - n + 1$ is zero. Whenever two hydrogen atoms are removed from an alkane to form a cyclic graph, a new bond is added, so that a hydrocarbon C_nH_{2k} will have $q = n - 1 + (2n + 2 - 2k)/2 = 2n - k$ carbon-carbon bonds. If the corresponding graph is a simple graph, then this hydrocarbon C_nH_{2k} will have $\mu = n - k + 1$ rings; if the hydrocarbon has multiple bonds, then μ will be the number of rings plus double bonds (a triple bond being counted as two double bonds). In particular for C_4H_4, $q = 8 - 2 = 6$ edges and $\mu = 3$ rings and/or double bonds.

One can see that valence isomers of [n]annulenes $(CH)_n$ are cubic graphs with n vertices,[28] where according to the corrollary of Euler's theorem, n is an even number.

V. FINDING ALL POSSIBLE VALENCE ISOMERS OF ANNULENES $(CH)_n$ AND OF HOMOANNULENES

The enumeration and construction of cubic graphs of order n, i.e., of the constitutional valence isomers of [n]annulenes, was an unsolved mathematical problem in 1966 when a constructive algorithm was developed.[28] A formula (1) has been found[22,31] for the number K of general cubic graphs with $n = 2k$ points, including disconnected general graphs:

$$K = Z(S_{2k}[S_3]) \cap Z(S_{3k}[S_2])   \tag{1}$$

In this formula, Z is the cycle index according to Pólya's theory for the permutation groups, thus

$$Z(S_3) = (s_1^3 + 3s_1s_2 + 2s_3)/6$$

for the symmetric permutation group S_3; similarly, $Z(S_2) = (s_1^2 + s_2)/2$ for the symmetric permutation group S_2; in general,[32]

$$Z(S_p) = \frac{1}{p!} \sum_{\kappa} \frac{p!}{\prod_k k^{j_k} j_k!} s_1^{j_1} s_2^{j_2} \ldots s_p^{j_p}$$

where the sum is over all partitions (j) of p.

Redfield's operation cap $\cap$ indicates that whenever two identical monomials appear in the two Z terms of Formula 1, their coefficients are to be multiplied and these monomials are to be combined so as to afford a number.

For instance it can be shown that

$$Z(S_2[S_3]) \cap Z(S_3[S_2]) = 2$$

in agreement with the two possible general cubic graphs on two points **17** and **18**.

17   18

In order to apply Formula 1 to higher k values, one needs a computer program,[33] and one has also to subtract the disconnected graphs for which a recurrent formula can be devised. This approach serves as a check-up for the correctness of the numbers obtained from the constructive algorithm.

For chemists, a more useful approach is the constructive algorithm developed by Balaban.[28,29] In order to obtain all possible cubic multigraphs on $2k$ vertices,[28] one has to obtain all connected general graphs with $2k - 2$ vertices.[29] Then one applies Operations 1 and 2 which increase by two the number of vertices.

Operation 1: On any line(s) of the graph, mark two new vertices and join them by a new line.

Operation 2: on any line of the graph mark a new vertex and join it to a loop.

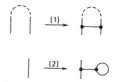

These operations must be performed for every graph with $n - 2$ vertices, testing each line for Operation 2 and each line or pair of lines for Operation 1. Redundancies, i.e., isomorphic graphs, must be discarded, and one obtains the complete set of general cubic graphs of order $2k$. In order to assist in spotting isomorphic graphs and to arrive at a simple notation for general cubic graphs, it was proposed to denote each graph by a sequence of numbers separated by hyphens, indicating in turn:

- The number $2k$ of vertices
- The number λ of loops (one-membered rings)
- The number β of double bonds (two-membered rings)
- The number t of three-membered rings
- The number f of four-membered rings
- A serial number which is one, unless two or more graphs have identical sequences of all preceding numbers

In the latter case this serial number takes integer values starting with one, in the order of increasing numbers of five-membered rings (or, if the sequence is still identical, six-membered rings, etc.). A computer program was developed for providing all circuits in a given graph.[33] Additional conventions were proposed for general cubic graphs where the above criteria fail to provide a nonsubjective assignment of serial numbers.[29]

Table 2 indicates the numbers of general cubic graphs of orders 2 to 10. Figure 2 presents structures of general cubic graphs with $n = 2, 4,$ or 6 vertices. Figure 3 presents structures of general cubic graphs with n = 8 vertices.

It can be seen from Table 2, and it can be demonstrated for any general cubic graphs on n vertices, that the number of general graphs with exactly one double bond and no loops equals the number of general graphs with exactly two loops and no double bond, and that

Table 2
NUMBERS OF GENERAL CUBIC
GRAPHS OF ORDERS n = 2—10
(INCLUDING NONPLANAR GRAPHS)

No. of points (n)	Double bonds β Loops λ	Double bonds β						Total
		0	1	2	3	4	5	
2	0	0	1					**1**
	1	0	0					0
	2	1	0					1
	Total	**1**	1					2
4	0	1	0	1				**2**
	1	0	1	0				1
	2	0	1	0				1
	3	1	0	0				1
	Total	**2**	2	1				5
6	0	2	1	2	1			**6**
	1	1	1	2	0			4
	2	1	2	1	0			4
	3	1	1	0	0			2
	4	1	0	0	0			1
	Total	**6**	5	5	1			17
8	0	5	5	5	4	1		**20**
	1	4	6	6	3	0		19
	2	5	6	6	1	0		18
	3	3	4	2	0	0		9
	4	2	2	0	0	0		4
	5	1	0	0	0	0		1
	Total	**20**	23	19	8	1		17
10	0	19	23	25	16	7	1	**91**
	1	19	33	32	18	4	0	106
	2	23	34	31	13	1	0	102
	3	15	21	16	3	0	0	55
	4	9	11	5	0	0	0	25
	5	4	3	0	0	0	0	7
	6	2	0	0	0	0	0	2
	Total	**91**	125	109	50	12	1	388

the number of all multigraphs (no loops) equals the number of all loop-graphs (no double bonds): these numbers are boldface in Table 2.

The former observation is due to the one-to-one correspondence (i), and the second observation to the similar correspondences (ii) to (v) which conserve the order of the graph but change both numbers λ and β of loops and double bonds, respectively.

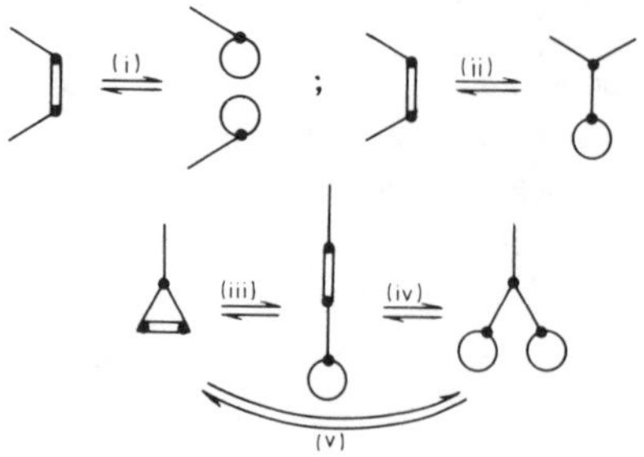

Operations 1 to 5 serve as internal check-up after obtaining the table of general cubic graphs with $n + 2$ vertices from that of those with $n - 2$ vertices.

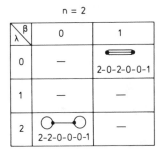

n = 2

λ \ β	0	1
0	—	2-0-2-0-0-1
1	—	—
2	2-2-0-0-0-1	—

n = 4

λ \ β	0	1	2
0	4-0-0-4-3-1	—	4-0-2-0-1-1
1	—	4-1-1-1-0-1	—
2	—	4-2-1-0-0-1	—
3	4-3-0-0-0-1	—	—

n = 6

λ \ β	0	1	2	3
0	((6-0-0-0-9-1)) 6-0-0-2-3-1	6-0-1-2-1-1	6-0-2-0-2-1 (6-0-2-2-0-1)	6-0-3-0-0-1
1	6-1-0-2-3-1	6-1-1-1-1-1	6-1-2-0-0-1 6-1-2-1-0-1	—
2	6-2-0-2-1-1	6-2-1-0-1-1 6-2-1-1-0-1	6-2-2-0-0-1	—
3	6-3-0-1-0-1	6-3-1-0-0-1	—	—
4	6-4-0-0-0-1	—	—	—

FIGURE 2. General cubic graphs with n = 2, 4, or 6 vertices, λ loops, and β double bonds. Brackets around codes indicate separable graphs without loops, and the code in double brackets indicates a nonplanar graph.

The general cubic graphs on *n* vertices with no loops (λ = 0), i.e., the multigraphs, are constitutional formulas for valence isomers of [*n*]annulenes, which will be discussed in detail below. The general cubic graphs with one loop[34] represent constitutional formulas for valence isomers of homo-[*n* − 2]annulenes, if the loop is deleted and the vertex to which it was attached (denoted by a black point in Figure 4) is considered to represent a CH$_2$ group; if it is considered to represent a CO group, one obtains valence isomers of annulenones (e.g., of tropone from general cubic graphs with eight vertices and one loop); if finally it is considered to represent a heteroatom one obtains valence isomers of heterocycles with one heteroatom, e.g., the four valence-isomers **19** to **22** (see Figure 5) of furan (from the graphs of order six with one loop). The vertex of degree two is denoted in **19** to **22** by a black point (in the parent general cubic graph it had a loop attached to it). Valence isomers of bis-homo[*n*]annulenes are among the graphs with two loops with *n* + 4 vertices, according

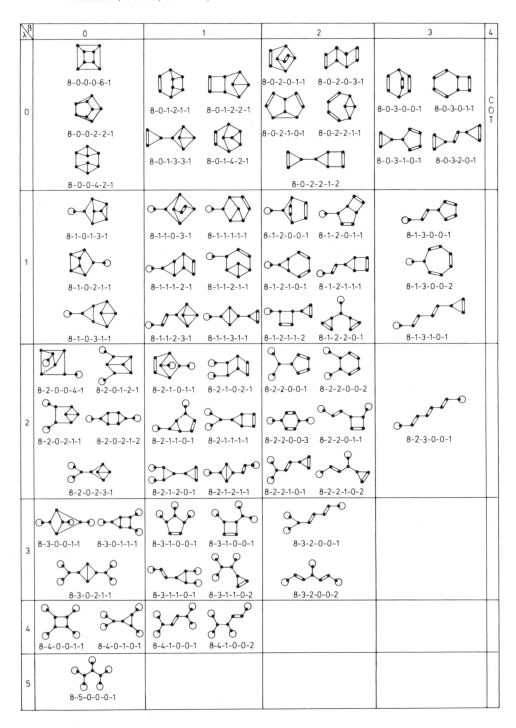

FIGURE 3.   General planar cubic graphs with n = 8 vertices, λ loops, and β double bonds. Brackets around codes indicate separable graphs without loops. The column headed β = 4 contains only cyclooctatetraene with λ = 0, symbolized by COT, whose full code *n-λ-β-t-f-s* is 8-0-4-0-0-1.

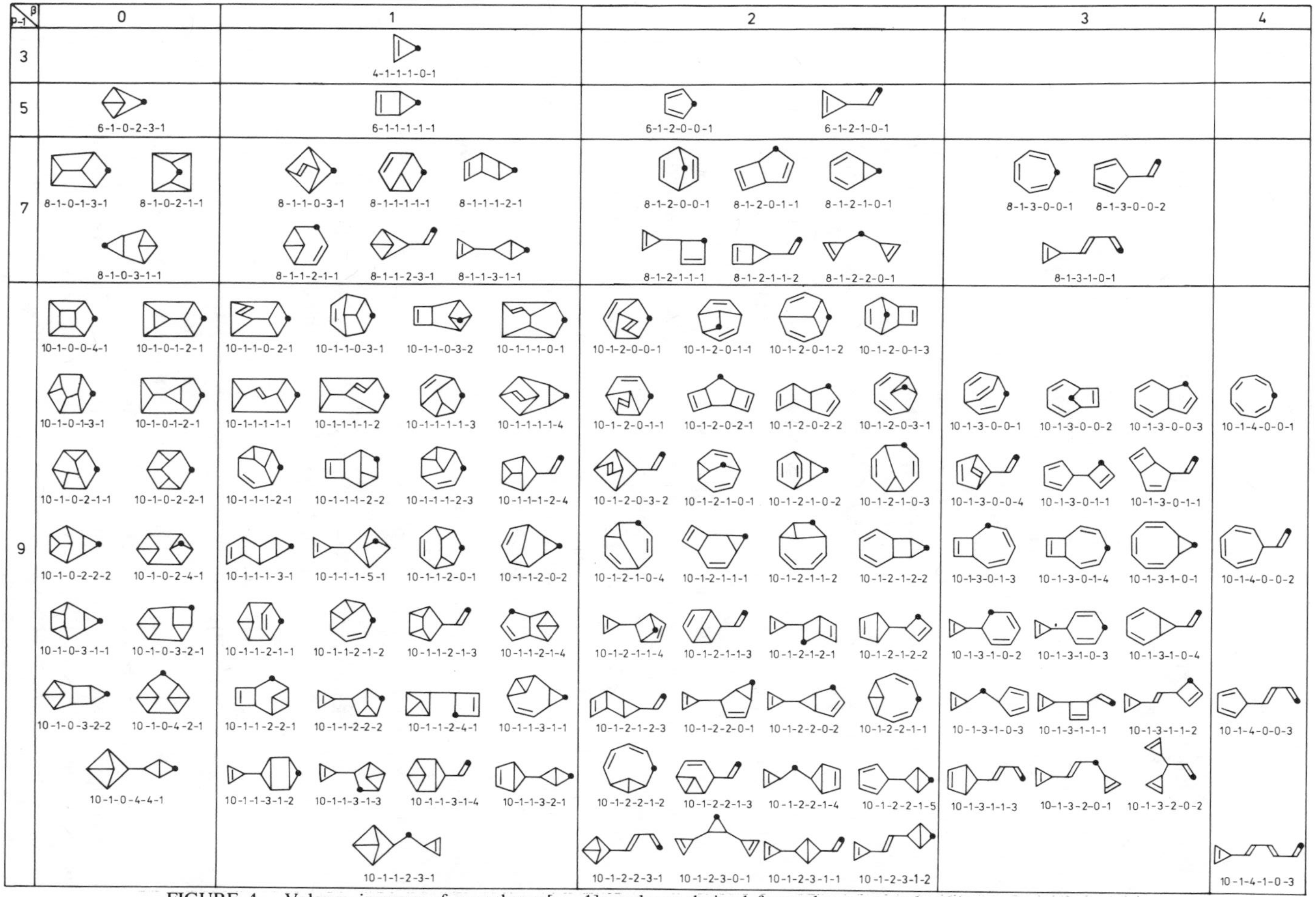

FIGURE 4.   Valence isomers of monohomo[p − 1]annulenes derived from planar general cubic graphs with p vertices and one loop, by deleting the loop and obtaining thereby a vertex of degree 2, indicated by a black dot. The full notation of general cubic graphs is conserved.

FIGURE 5.   The four valence isomers of cyclopentadiene, cyclopenta-
dienyl anion or cation, furan, thiophene, pyrrole, etc.; cf. Figure 4, graphs
with $p - 1 = 5$ vertices.

to the above deletion of both loops; however, in this case, one must ignore graphs with two
loops attached to the same vertex.

Table 3 presents, from all possible cubic multigraphs with 4, 6, 8, and 10 vertices, only
the planar ones.[28,29] It will be observed that there exist nonplanar cubic multigraphs with
$n \geq 6$ (in Figure 2 their notation was in double brackets), e.g., **12**, which cannot correspond
to chemical compounds if the usual assumption holds: a vertex symbolizes an atom and an
edge symbolizes a covalent bond. However, if homeomorphic reductions and expansions of
graph edges are allowed, recently several research groups succeeded in preparing compounds
whose constitutional formulas are graphs which are homeomorphically reducible to nonplanar
graphs: **11**,[35a,b] or a molecular Möbius strip[35c] (it will be observed that **12** is also a Möbius
strip corresponding to crossing two edges in the benzprismane formula). Nevertheless, in
the following chapters nonplanar graphs will be of no use and thus are not included in Table
3.

Walba succeeded in synthesizing the first molecular Möbius strip, and reviewed recently
topological stereochemistry,[63] a topic which includes also catenanes, rotaxanes, knots, to-
pologically chiral compounds, and molecules homeomorphic with nonplanar graphs.

Since the numbers of possible cubic multigraphs increase practically exponentially with
the number n of vertices, it is easily seen that it makes no sense to coin trivial names for
valence isomers such as those which proliferated in the last 10 years (cuneane, snoutene,
hypostrophene, etc.); therefore the five-digit notation n- β-t-f-s (or if n is specified otherwise,
the remaining four-digit notation) is a reasonable alternative for valence isomers of annulenes,
and the 6-number notation, including λ, for homoderivatives. For displaying multigraphs
corresponding to valence isomers of [n]annulenes with $n \geq 12$ it will be necessary to reduce
further the numbers of graphs. This will be done by indicating nonseparable graphs (for
cubic graphs this means graphs devoid of cut edges), or by including numbers of 5- and 6-
circuits.

In Table 4 the separable cubic multigraphs on 12 vertices[36] are indicated with their notation
in brackets.

Table 5 presents[36] the numbers of cubic graphs with $n \leq 14$. It should be noted that
multigraphs include simple graphs, and that general graphs include multigraphs. From Table
5, the numbers which are important in the present context are those of planar multigraphs,
and especially of the nonseparable ones. Indeed, among valence isomers of annulenes those
whose formulas are separable cubic multigraphs are often less interesting, and their chemistry
is less developed. For instance, there exist for $n = 6$ five planar cubic multigraphs: benzene
or [6]annulene, Dewar benzene, benzvalene, and benzprismane whose formulas are nonseparable graphs, as well as bicyclopropenyl whose formula is a separable graph.

The last compound was not recognized in early reviews on valence isomers of benzene[38]
as a valence-isomer $(CH)_6$ until the graph-theoretical analysis was published;[28] it is even
less investigated today. Separable valence isomers are easy to synthesize and may serve as
entry points for the "energy hypersurface" interconnecting the various valence isomers.

VI. PÓLYA'S THEOREM AND VALENCE ISOMERS OF HETEROANNULENES

It was mentioned earlier that Pólya's enumeration theorem[21] represented a landmark in

Table 3

PLANAR CUBIC MULTIGRAPHS WITH N = 4 TO 10 VERTICES

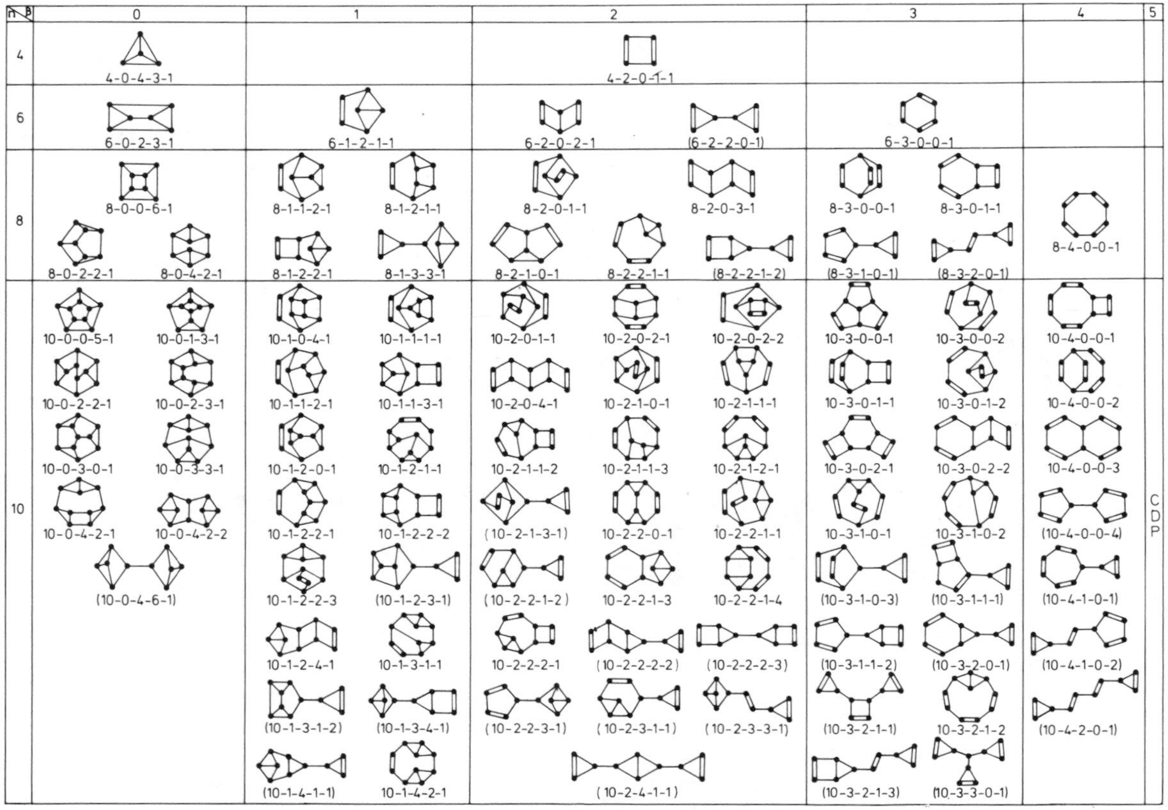

Note: Systems whose notation n-β-t-f-s is in brackets are separable graphs. For brevity, λ was omitted from the code in this table. The column headed $\beta = 5$ contains only cyclodecapentaene with $n = 10$, symbolized by CDP, whose code is 10-5-0-1.

Table 4A
PLANAR CUBIC MULTIGRAPHS WITH N = 12 VERTICES AND β = 0 OR 1 DOUBLE BONDS, ARRANGED ACCORDING TO THEIR T VALUES

β \ t	0	1	2	3	4	5	6
0	0-0-4-1, 0-0-6-1	0-1-3-1, 0-1-4-1	0-2-0-1, 0-2-1-1, 0-2-2-1, 0-2-2-2, 0-2-2-3, 0-2-3-1, 0-2-4-1, 0-2-4-2, 0-2-5-1	0-3-1-1, 0-3-1-2, 0-3-2-1, 0-3-3-1, 0-3-3-2, 0-3-3-3, (0-3-6-1)	0-4-0-1, 0-4-1-1, 0-4-2-1, 0-4-2-2, 0-4-2-3, 0-4-3-1, 0-4-3-2, 0-4-3-3, (0-4-4-1)	0-5-2-1, (0-5-4-1)	0-6-3-1
1	1-0-3-1, 1-0-3-2, 1-0-4-1, 1-0-5-1	1-1-1-1, 1-1-1-2, 1-1-2-1, 1-1-2-2, 1-1-2-3, 1-1-2-4, 1-1-2-5, 1-1-3-1, 1-1-3-2, 1-1-3-3, 1-1-3-4, 1-1-4-1, (1-1-4-2)	1-2-0-1, 1-2-0-2, 1-2-1-1, 1-2-1-2, 1-2-1-3, 1-2-1-4, 1-2-1-5, 1-2-2-1, 1-2-2-2, 1-2-2-3, 1-2-2-4, 1-2-2-5, 1-2-2-6, 1-2-2-7, 1-2-2-8, 1-2-2-9, (1-2-2-10), 1-2-3-1, 1-2-3-2, 1-2-3-3, 1-2-3-4, 1-2-3-5, (1-2-3-6), 1-2-4-1, (1-2-4-2), (1-2-6-1)	1-3-0-1, 1-3-0-2, (1-3-0-3), 1-3-1-1, 1-3-1-2, 1-3-1-3, 1-3-1-4, (1-3-1-5), 1-3-2-1, 1-3-2-2, 1-3-2-3, 1-3-2-4, (1-3-2-5), (1-3-2-6), (1-3-2-7), 1-3-3-1, (1-3-4-1), (1-3-4-2), (1-3-5-1)	1-4-1-1, (1-4-1-2), 1-4-2-1, 1-4-2-2, (1-4-2-3), (1-4-2-4), (1-4-2-5), 1-4-2-6, 1-4-3-1, (1-4-4-1), (1-4-6-1)	(1-5-2-1), (1-5-4-1)	

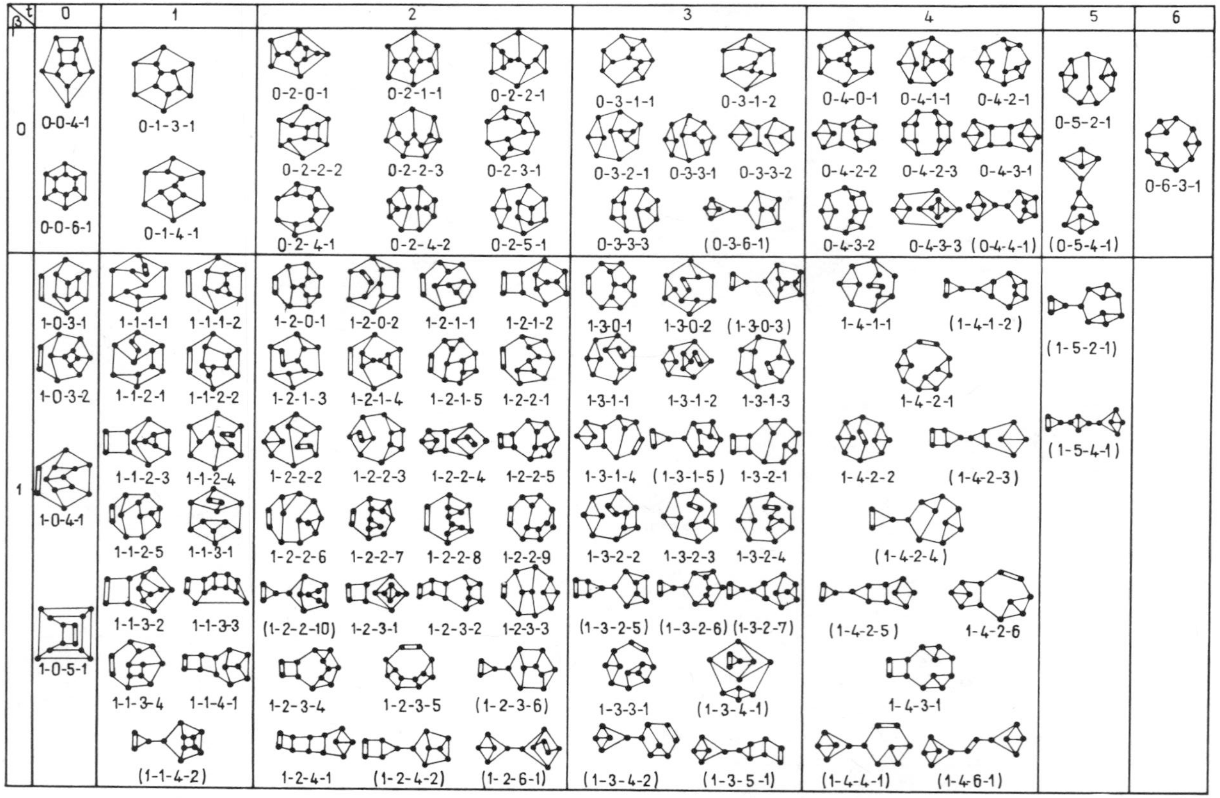

Note: For brevity, from their codes n and λ were omitted. Bracketed codes indicate separable graphs.

(From Banciu, M., Popa, E., and Balaban, A. T., *Chem. Scripta*, 1984. With permission.)

Table 4B
PLANAR CUBIC MULTIGRAPHS WITH N = 12 VERTICES AND β = 2 DOUBLE BONDS; OTHER EXPLANATIONS AS IN TABLE 4A

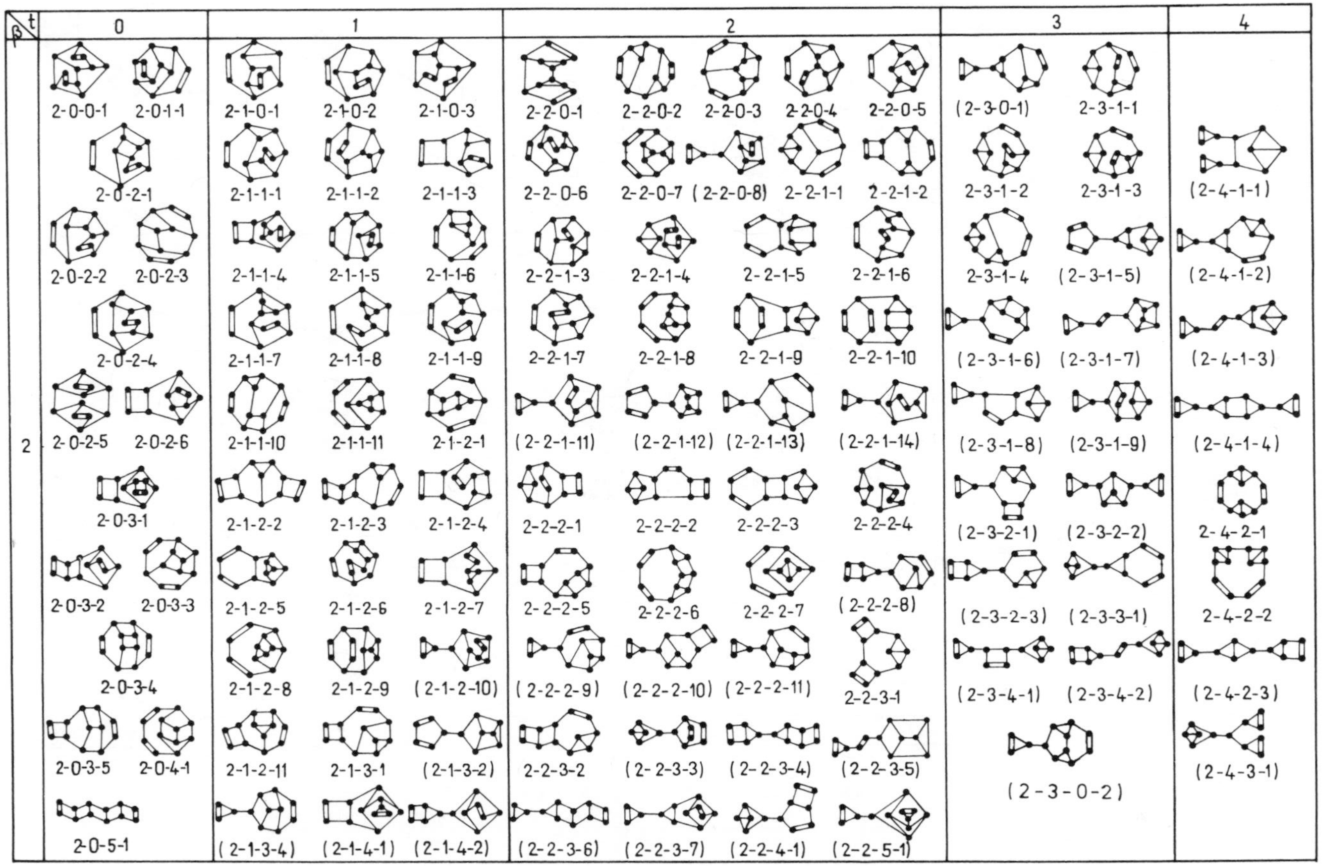

Table 4C

PLANAR CUBIC MULTIGRAPHS WITH N = 12 VERTICES AND β = 3 DOUBLE BONDS; OTHER EXPLANATIONS AS IN TABLE 4A

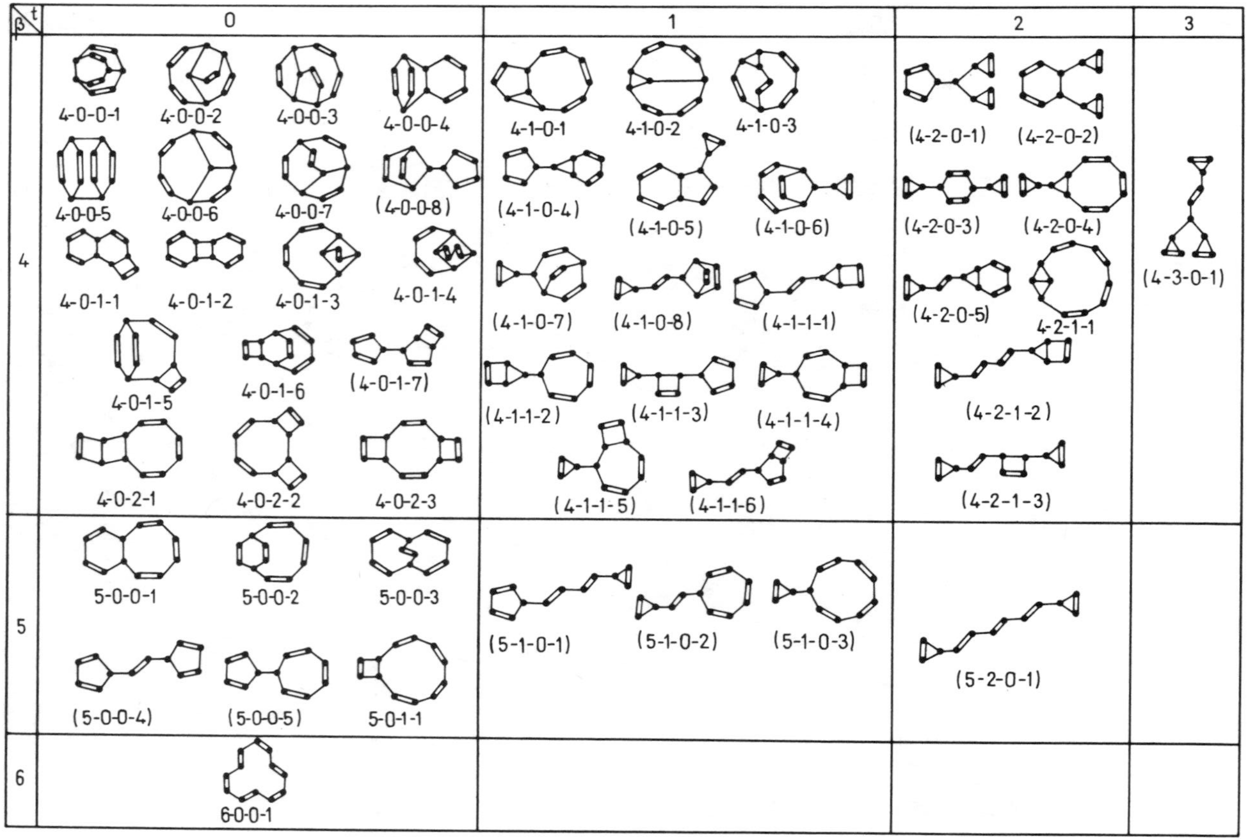

Table 4D
PLANAR CUBIC MULTIGRAPHS WITH N = 12 VERTICES AND β = 4, 5, OR 6 DOUBLE
BONDS; OTHER EXPLANATIONS AS IN TABLE 4A

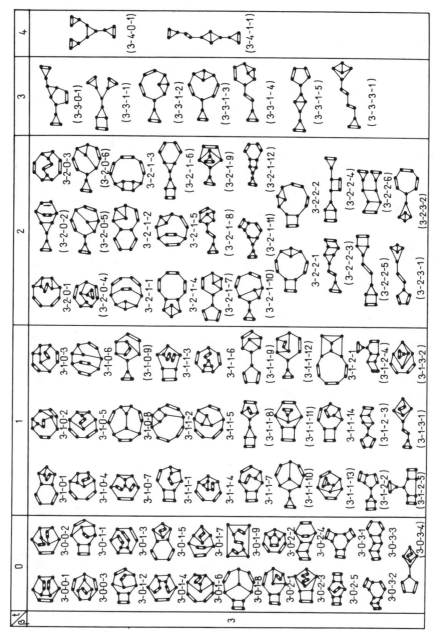

Table 5
NUMBERS OF CUBIC GRAPHS

Order n	2	4	6	8	10	12	14
Simple graphs*	0	1	2	5	19	85	509
Simple planar graphs*	0	1	1	3	9	32	133
Multigraphs*	1	2	6	20	91	506	—
Planar multigraphs*	1	2	5	17	71	357	2140
Nonseparable planar graphs*	1	2	4	13	47	226	1316
General graphs*	2	5	17	71	388	—	—
Disconnected general graphs	0	3	14	69	334	—	—
Total general graphs	2	8	31	140	722	—	

Note: Asterisk means connected.

graph theory and in mathematics as a whole. Burnside,[39] Lunn and Senior,[17] and Henze and Blair[18-20] developed approaches for applying permutation groups to graph enumerations; the latter authors applied Cayley's method to the enumeration of constitutionally isomeric alkanes and alkyls and corrected a few arithmetical errors. However, a radically new approach was prefigurated in the only article ever published by Redfield;[16] unfortunately, his paper was completely overlooked for many years, though his results closely adumbrated Pólya's contribution, and were recognized after Pólya's theorem became widely known.

The basic idea on which Pólya's theorem rests[21] is that the numbers of substitution isomers with $0, 1, 2, \ldots, n$ substituents in place of, or attached to, the n atoms of a given molecule, can be expressed as a polynomial (configuration counting series)

$$C(x) = a_0 x^0 + a_1 x^1 + a_2 x^2 + \ldots + a_n x^n$$

In order to find this polynomial, one has to describe the symmetry of the molecule (only rotation axes are taken into account) in terms of the cycle index $Z(s)$ and subsequently one substitutes for the variable s a figure-counting series according to the type of attachment or substitution. For instance, for heteroatom-substituted benzenes, e.g., azabenzenes, the cycle index of benzene (whose symmetry group is D_{6h}) is

$$Z(s_i) = (s_1^6 + 3s_1^2 s_2^2 + 4s_2^3 + 2s_3^2 + 2s_6)/12$$

and the figure-counting series is simply

$$s_i = 1 + x^i$$

The meaning of the 12 terms in $Z(s)$ is the following: s_1^6 indicates the identity permutation, i.e., each of the six CH groups is unchanged; $3s_1^2 s_2^2$ indicates that there exist three axes C_2 passing through two CH groups (leaving them unchanged s_1^2) which permute two pairs of CH groups, s_2^2; $4s_2^3$ indicates that there exist four rotations which permute three pairs of CH groups: three around the in-plane axes which bisect the molecule without encountering atoms, and a fourth rotation with 180° around the sixfold axis; $2s_3$ and $2s_6$ indicate rotations with 120 and 60° around the sixfold axis, respectively. By substituting s_i into $Z(s_i)$ one obtains the configuration-counting series:

$$C(x) = (1 + x)^6 + 3(1 + x)^2 (1 + x^2)^2 + 4(1 + x^2)^3 + 2(1 + x^3)^2 + 2(1 + x^6)/12$$

$$= 1 + x + 3x^2 + 3x^3 + 3x^4 + x^5 + x^6$$

This polynomial indicates that there may exist three di-, tri-, and tetra-azabenzenes and one

Table 6
NUMBER OF ISOMERIC HETEROATOM-SUBSTITUTED VALENCE ISOMERS OF BENZENE

Substituent pattern[a]	x^5	x^4	x^3	x^4y	x^3y^2	x^2y^2	x^3yz	x^2y^2z	x^2yzu	xyzuv
Benzvalene	3	7	8	11	20	29	36	52	56	180
Dewar-benzene	2	6	6	8	16	26	30	46	90	180
Bicyclopropenyl	2	5	5	7	12	20	21	29	51	90
Benzprismane[b]	1	4	4	5	10	18	20	30	60	120
Benzene[c]	1	3	3	3	6	11	10	16	30	60

[a] Penta- or monosubstituted denoted by x^5; di- or tetra- with one type of substituent by x^4; substituted with two different substituents by x^4y, x^3y^2, etc.

[b] Including stereoisomerism.

[c] Or benzprismane, excluding stereoisomerism.

benzene, azabenzene, penta- and hexa-azabenzene, as it is well known. It is for more complicated structures and for more complex substitutions that the power of Pólya's theorem becomes apparent allowing rapid, rigorous, and error-proof computation of numbers of isomers. For example, when two different heteroatoms are present, the figure-counting series is[40]

$$s_i = 1 + x^i + y^i$$

and this series can be generalized easily.[41-47]

Table 6 indicates the numbers of isomers for heteroatom-substituted valence isomers of benzene such as aza-benzvalenes, etc.

Interestingly, if stereoisomerism is not included in calculating the cycle index, benzprismane is found to possess the same cycle index as benzene, and leads therefore to the same numbers of constitutional isomers. This explains why the dispute between Kekulé and Ladenburg (the latter advocating the benzprismane formula) lasted so long. This is by no means the only pair of co-isomeric valence isomers; many other pairs, triplets, etc. were reviewed.[45] Three such pairs with $n = 8$ CH groups are **23** and **24**; **25** and **26**; **27** and **28** (see also Chapter 2, Section I.A.).

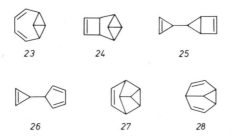

It is simple to find via Figures 2 to 4 the valence isomers of monohomo-annulenes (e.g., of tropylidene, etc. as shown in Figure 6), and slightly more complicated to find those for bishomo-annulenes (e.g., Figure 7). In the latter case, however, sometimes in addition to the normal (neutral) systems, supplementary valence isomers with charged heteroatoms may appear, resulting in dipolar structures. Thus, if X = BH in Figure 7 then, in addition to the 18 valence isomers from Figure 7 which appear on the right-hand side of Figure 8, supplementary structures appear (left-hand side of Figure 8); some structures on corresponding lines of the two sides are just resonance structures of the same compound, but other structures on the left-hand side have no counterpart on the right-hand side.

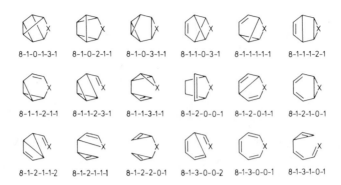

FIGURE 6.   Valence isomers of tropylidene, tropone, cycloheptatrienyl cation or anion (X = CH$_2$, CO, CH$^\pm$, O, S, NR, etc.); cf. Figure 4, graphs with p − 1 = 7 vertices whose full notation is conserved.

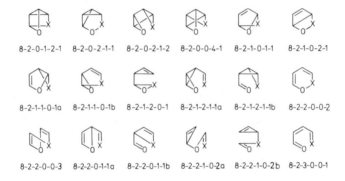

FIGURE 7.   The 18 valence isomers of bishomo(CH)$_4$ systems, with two kinds of different vertices of degree two, e.g., pyrones (X = CO), pyrans (X = CH$_2$), etc.; cf. general cubic multigraphs with λ = 2 loops from Figure 3, having 8 vertices. The full notation is conserved, and letters a, b, . . . are added when two or more nonisomorphic structures result on interchanging the heteroatoms.

VII. FINDING BENZODERIVATIVES OF VALENCE ISOMERS OF [n]ANNULENES

According to the graph-theoretical definition, valence isomers of benzo[*n*]annulenes occur in enormous numbers, even for the simple cases. Thus benzocyclobutene (old name: benzocyclobutadiene) has 355 valence isomers, and naphthalene (benzobenzene) has 3838 valence isomers (only constitutional isomerism is taken into account).[45]

We are interested primarily, however, in a very small subset of these compounds, namely benzoderivatives of valence isomers of [*n*]annulenes which will be called henceforth benzo(CH)$_n$ valence isomers, dibenzo(CH)$_n$ valence isomers, etc. A benzoderivative of a valence isomer of [*n*]annulene having *b* benzenoid rings will be called [*b*]benzo(CH)$_n$ valence isomer.[46,47] The distinction originates from the fact that graphs of the subset must have adjacent pairs of vertices of degree four, while for the large set no such restriction exists.

There exist two possible ways for finding all possible [*b*]benzo(CH)$_n$ valence isomers. The trivial one is to look in the table of cubic multigraphs (valence isomers of [*n* + 4*b*]annulenes) for structures having *b* condensed 1,3-cyclohexadienic rings; by removing 2*b* hydrogen atoms, i.e., by converting the above rings into benzenic ones, one obtains the result. This approach requires, however, tables with cubic multigraphs having *n* > 14 vertices

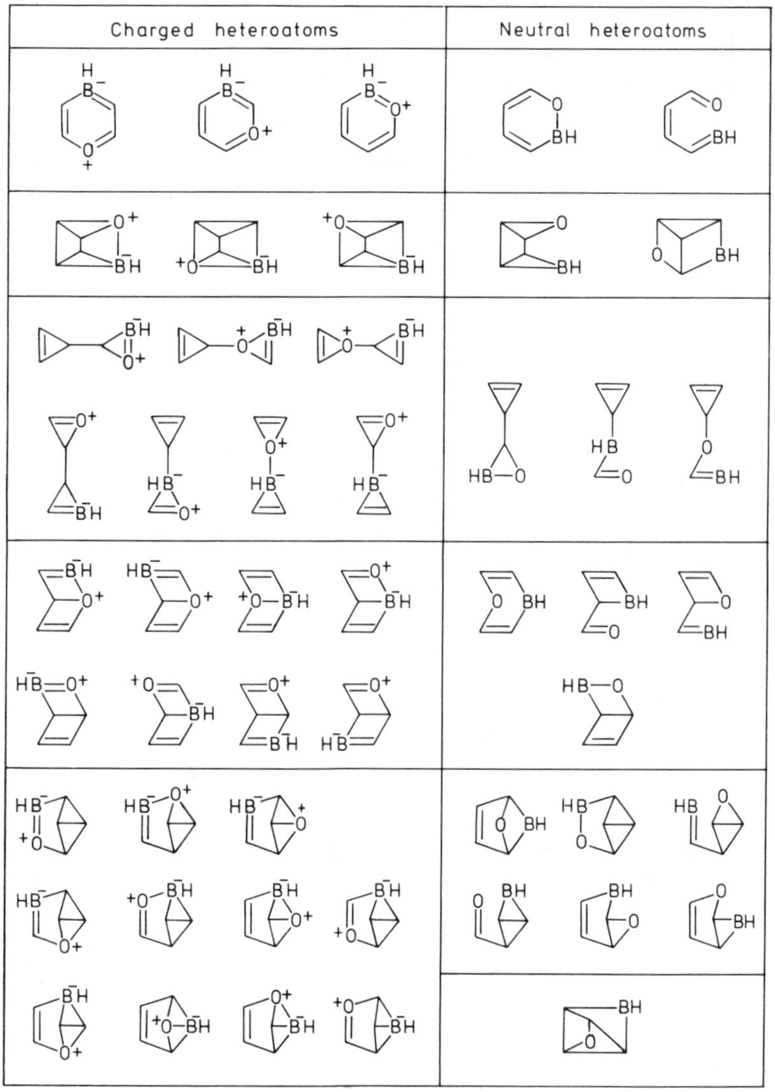

FIGURE 8. Valence isomers of $(CH)_4(BH)O$ illustrating the possibility for the apparition of dipolar structures. The 18 graphs on the right-hand side are bishomo-derivatives derived from general cubic multigraphs with $n = 8$ and $\lambda = 2$ (cf. Figure 7); those on the left-hand side are heteroatom-substituted cubic multigraphs with $n = 6$ and $\lambda = 0$.

which are not yet available; the only systems which can be found from the available tables are monobenzeno$(CH)_n$ valence isomers with $n = 4, 6, 8$, and dibenzo$(CH)_4$ valence isomers.

Therefore, we shall use the second approach, annelation of cubic multigraphs at their double bond(s), or when at least two conjugated double bonds exist, at any bonds of the whole conjugated system. In the latter case one can obtain *ortho*-quinonoid structures, which are expected to be less stable than isomeric structures which are not *ortho*-quinonoid. The simplest examples are of the two dibenzo$(CH)_6$ valence isomers, phenanthrene and anthracene. The latter is *ortho*-quinonoid, i.e., the two added benzo rings cannot simultaneously have three inscribed double bonds, and indeed anthracene has a lower resonance energy than phenanthrene. We use double bonds in formulas of *ortho*-quinonoid systems and asterisks in their notation, while non-*ortho*-quinonoid (NOQ) systems will have circles inscribed in their benzenoid rings (these are not meant to have the significance of Clar's formulas but only to serve as distinction that they are not *ortho*-quinonoid).

Table 7
NUMBERS OF PLANAR VALENCE ISOMERS OF [n] ANNULENES ACCORDING TO THE NUMBER β OF DOUBLE BONDS

β	Exactly β double bonds (n =)						At least β double bonds (n =)					
	4	6	8	10	12	14	4	6	8	10	12	14
0	1	1	3	9	32	133	2	5	17	71	357	2140
1	0	1	4	16	75	404	1	4	14	62	325	2007
2	1	2	5	22	112	697	1	3	10	46	250	1603
3	—	1	4	16	86	556	—	1	5	24	138	956
4	—	—	1	7	41	294	—	—	1	8	52	400
5	—	—	—	1	10	90	—	—	—	1	11	106
6	—	—	—	—	1	15	—	—	—	—	1	16
7	—	—	—	—	—	1	—	—	—	—	—	1
Total	2	5	17	71	357	2140						

Table 8
NUMBERS OF PLANAR [b]BENZO (CH)$_n$ VALENCE ISOMERS

b	Total (n =)				NOQ (n =)				NOQ and nonseparable (n =)			
	4	6	8	10	4	6	8	10	4	6	8	10
1	1	4	21	113	1	4	18	96	1	3	11	55
2	1	4	17	102	1	3	14	79	1	2	9	43
3	—	1	6	44	—	1	5	33	—	1	3	17
4	—	—	1	10	—	—	1	8	—	—	1	4
5	—	—	—	1	—	—	—	1	—	—	—	1
Total	2	9	45	270	2	8	38	217	2	6	24	120

We illustrate the annelation of 8-0-2-2-1-1 (**29**) to an *ortho*-quinonoid (**30**) or a NOQ benzo[8]valence isomer (**31**). Since so far no *ortho*-quinonoid mono- or polybenzo(CH)$_n$ valence isomer has yet been reported (excepting the benzo[n]annulenes themselves) we shall not display *ortho*-quinonoid formulas but only NOQ systems. We shall, however, indicate separable graphs although at first sight it seems surprising to find in the table of benzo(CH)$_8$ valence isomers *ortho*-dicyclopropenylbenzene, while its *meta*- and *para*-isomers are excluded (they are members of the larger set, i.e., valence isomers of benzo[8]annulene). Of course, separable graphs are, as before, less interesting. Only planar, connected multigraphs representing constitutional isomers are discussed, ignoring stereoisomerism.[46,47]

30   29   31

We start from the cubic multigraphs with β double bonds (β > 0) whose numbers are indicated in Table 7. From these we obtain by annelation with b benzo rings the [b]benzo(CH)$_n$ valence isomers whose numbers are indicated in Table 8.[46,47] We display structures in Table 9.[47]

The proposed[46,47] notation for these systems is derived from that of the [n]annulene which has undergone the annelation. Its notation according to Table 9 is preceded by a symbol bB or bB$_i$, where b indicates the number of annelated benzenoid rings, B stands for benzo, and the index i is a serial number which is absent if one benzoderivative can result from

Table 9A
MONOBENZO(CH)$_N$ VALENCE ISOMERS WITH N = 4, 6, AND 8

n \ b	1		2		3	4
4	1B-4-2-0-1-1		2B-4-2-0-1-1			
6	1B-6-1-2-1-1	1B-6-2-0-2-1	2B-6-2-0-2-1	2B-6-2-2-0-1		
	1B-6-2-2-0-1	1B-6-3-0-1-1	$2B_1$-6-3-0-1-1	$2B_2$-6-3-0-1-1	3B-6-3-0-1-1	
8	1B-8-1-1-2-1	1B-8-1-2-1-1	2B-8-2-0-1-1	2B-8-2-0-3-1	3B-8-3-0-0-1	
	1B-8-1-2-2-1	1B-8-1-3-3-1	2B-8-2-1-0-1	2B-8-2-2-1-1		
	1B-8-2-1-0-1	1B-8-2-0-3-1	2B-8-2-2-1-2	2B-8-3-0-0-1	3B-8-3-0-1-1	
	1B-8-2-1-0-1	$1B_1$-8-2-2-1-1	$2B_1$-8-3-0-1-1	$2B_2$-8-3-0-1-1		
	$1B_2$-8-2-2-1-1	$1B_1$-8-2-2-1-2	$2B_3$-8-3-0-1-1	$2B_1$-8-3-1-0-1	3B-8-3-1-0-1	4B-COT
	$1B_2$-8-2-2-1-2	1B-8-3-0-0-1	$2B_2$-8-3-1-0-1	$2B_3$-8-3-1-0-1		
	$1B_1$-8-3-0-1-1	$1B_2$-8-3-0-1-1	$2B_1$-8-3-2-0-1	$2B_2$-8-3-2-0-1	3B-8-3-2-0-1	
	$1B_3$-8-3-0-1-1	$1B_1$-8-3-1-0-1	$2B_1$-8-4-0-0-1	$2B_2$-8-4-0-0-1		
	$1B_2$-8-3-1-0-1	$1B_3$-8-3-1-0-1			$3B_1$-8-4-0-0-1	
	$1B_1$-8-3-2-0-1	$1B_2$-8-3-2-0-1	$2B_3$-8-4-0-0-1			
	1B-8-4-0-0-1				$3B_2$-8-4-0-0-1	

Note: From all codes the symbol λ = 0 has been omitted for brevity. The column with b = 4 has just tetrabenzo-cyclooctene, symbolized by 4B-COT (code 4B-8-4-0-0-1).

Table 9B
MONOBENZO(CH)$_{10}$ VALENCE ISOMERS; CODES AS IN TABLE 9A

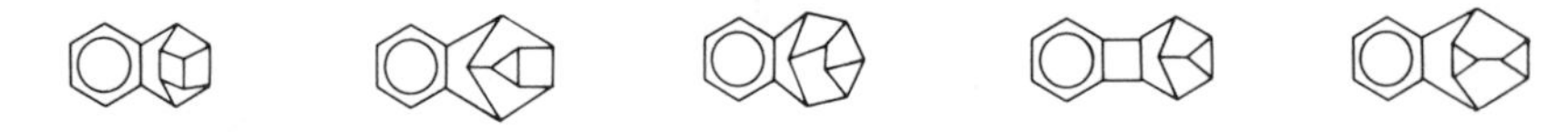

1B-10-1-0-4-1	1B-10-1-1-1-1	1B-10-1-1-2-1	1B-10-1-1-3-1	1B-10-1-2-0-1
1B-10-1-2-1-1	1B-10-1-2-2-1	1B-10-1-2-2-2	1B-10-1-2-2-3	1B-10-1-2-3-1
1B-10-1-3-1-1	1B-10-1-4-2-1	1B-10-2-0-1-1	1B-10-2-0-2-1	1B$_1$-10-2-0-2-2
1B$_2$-10-2-0-2-2	1B-10-2-0-4-1	1B$_1$-10-2-1-0-1	1B$_2$-10-2-1-0-1	1B-10-2-1-1-1
1B$_1$-10-2-1-1-2	1B$_2$-10-2-1-1-2	1B$_1$-10-2-1-1-3	1B$_2$-10-2-1-1-3	1B$_1$-10-2-1-2-1
1B-10-2-2-0-1	1B-10-2-2-1-1	1B$_1$-10-2-2-1-2	1B$_1$-10-2-2-1-4	1B$_1$-10-2-2-2-1
1B$_2$-10-2-2-2-1	1B$_1$-10-3-0-0-1	1B$_2$-10-3-0-0-1	1B-10-3-0-0-2	1B$_1$-10-3-0-1-1
1B$_2$-10-3-0-1-1	1B$_1$-10-3-0-1-2	1B$_2$-10-3-0-1-2	1B$_1$-10-3-0-2-1	1B$_2$-10-3-0-2-1
1B$_1$-10-3-0-2-2	1B$_2$-10-3-0-2-2	1B-10-3-1-0-1	1B$_1$-10-3-1-0-2	1B$_2$-10-3-1-0-2
1B$_3$-10-3-1-0-2	1B$_1$-10-3-2-1-2	1B$_2$-10-3-2-1-2	1B$_1$-10-4-0-0-1	1B$_2$-10-4-0-0-1
1B$_1$-10-4-0-0-2	1B$_1$-10-4-0-1-1	1B$_2$-10-4-0-1-1	1B$_3$-10-4-0-1-1	1B-10-5-0-0-1

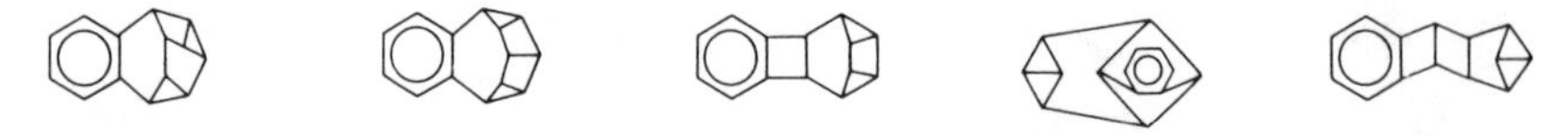

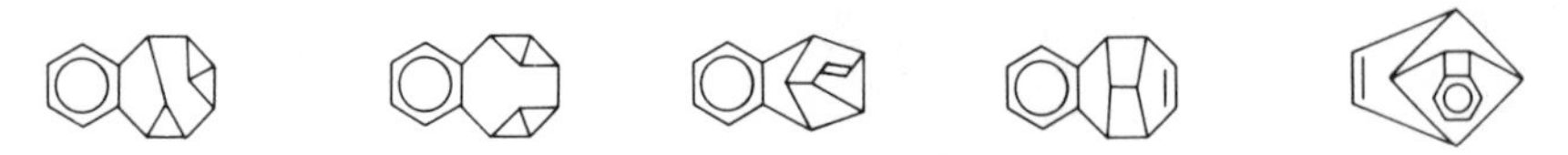

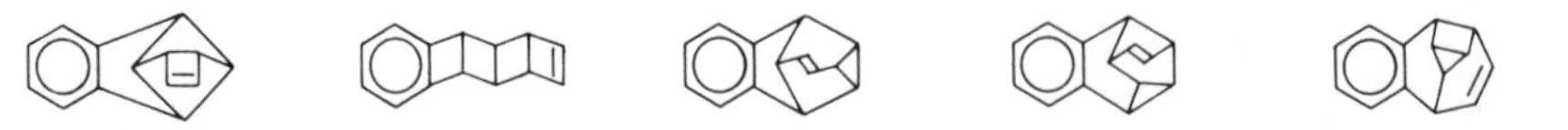

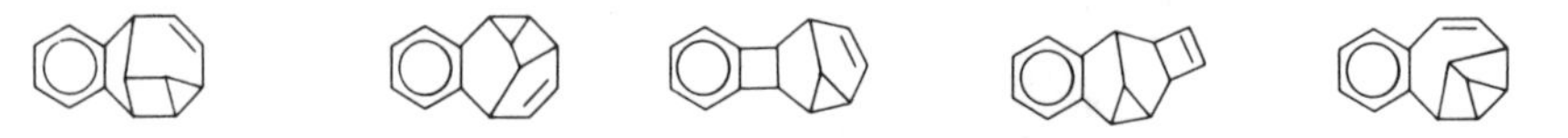

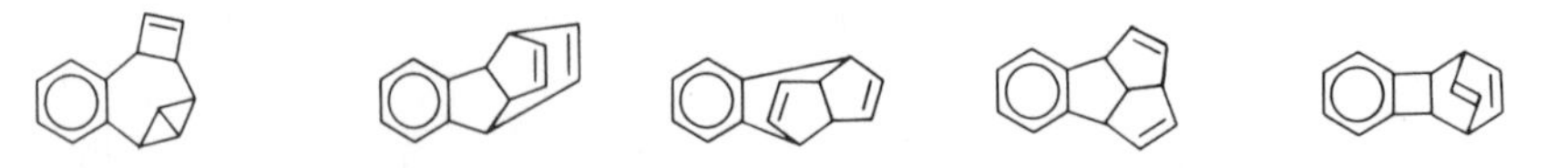

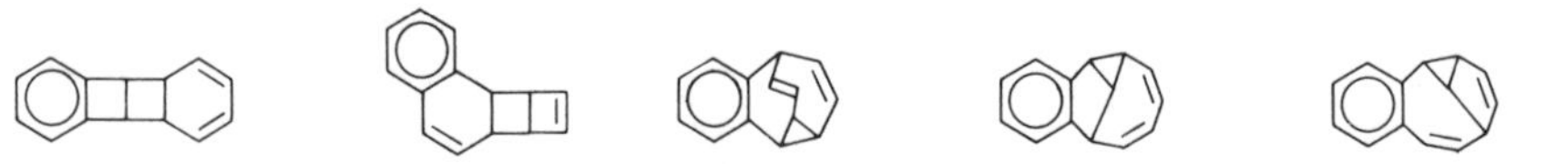

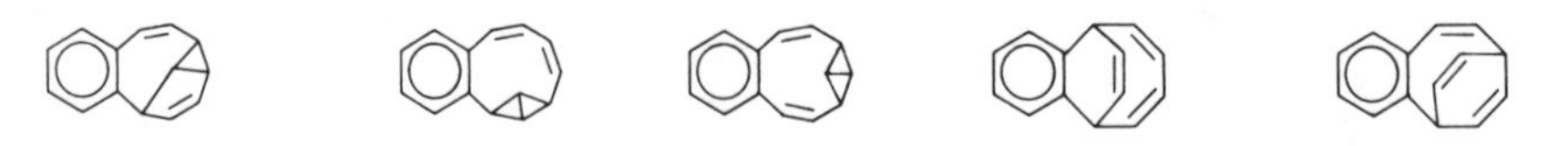

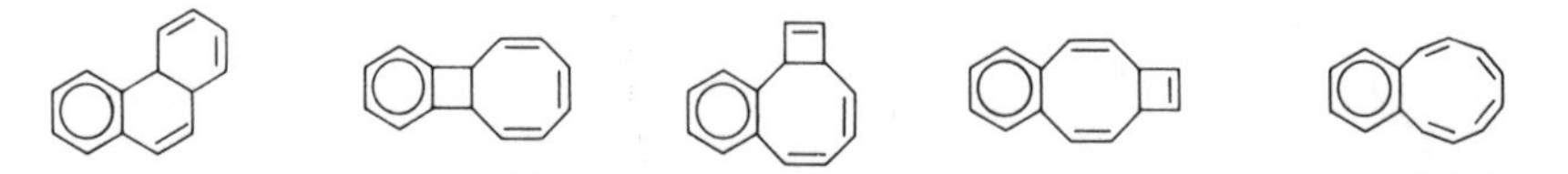

Table 9C
DIBENZO(CH)$_{10}$ VALENCE ISOMERS; CODES AS IN TABLE 9A

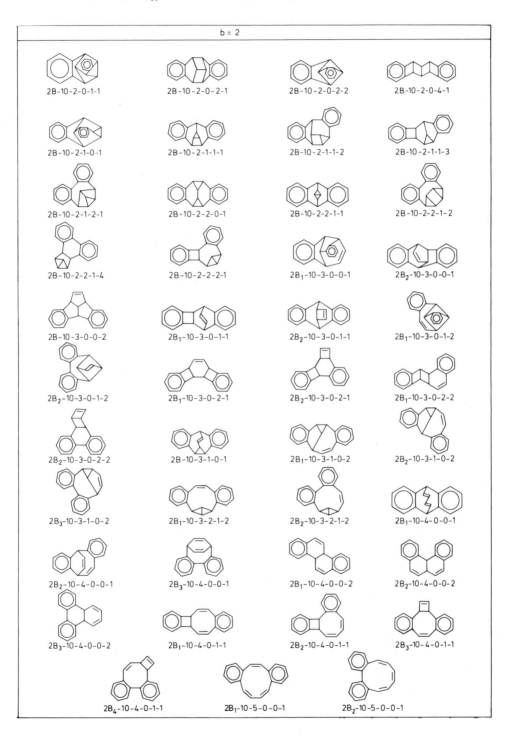

2B-10-2-0-1-1	2B-10-2-0-2-1	2B-10-2-0-2-2	2B-10-2-0-4-1
2B-10-2-1-0-1	2B-10-2-1-1-1	2B-10-2-1-1-2	2B-10-2-1-1-3
2B-10-2-1-2-1	2B-10-2-2-0-1	2B-10-2-2-1-1	2B-10-2-2-1-2
2B-10-2-2-1-4	2B-10-2-2-2-1	2B$_1$-10-3-0-0-1	2B$_2$-10-3-0-0-1
2B-10-3-0-0-2	2B$_1$-10-3-0-1-1	2B$_2$-10-3-0-1-1	2B$_1$-10-3-0-1-2
2B$_2$-10-3-0-1-2	2B$_1$-10-3-0-2-1	2B$_2$-10-3-0-2-1	2B$_1$-10-3-0-2-2
2B$_2$-10-3-0-2-2	2B-10-3-1-0-1	2B$_1$-10-3-1-0-2	2B$_2$-10-3-1-0-2
2B$_3$-10-3-1-0-2	2B$_1$-10-3-2-1-2	2B$_2$-10-3-2-1-2	2B$_1$-10-4-0-0-1
2B$_2$-10-4-0-0-1	2B$_3$-10-4-0-0-1	2B$_1$-10-4-0-0-2	2B$_2$-10-4-0-0-2
2B$_3$-10-4-0-0-2	2B$_1$-10-4-0-1-1	2B$_2$-10-4-0-1-1	2B$_3$-10-4-0-1-1
	2B$_4$-10-4-0-1-1	2B$_1$-10-5-0-0-1	2B$_2$-10-5-0-0-1

Table 9D

TRI-, TETRA-, AND PENTABENZO(CH)$_{10}$ VALENCE ISOMERS; CODES AS IN TABLE 9A

b = 3		b = 4	b = 5
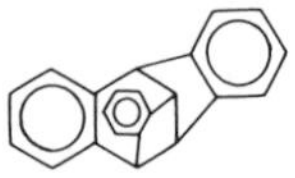 3B-10-3-0-0-1	 3B-10-3-0-0-2		
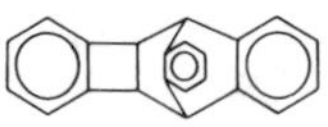 3B-10-3-0-1-1	 3B-10-3-0-1-2	 4B-10-4-0-0-1	
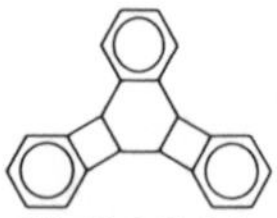 3B-10-3-0-2-1	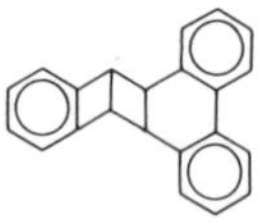 3B-10-3-0-2-2		
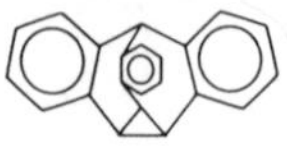 3B-10-3-1-0-1	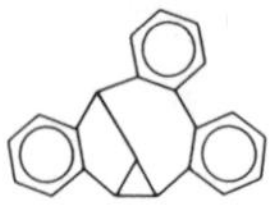 3B-10-3-1-0-2	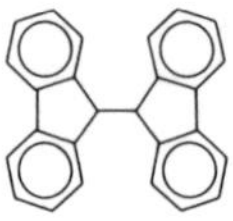 4B-10-4-0-0-2	
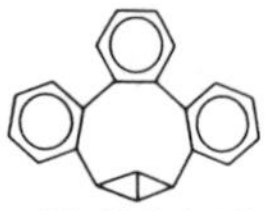 3B-10-3-2-1-2	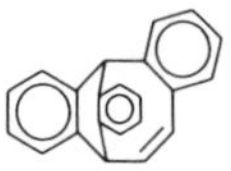 3B$_1$-10-4-0-0-1		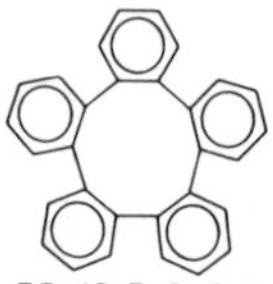 5B-10-5-0-0-1
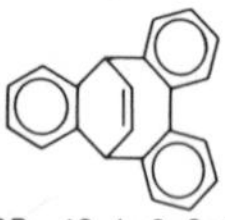 3B$_2$-10-4-0-0-1	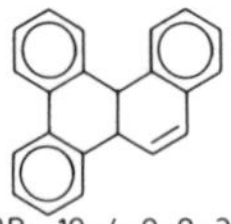 3B$_1$-10-4-0-0-2	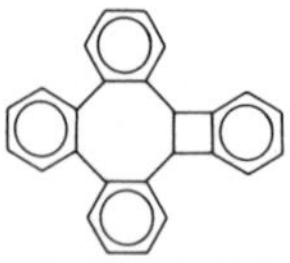 4B-10-4-0-1-1	
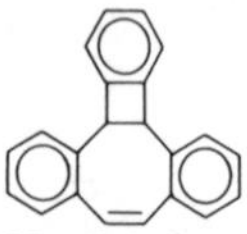 3B$_1$-10-4-0-1-1	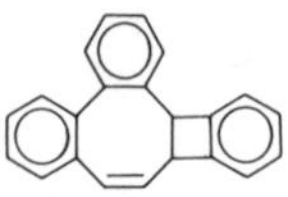 3B$_2$-10-4-0-1-1		
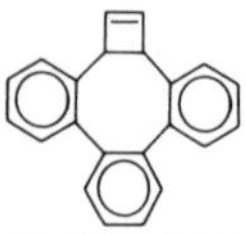 3B$_3$-10-4-0-1-1	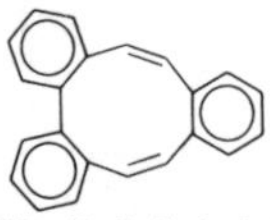 3B$_1$-10-0-5-0-0-1	 4B$_1$-10-5-0-0-1	
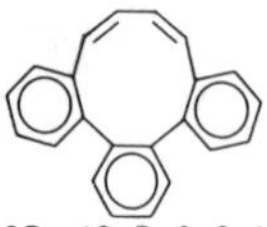 3B$_2$-10-5-0-0-1			

the given [*n*]annulene valence isomer, and takes integer values starting with 1 when the [*n*]annulene valence isomer can be [*b*]annelated in several ways. Thus, phenanthrene is $2B_1$-6-3-0-1-1 and anthracene is $2B_2$-6-3-0-1-1.

Interestingly, biphenylene appears as dibenzocyclobutene 2B-4-2-0-1-1, while dihydro-biphenylene $1B_1$-8-3-0-1-1 is a monobenzoderivative of a valence isomer of cyclooctate-traene. This is a general feature of Table 9, where phenanthrene appears as dibenzo[6]annulene, while dihydrophenanthrene $1B_1$-10-4-0-0-2 is a monobenzo$(CH)_{10}$ valence isomer.

Tables of benzo-homovalenes were compiled and are available.[48] Again, such isomers can be *ortho*-quinonoid (**32**) or NOQ (**33**); the latter are much more stable than the former ones.

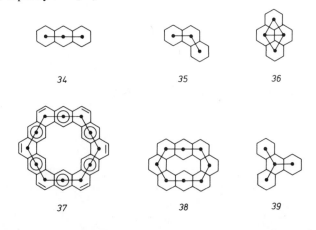

VIII. FINDING ALL POSSIBLE ISOMERS OF BENZENOID POLYCYCLIC CONDENSED HYDROCARBONS

Soon after Hückel's theory was formulated, it was recognized that formally some ben-zenoid-conjugated hydrocarbons (naphthalene, anthracene, phenanthrene etc.) can be viewed as bridged [4*n* + 2]annulenes; moreover, even nonbenzenoid systems such as azulenes fit this idea. Actually, Hückel's theory was formulated strictly for monocyclic systems and such an extension to polycyclics is bound to have limitations due to the perturbation induced by bridges. This problem was examined in the chapter on aromaticity. We shall discuss here only problems of definition, classification, and enumeration of isomers.

Polycyclic benzenoid aromatic hydrocarbons will be called for brevity "polyhexes" since they are composed of condensed fused hexagons which may share their edge(s). A classical dichotomy of polyhexes into *cata*- and *peri*-condensed systems (which will be called for brevity "catafusenes" and "perifusenes", respectively) is based on the number of hexagons surrounding each vertex: in catafusenes no carbon atom is common to more than three hexagons, while in perifusenes there exist carbon atoms common to three condensed hex-agons. A more versatile definition[49] is based on the graph obtained by joining the centers of condensed hexagons; if this graph is acyclic (a tree), the polyhex is a catafusene; if it has three-membered rings, the polyhex is a perifusene; if it also has larger rings, the polyhex is a coronafusene. Benzene, naphthalene, phenanthrene (**34**), anthracene (**35**), chrysene, picene, and all helicenes are catafusenes; pyrene (**36**), perylene, benzo[*a*]pyrene are peri-fusenes; the known Kekulene (**37**) and the unknown hydrocarbon (**38**) are coronafusenes. In the new definition, all catafusenes with the same number *r* of benzenoid rings have formula $C_{4r + 2}H_{2r + 4}$, and they may be nonbranched as all examples above, or may have branches, e.g., triphenylene (**39**).

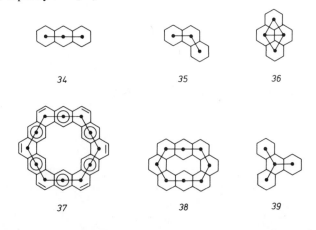

Table 10
NUMBERS OF POLYHEXES WITH r = 1 TO 7
BENZENOID RINGS

| | Catafusenes | | | | Total |
r	Linear	Branched	Total	Perifusenes	polyhexes
1	1	0	1	0	1
2	1	0	1	0	1
3	2	0	2	1	3
4	4	1	5	2	7
5	10	2	12	10	22
6	25	12	37	45	82
7	70	53	123	210	333

The graph whose vertices are ring centers and whose edges indicate condensed rings is called dualist graph;[49-51] unlike normal graphs, in this graph angles are important, because they distinguish for instance anthracene **34** (angle 180°) from phenanthrene **35** (angle 120°). By means of these dualist graphs one can enumerate and code all possible nonbranched catafusenes. For the other polyhexes, computer algorithms exist[53] but no simple enumeration procedure has yet been devised. Table 10 presents the numbers of polyhexes with r ≤ 7 benzenoid rings.

The coding of nonbranched catafusenes starts from one end of the dualist graph and indicates the angles at each successive vertex by digit 0 for 180°, and digits 1 or 2 for 120°/240°; a left-kink on following the dualist graph may be denoted either by 1 or 2, but once one has made the choice, this should be conserved. Among several possible codings (according to the selected end, and to the choice 1/2) the canonical one must correspond to the smallest number resulted on reading the digits sequentially.

Branched catafusenes have more complicated enumeration formulas[52] and codes[49] than nonbranched ones. Other uses of dualist graphs were reviewed.[54]

If the distinction between digits 1 and 2 is abandoned and both are replaced by letter *l*, one converts the three-digit code (3DC) into a two-digit code (2DC) for catafusenes.[55] One may read this two-digit notation as a binary number. Interestingly, catafusenes with the same 2DC have the same number of Kekulé structures, i.e., the same structure count; such catafusenes are called isoarithmic,[56] e.g., **40** and **41**, or **42** and **43**.

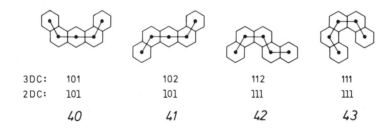

3DC:	101	102	112	111
2DC:	101	101	1l1	1l1
	40	*41*	*42*	*43*

Many properties of isoarithmatic catafusenes are very similar: resonance energy,[57] rate of Diels-Alder cycloaddition with dienophiles,[58] spectral properties,[58] biochemical properties,[59] etc.

Counting nonbenzenoid fully conjugated systems, even for *cata*-condensed systems, is more difficult, but the problem was solved by means of similar reasoning.[60]

Clar's observations about the effect of annelation on electronic absorption spectra of catafusenes, especially acenes, led him to postulate that in acenes, e.g., **35**, only one ring can have an aromatic sextet, whereas in kinked catafusenes such as phenanthrene (**34**), more than one ring has an aromatic sextet.[61] Clar's structures **34A**, **35A**, with circles for sextets and double bonds for remaining rings, express particular features of electronic absorption spectra and also of ¹H-NMR spectra (e.g., for Kekulene **37**, which does not have shielded internal protons like [18]annulene, but behaves rather like a normal catafusene). There exists a close link between Clar's postulate and graph theory: it can be shown that in catafusenes the numbers of rings in rectilinear portions (in each portion only one sextet can be written) are important constitutional parameters for various properties: number of Kekulé structures,[56] resonance energy (linearly correlated to the number of zeros in the 2DC or 3DC codes),[57] kinetics of Diels-Alder addition (where only the longest linear portion matters),[58] etc. The 3DC can also serve for denoting geometrical isomerism, e.g., configurations of annulenes, provided that these configurations are superimposable on the graphite lattice. This lattice is composed of lines which may have one from three possible orientations (hence 3DC). To make this specification of configurations unique, one has to test all starting points around the periphery and the two senses of rotation, and then to choose from the resulting codes that one which leads to the smallest number when read sequentially.[62] For instance, the two geometrical isomers of [16]annulene **44** and **46** have the three-digit notation (3DC) and two-digit notation (2DC) specified under the formulas (middle and bottom lines, respectively); the starting point is marked by a black dot. The upper line indicates the usual notation (UN) resulted by indicating *cis* and *trans* double bond configurations by digits 0 and 1, respectively (also using the convention of the minimal number); in this case, the starting point and the sense of rotation are indicated by the curved arrow. It may be seen that the two Kekulé structures of the former isomer (**44** and **45**) differ in this notation. This may be useful in cases when there is double bond fluctuation, but can be misleading when the two formulas are limiting structures.

UN	00001011	01011011	01011011
3DC	1212123121312323		1212123131212323
2DC	00001110101100		00001100100100

44   **45**   **46**

REFERENCES

1. **Balaban, A. T., Ed.,** *Chemical Applications of Graph Theory,* Academic Press, London, 1976; (a) chap. 1; (b) chap. 2; (c) chap. 5; (d) **Balaban, A. T.,** *Math. Chem.,* 1, 33, 1975; *J. Mol. Struct. Theochem.,* 120, 117, 1985; *J. Chem. Inf. Comput. Sci.,* 25, 334, 1985.
2. **Trinajstić, N., Ed.,** *Chemical Graph Theory,* CRC Press, Boca Raton, Fla., 1983.
3. **Harary, F., Ed.,** *Graph Theory,* Addison-Wesley, Reading, Mass., 1969.
4. **Wilson, R. J., Ed.,** *Introduction to Graph Theory,* Academic Press, New York, 1972.

5. **Essam, J. W. and Fisher, M. E.,** *Rev. Mod. Phys.,* 42, 272, 1970.
6. **Rouvray, D. H.,** *R. Inst. Chem. Rev.,* 4, 173, 1971.
7. **Rouvray, D. H. and Balaban, A. T.,** in *Applications of Graph Theory,* Beineke, L. W. and Wilson, R. J., Eds., Academic Press, New York, 1979, 177.
8. **Sylvester, J. J.,** *Proc. R. Soc. London,* 7, 219, 1854—5.
9. **Biggs, N. L., Lloyd, E. R., and Wilson, R. J., Eds.,** *Graph Theory 1736 — 1936,* Clarendon Press, Oxford, 1976.
10. **Cayley, A.,** *Phil. Mag.,* (4), 13, 172, 1857; (4) 18, 374, 1859.
11. **Cayley, A.,** *Phil. Mag.,* (4), 47, 444, 1874; *Rep. Br. Assoc. Adv. Sci.,* 45, 257, 1875; *Ber. Dtsch. Chem. Ges.,* 8, 1056, 1857.
12. **Cayley, A.,** *Q. J. Pure Appl. Math.,* 23, 376, 1889.
13. **Schiff, H.,** *Ber. Dtsch. Chem. Ges.,* 8, 1542, 1875.
14. **Losanitsch, S. M.,** *Ber. Dtsch. Chem. Ges.,* 30, 1917, 3059, 1897.
15. **Hermann, F.,** *Ber. Dtsch. Chem. Ges.,* 13, 792, 1880; 30, 2423, 1897; 31, 91, 1898.
16. **Redfield, J. H.,** *Am. J. Math.,* 49, 433, 1927.
17. **Lunn, A. C. and Senior, J. K.,** *J. Phys. Chem.,* 33, 1027, 1929.
18. **Henze, H. R. and Blair, C. M.,** *J. Am. Chem. Soc.,* 53, 3042, 3077, 1931; 55, 680, 1933.
19. **Blair, C. M. and Henze, H. R.,** *J. Am. Chem. Soc.,* 54, 1098, 1538, 1932.
20. **Coffman, D. D., Blair, C. M., and Henze, H. R.,** *J. Am. Chem. Soc.,* 55, 252, 1933.
21. **Pólya, G.,** *Acta Math.,* 68, 145, 1937; *Z. Kristallogr.,* (A), 93, 414, 1936; *C. R. Acad. Sci. Paris,* 201, 1167, 1935; 202, 1534, 1936.
22. **Harary, F. and Palmer, E. M.,** *Graphical Enumeration,* Academic Press, New York, 1973; **Balaban, A. T.,** in press.
23. **Otter, R.,** *Ann. Math.,* 49, 583, 1948.
24. **Ruch, E., Hässelbarth, W., and Richter, B.,** *Theor. Chim. Acta,* 19, 225, 1970; **Hässelbarth, W. and Ruch, E.,** *Theor. Chim. Acta,* 29, 259, 1973.
25. **Brocas, J., Gielen, M., and Willem, R.,** *The Permutational Approach to Dynamic Stereochemistry,* McGraw-Hill, New York, 1983, 240.
26. **Robinson, R. W., Harary, F., and Balaban, A. T.,** *Tetrahedron,* 32, 355, 1976.
27a. **Cope, A. C., Haven, A. C., Jr., Ramp, F. L., and Trumbull, E. R.,** *J. Am. Chem. Soc.,* 74, 4867, 1952.
27. **Maier, G.,** *Valenzisomerisierungen,* Verlag-Chemie, Weinheim, 1972, 2.
28. **Balaban, A. T.,** *Rev. Roum. Chim.,* 11, 1097, 1966; erratum, *Rev. Roum. Chim.,* 12 (1), last page, 1967.
29. **Balaban, A. T.,** *Rev. Roum. Chim.,* 15, 463, 1970; **Balaban, A. T. and Deleanu, C.,** *Rev. Roum. Chim.,* in press; **Balaban, A. T. and Banciu, M.,** *J. Chem. Educ.,* 61, 766, 1984.
30. **Balaban, A. T.,** *Rev. Roum. Chim.,* 18, 635, 1973; erratum, *Rev. Roum. Chim.,* 19, 338, 1974.
31. **Read, R. C.,** *J. London Math. Soc.,* 34, 417, 1959; 35, 344, 1960.
32. **Harary, F., Ed.,** *A Seminar on Graph Theory,* Holt, Rinehart & Winston, New York, 1967, 25.
33. **Balaban, A. T., Vancea, R., Motoc, I., and Holban, S.,** *J. Chem. Inf. Comput. Sci.,* 26, 72, 1986; **Balaban, A. T., Filip, P., and Balaban, T. S.,** *J. Comput. Chem.,* 6, 316, 1985.
34. **Balaban, A. T.,** *Rev. Roum. Chim.,* 19, 1611, 1974; erratum, *Rev. Roum. Chim.,* 23, 311, 1978; **Carhart, R. E., Smith, D. H., Brown, H., and Sridharan, N. S.,** *J. Chem. Inf. Comput. Sci.,* 15, 124, 1975.
35. (a)**Simmons, H. E., III and Maggio, J. E.,** *Tetrahed. Lett.,* 22, 287, 1981; (b) **Paquette, L. A. and Vazeux, M.,** *Tetrahed. Lett.,* 22, 291, 1981; (c) **Walba, D. M., Richards, R. M., and Haltinwanger, R. C.,** *J. Am. Chem. Soc.,* 104, 3219, 1982.
36. **Balaban, A. T.,** *Rev. Roum. Chim.,* 17, 865, 1972; **Banciu, M., Popa, E., and Balaban, A. T.,,** *Chem. Scripta,* 24, 78, 1984.
38. **Schäfer, W. and Hellmann, H.,** *Angew. Chem.,* 79, 566, 1967; *Angew. Chem. Int. Ed. Engl.,* 6, 518, 1967.
39. **Burnside, W.,** *Theory of Groups of Finite Order,* 2nd ed., Theorem VII, 1911, 191; reprinted, Dover, New York, 1959.
40. **Balaban, A. T., Farcasiu, D., and Harary, F.,** *J. Label. Comp.,* 6, 211, 1970.
41. **Balaban, A. T. and Harary, F.,** *Rev. Roum. Chim.,* 12, 1511, 1967.
42. **Balaban, A. T., Palmer, E. M., and Harary, F.,** *Rev. Roum. Chim.,* 22, 517, 1977.
43. **Balaban, A. T.,** *Rev. Roum. Chim.,* 20, 239, 1975; **Balaban, A. T. and Baciu, V.,** *Math. Chem.,* 4, 131, 1978.
44. **Balaban, A. T.,** *Rev. Roum. Chim.,* 19, 1323, 1974.
45. **Banciu, V. and Balaban, A. T.,** *Rev. Roum. Chim.,* 25, 1213, 1980.
46. **Balaban, A. T.,** *Rev. Roum. Chim.,* 19, 1185, 1974.
47. **Banciu, M. and Balaban, A. T.,** *Chem. Scripta,* 22, 188, 1983.
48. **Balaban, A. T.,** *Rev. Roum. Chim.,* 22, 987, 1977.
49. **Balaban, A. T. and Harary, F.,** *Tetrahedron,* 24, 2505, 1968.

50. **Balaban, A. T.,** *Tetrahedron,* 25, 2949, 1969.
51. **Balaban, A. T.,** *Math. Chem.,* 2, 51, 1976.
52. **Read, R. C. and Harary, F.,** *Proc. Edinburgh Math. Soc.,* (2) 17, 1, 1970.
53. **Lunnon, W. F.** *Graph Theory and Computing,* Read, R. C., Ed., Academic Press, New York, 1972, 87; **Knop, J. V., Müller, W. R., Saymanski, K., and Trinajstić, N.,** *Computer Generation of Certain Classes of Molecules,* Editions Kemija u Industriji, Zagreb, 1985.
54. **Balaban, A. T.,** *Pure Appl. Chem.,* 52, 1409, 1980.
55. **Balaban, A. T.,** *Rev. Roum. Chim.,* 22, 45, 1977.
56. **Balaban, A. T. and Tomescu, I.,** *Math. Chem.,* 14, 155, 1983.
57. **Balaban, A. T.,** *Rev. Roum. Chim.,* 15, 1243, 1970.
58. **Biermann, D. and Schmidt, W.,** *J. Am. Chem. Soc.,* 102, 3163, 3173, 1980.
59. **Balaban, A. T., Biermann, D., and Schmidt, W.,** *Nouv. J. Chim.,* 9, 443, 1985.
60. **Balaban, A. T.,** *Rev. Roum. Chim.,* 15, 1251, 1970.
61. **Clar, E.,** *Polycyclic Hydrocarbons,* Academic Press, London, 1964; *The Aromatic Sextet,* John Wiley & Sons, New York, 1972.
62. **Balaban, A. T.,** *Tetrahedron,* 27, 6115, 1971; *Rev. Roum. Chim.,* 21, 1045, 1976.
63. **Walba, D. M.,** *Tetrahedron,* 41, 3161, 1985; **Walba, D. M.,** in *Chemical Applications of Topology and Graph Theory,* Elsevier, Amsterdam, 1983, 17.

Chapter 4

ANNULENES

I. INTRODUCTION

Sondheimer and Wolovsky[1] introduced the term [*n*]annulenes for denoting monocyclic conjugated systems (CH)$_n$ where *n* is an even number. We should recall the theorem stating that cubic graphs (corresponding to annulenes and their valence isomers) must have an even number of vertices (corresponding to CH groups). Thus cyclobutadiene, benzene, and cyclooctatetraene are [4]-, [6], and [8]annulene, respectively. Actually, Thiele[2] is responsible for having first tried to probe the mystery of aromatic stabilization; he coined the term "conjugation" and postulated the "partial valence hypothesis" to explain 1,4-additions of conjugated dienes and to account for the strange behavior of benzene (no polymerization tendency, predominance of substitutions over additions). When Willstätter could not synthesize cyclobutadiene[3] and showed[4] that cyclooctatetraene had typical olefinic character, although according to Thiele's hypothesis they were expected to resemble benzene, this hypothesis was abandoned; its merit was to adumbrate electronic delocalization.

According to Hückel's (4*m* + 2) π-electron rule, annulenes are grouped into two series: (1) the Hückel [*n*]annulenes which fulfill the Hückel rule, *n* = 4*m* + 2, having an odd number of cyclically conjugated double bonds; whenever there is a satisfactory planar geometry these annulenes are aromatic; (2) the anti-Hückel [*n*]annulenes, *n* = 4*m*, with an even number of double bonds; the negative value of the DRE makes the planar delocalized systems antiaromatic, i.e., destabilized relative to the systems with noninteracting π-electrons. It was believed that, whenever possible, such systems avoid the planar geometry, e.g., [8]annulene assumes a tub-shape fluxional geometry.

Calculations by Allinger and Yuh[5] using molecular mechanics (MMP2 program) gave good agreement with experiment for the heat of formation and for the barrier to planarity ($\Delta G^{\ddagger}_{exper}$ = 13.7 kcal/mol, while the difference in *calculated* energy between the planar and nonplanar forms is 13.67 kcal/mol). Allinger et al.[5] pointed out that comments such as "planar cyclooctatetraene puckers because it is antiaromatic" are completely erroneous. Puckering serves only to raise its pi energy still further, from −2.8 to −21.4 kcal/mol. The ring does not pucker to relieve its antiaromaticity, but to relax the strain in the σ system from 25.5 to 12.2 kcal/mol, even though it raises thereby the energy of the pi system. The only way it might relieve the antiaromatic character would be to change its bonding, e.g., to break a bond and form an open chain. On the other hand, localized double bonds are possible even for planar [4*m*]annulenes, as proved by cyclobutadiene.

The DRE values and the steric strain decrease with increasing *n*; therefore, the larger annulenes have lower energy constraints than the smaller annulenes.

The chemistry of annulenes was reviewed by several authors both *in toto*[6-9] and by classes (*vide infra*). In the present chapter we shall discuss the monocyclic (nonbridged) systems; dehydroannulenes will be dealt with at the end of this chapter; the following chapter discusses the bridged systems.

This chapter will discuss first the small- and medium-sized annulenes in individual sections ([4]-, [8]-, and [10]annulene). Benzene, i.e., [6]annulene is too well known to be mentioned here other than as a reference compound. Then two general methods for the synthesis of macrocyclic annulenes ([12]- to [30]annulenes) will be discussed and the remaining individual annulenes will be reviewed.

Since the C–C–C bond angles for sp^2-hybridized systems are 120°, the least strained configurations of annulenes are those which (1) are superimposable on the graphite lattice and (2) have the lowest van der Waals interactions between "inner" hydrogens. By means of dualist graphs for *peri*-condensed polyhexes, the perimeters of such systems (hence all possible configurations of annulenes) may be easily obtained.[11] There are thus 12 possible configurations of [18]annulene with angles equal to 120° from which two are indicated by formulas **1** and **2,** and two possible such configurations of [16]annulene, namely **3** and **4.** In addition, there exist (1) configurations with angles of 120° where an edge is used several times (e.g., twice the benzene perimeter leads to a [12]annulene, three times to [18]annulene, etc.), but such configurations were never observed and (2) configurations with angles differing from 120°, e.g., for [16]annulene the configuration **5** with fourfold symmetry.

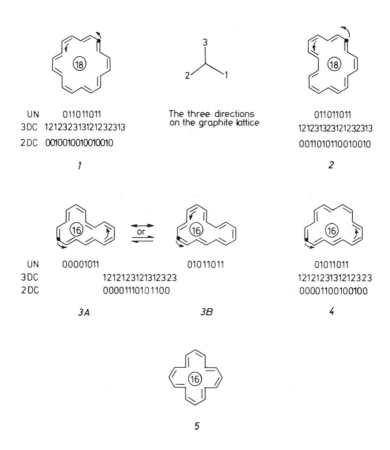

UN   011011011
3DC   121232313121232313
2DC   0010010010010010

1

The three directions
on the graphite lattice

011011011
121231323121232313
0011010110010010

2

UN   00001011
3DC
2DC

01011011
1212123121312323
0000111010 1100

3A   *3B*

01011011
1212123131212323
00001100100100

4

5

For specifying and naming configurations, the oldest proposal derived from usual notation (UN) was to denote by *cis/trans* or ZE the double bond configurations (sometimes these prefixes are abridged as C/T for brevity when there are many such bonds, e.g., *all-cis* [10]annulene is CCCCC-[10]annulene). A newer variant assigns digit 0 to *cis* and 1 to *trans*, selects the lowest resulting number read sequentially among all possible ones for a given system, and converts this binary number into a decimal.[10] However, it was pointed out[11] that this system presents ambiguities both for localized and for delocalized systems, as demonstrated by the fact that the localized **3A** and **3B** formulas have different ZE notations, while the different configurations **3B** and **4** have the same UN(ZE) notation (the latter configuration has twofold symmetry; therefore, its UN symbol is the same for a localized or delocalized electronic structure). The two different configurations **1** and **2** also have the same UN notation. In Chapter 3, structures **3** and **4** have already been mentioned as **44** to **46**.

For configurations having only angles of 120° a different notation was proposed,[11] taking advantage of the fact that there exist only three directions for C–C bonds in a given graphite lattice[12] we may denote these three directions by digits 1 to 3 and we obtain thus a three-digit code (3DC) as indicated by formulas **1** to **4**. Again, for uniqueness we select from all initial points and from the two senses of rotation the one which leads to the smallest number on sequential reading of the digits. As a simplification, the 3DC may be converted into a two-digit code (2DC) by converting overlapping triads of digits *aba* (transoid triad) into 0 and *abc* (cisoid triad) into 1. The 3DC and 2DC are in one-to-one correspondence, and have no ambiguities, but are applicable only to systems superimposable on the graphite lattice, whereas the ZE system may be applied to any configuration, including **5**. On formulas **1** to **4**, inner arrows denote the start for the ZE code, while the dot and outer arrow are for the 2DC and 3DC. It may be observed that for a given [*n*]annulene, the ZE code has *n*/2 digits in the binary system, the 3DC has *n* digits, and the 2DC has *n* − 2 digits. Like the ZE system, the 2DC binary number can be converted into decimal notation.

We shall now start to review the various annulenes in the order of increasing ring size.

Haddon et al.[302] as well as Ichikawa[303] calculated the stabilities of hydrogen chain systems for obtaining simple models of aromaticity and antiaromaticity.

Other theoretical approaches for aromaticity were published by Ichikawa and Ebisawa[304] using Hartree-Fock MO theory, by Haddon and Raghavachari[305] for [18]annulenes and bridged [10]-annulenes, by Herndon and Hosoya[306] using Clar structures, by Glidewell and Lloyd[307] using the MNDO method for bi- and tricyclic hydrocarbons, and by Norrinder et al.[308] using HMO methods for various π-systems.

Michl[310] reviewed the magnetic circular dichroism of aromatic molecules. Theoretical models and calculations for para- and diatropicity of anti-aromatics were presented by Stollenwerk et al.,[311] for dehydro[12]annulenes, Minsky et al.[312] for antiaromatics, and Aihara[313] for biphenylene and related hydrocarbons.

II. CYCLOBUTADIENE (CBD)

A book[13] published in 1967 and two more recent excellent reviews[14,15] discuss in detail the chemistry of cyclobutadiene (CBD). Attempts to synthesize cyclobutadiene were first performed by Kekulé.[16] Perkin[17] attempted to obtain 1,2-cyclobutadienedicarboxylic acid from 1,2-dibromocyclobutane-1,2-dicarboxylic acid and bases but the reaction yielded only 2-bromo-1-cyclobutenecarboxylic acid (**6**). Willstätter and Schmädel[18] prepared *trans*-1,2-dibromocyclobutane **7**; on refluxation with aqueous KOH the product was 1-bromocyclobutene (**10**, X = Br); 2 mol of hydrogen bromide were eliminated by heating in quinoline, when 1,3-butadiene was formed, or with KOH at 210°C when acetylene resulted.

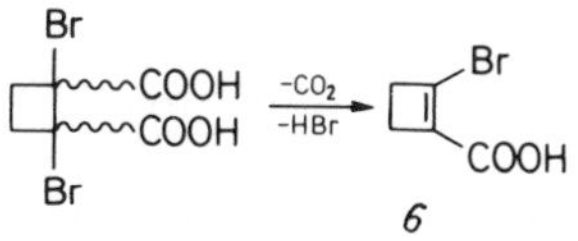

Later, Buchman et al.[19] studied the Hofmann degradation of the 1,2-diquarternary ammonium salt **8**, but no CBD was found. In these cases (**9**), monoelimination leads to an unreactive intermediate **10** with the second leaving group bonded to an olefinic cyclobutene carbon.

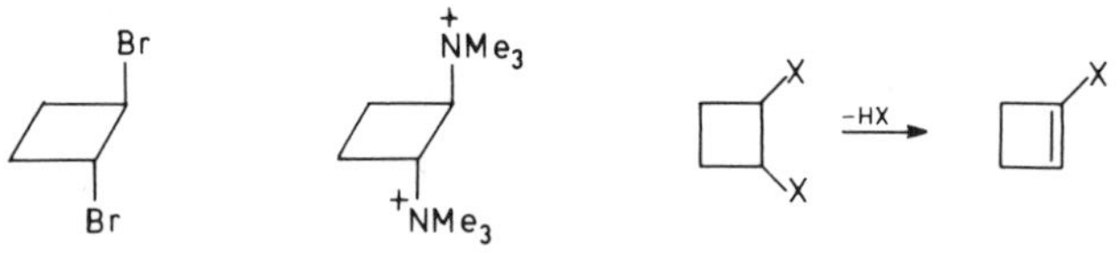

7   8   9   10

Nenitzescu and co-workers[10,21] realized that elimination from *trans*-1,2-disubstituted cyclobutanes led to 1-substituted cyclobutenes rather than to cyclobutadiene; they reasoned that these systems required too drastic conditions and that either *cis*-1,2-dibromocyclobutane or 1,3-disubstituted cyclobutanes had a better chance of leading to cyclobutadiene. However, Hofmann degradation of the 1,3-diquaternary base **11** afforded 1,3-butadiene **12**; the same product resulted from the Alder-Rickert degradation of various adducts, e.g., **13**; it appears that reactions performed at elevated temperature yield 1,3-butadiene instead of cyclobutadiene by a kind of disproportionation. Reactions proceeding under milder conditions were then tested, namely (1) elimination of HBr from 1,3-dibromocyclobutane when vinylacetylene was identified as the only reaction product; (2) halogen elimination from tetrahalocyclobutanes or dihalocyclobutenes with metals: stereoisomeric 1,2,3,4-tetrabromocyclobutanes **14** and *cis*-3,4-dichlorocyclobutene **15** were prepared starting from cyclooctatetraene (COT).[20]

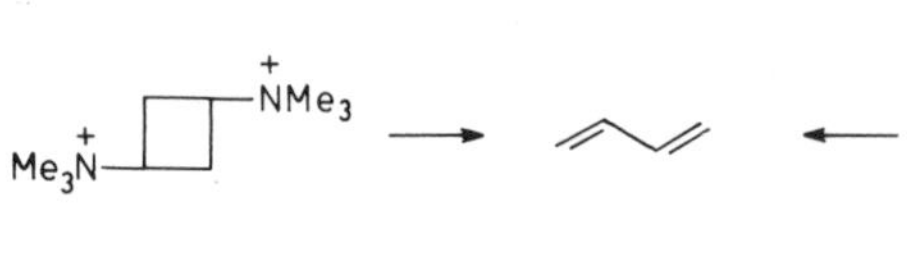

11   12   13

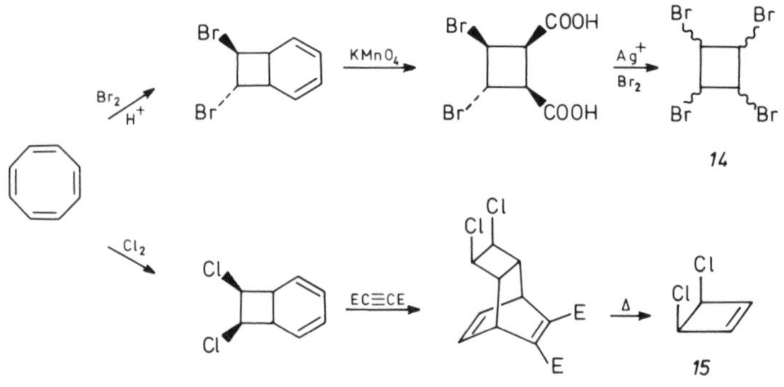

14

15

From *all-trans*-**14** with lithium amalgam, or from **15** with sodium amalgam, the *syn*-cyclobutadiene dimer **16** was obtained, but from **15** and lithium amalgam the *anti*-stereoisomer **17** was formed (see also Chapter 6). Both compounds rearranged thermally to COT.[21] The *syn*-dimer yields an AgNO₃ complex which was initially believed to be a CBD-complex.[21]

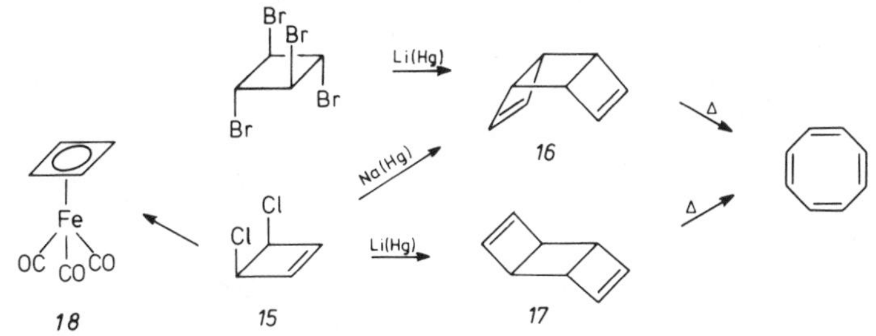

Pettit and co-workers[22,24] were the first to obtain compelling evidence for the fleeting existence of CBD; on treating Nenitzescu's dichloride **15** with $Fe_2(CO)_9$ they obtained a true CBD-iron tricarbonyl complex **18**.[22] The same complex results from dichloride **15** with lithium amalgam and iron pentacarbonyl. This complex **18** presents a remarkable stability despite its steric strain; for instance, it may be acylated under Friedel-Crafts conditions.[23] On treating this complex with ceric ion, CBD (**19**) is formed as a transient species which dimerizes to a mixture of **16** and **17** but may be trapped in stereospecific [4 + 2]-cycload-ditions with maleic or fumaric esters[24] yielding bicyclic adducts **20** and **21**, respectively.

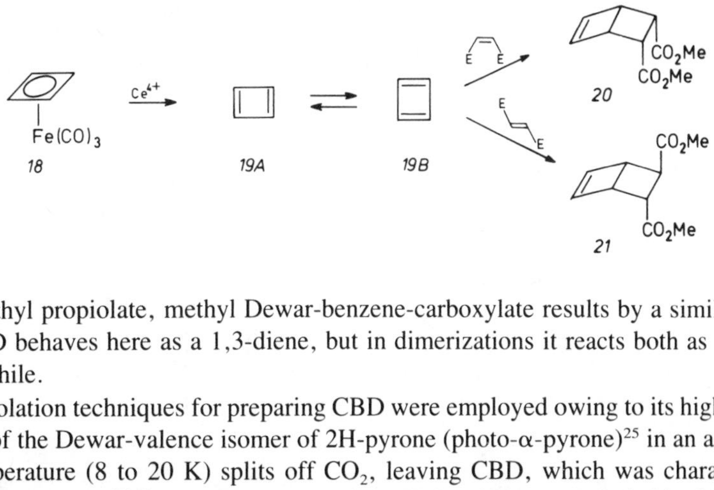

With methyl propiolate, methyl Dewar-benzene-carboxylate results by a similar cycload-dition. CBD behaves here as a 1,3-diene, but in dimerizations it reacts both as a diene and as a dienophile.

Matrix isolation techniques for preparing CBD were employed owing to its high reactivity: photolysis of the Dewar-valence isomer of 2H-pyrone (photo-α-pyrone)[25] in an argon matrix at low temperature (8 to 20 K) splits off CO_2, leaving CBD, which was characterized by IR spectrophotometry.[26,27] The debate around the IR bands was concluded by Maier,[28] who showed that the IR bands due to CBD were situated at 570 and 1240 cm^{-1} (low-intensity electronic absorption band around 300 nm)[15] and that other absorptions (IR and UV) were due to charge transfer complexes involving CBD and the other components present in the matrix (photofragments).

Alternatively, CBD may be produced by photolysis of other precursors **23** to **30** (some of them with a propellanic structure) in a low-temperature matrix. In some of these reactions, e.g., from **22** or **25**, cyclopentadienone is also formed. Alternatively, CBD may be also obtained in gas phase, by flash vacuum pyrolysis, or gas-phase flash photolysis; the half-life of CBD is 2 msec at a pressure of 0.1 Torr,[29] and 10 msec at 0.035 Torr.[30] Mild oxidation of the Dewar-dihydropyrazine **26** yields the diaza-Dewar-benzene **27** which loses nitrogen spontaneously, giving CBD.[31]

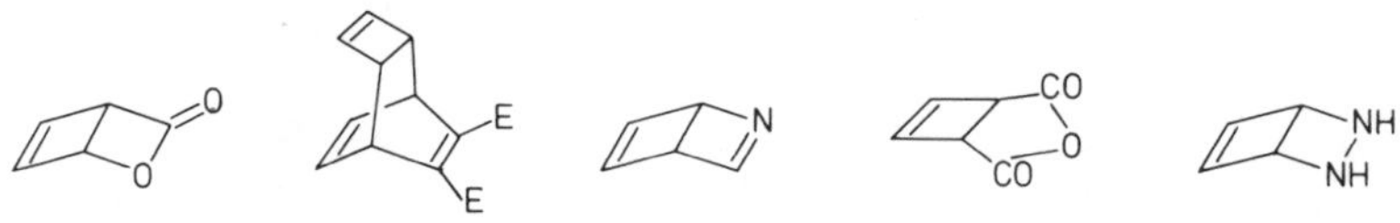

22   23   24   25   26

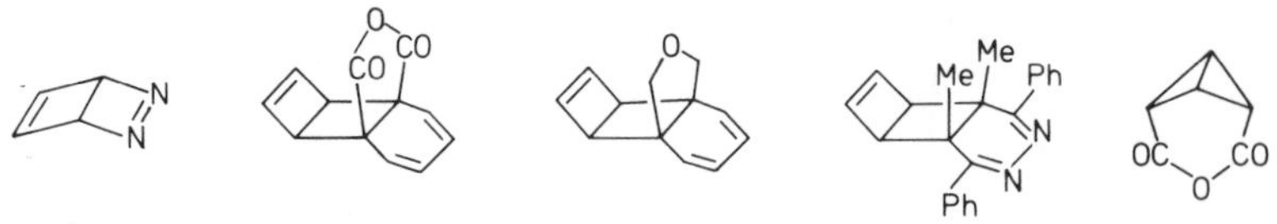

27   28   29   30   31

Photolysis of bicyclobutane-2,4-dicarboxylic anhydride **31** in an argon matrix at 10 K leads to CBD and not to tetrahedrane, as shown by IR and UV spectra;[32] thermolysis of **31** affords $\overset{.}{C}O_2$ and cyclopentadienone. On heating the matrix, CBD dimerizes to the *syn*-dimer **16** which may be isolated, but the monomeric CBD may be trapped by dienophiles (*p*-benzoquinone, cyclopentadiene, etc.).

When it was not yet clear whether the oxidation of the $(CH)_4Fe(CO)_3$ complex yielded free CBD or a complex of $(CH)_4$ with Ce ions, Rebek[33] applied the "three-phase test" for the detection of free CBD: a polymer-bound phenanthroline derivative of CBD-iron-carbonyl complex **32** was oxidized by ceric ions in the presence of a polymer-bound maleimide derivative **33**. The result (96% transfer of CBD between the two solid phases leading to **34**) clearly demonstrates the fact that Ce ions are not vehicles for CBD and is a proof for the intermediacy of free CBD. In the formulas, ℗ denotes polystyrene.

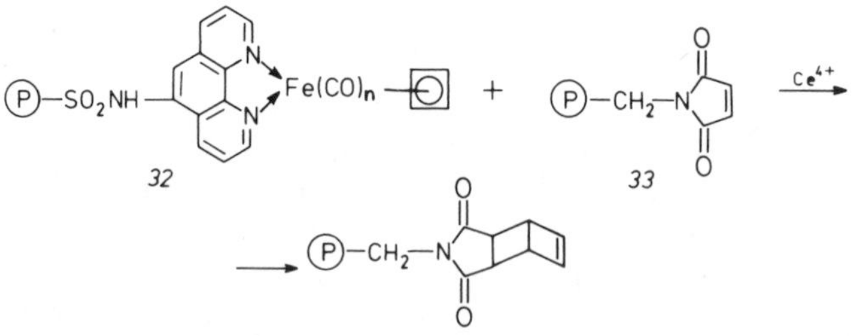

32   33

34

Another such proof involved optically active metallic complexes of substituted CBD (see below).

The structure of CBD was debated for a long time. Simple HMO theory predicted a triplet ground state, but it became apparent that such a square geometry was unstable relative to a second-order Jahn-Teller distortion leading to a singlet rectangular ground state.[34] With few exceptions, all types of theoretical calculations (*ab initio*[35a,b] or semiempirical[36a,b] result in a singlet rectangular ground state. The singlet square excited state is higher in energy (this resembles the transition state for the bond-shift process), while the singlet triplet state is probably still higher. It will be seen below that all substituted CBD derivatives whose molecular structure was established by X-ray diffraction have a rectangular geometry.

Chapman et al.[37] photolyzed in argon matrices at 8 K 5,6-[²H₂]- and 3,6-[²H₂]-2H-pyrone (**35** and **36**), observing the same IR spectrum in both cases. Their conclusion that CBD may have a square geometry was baffling for several years, but is contradicted by the number and position of the IR bands; *ab initio* calculations indicated[35b] that the IR spectra agree with a rectangular geometry undergoing fluctuation, so that the observed IR spectra are evidence for a mixture of two valence bond isomers **37A,B**.

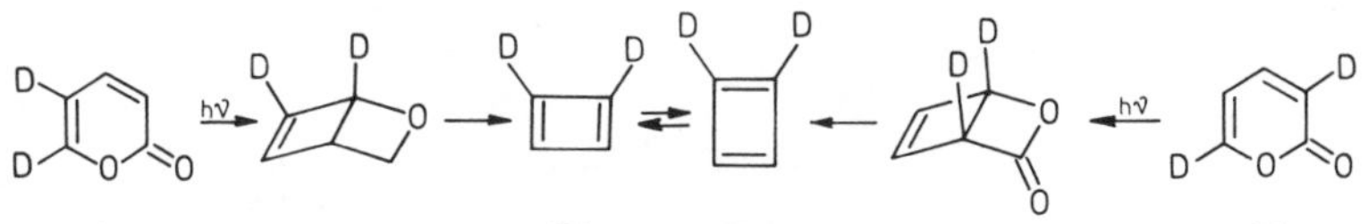

35   37A   37B   36

By preparing vicinal dideuteriated CBD via several different routes, Carpenter and co-workers[38] produced convincing evidence for the rectangular ground state. The bond-shift reaction activation parameters are indicative that the activation entropy is found to be negative at the low temperatures (-10 to $50°C$) of the trapping experiments.

Whereas the vicinally deuterated CBD-Fe(CO)$_3$ complex yields on oxidation with tBuOCl an equimolecular mixture of all adducts, the reaction of methyl (Z)-3-cyanoacrylate with the azo compound **38** gives adducts **39** to **42** (R = CN) with a distribution of deuterium labels which depends on concentrations and temperatures, i.e., on the ratio k_2/k_1. No such effect was found with methyl acrylate (k_2/k_1 too small), or with methyl azodicarboxylate (^{1}H-NMR coupling constants too small to be observable). The results were interpreted as indicating that the "negative entropy" for this intramolecular bond shift (rate constant k_1) is an artifact of the curvature in the Arrhenius plot due to quantum-mechanical tunneling; despite the fact that the CH or CD groups are heavy, they do travel a very short distance in the bond-shift reaction (0.2 Å). The above authors[38] argued that such tunneling might be operative in tri- and tetra-*t*-butyl-CBD (see below), and also in other bond-shift fluctuating structures such as cyclooctatetraene or semibullvalene. Newer calculations (Huang, M.-J. and Wolfsberg, M., *J. Am. Chem. Soc.*, 106, 4039, 1984; Dewar, M. J. S., Merz, Jr., K. M., and Stewart, J. J. P., *J. Am. Chem. Soc.*, 106, 4040, 1984) confirm the negative activation entropy and indicate that microwave spectra of cyclobutadiene will provide interesting information.

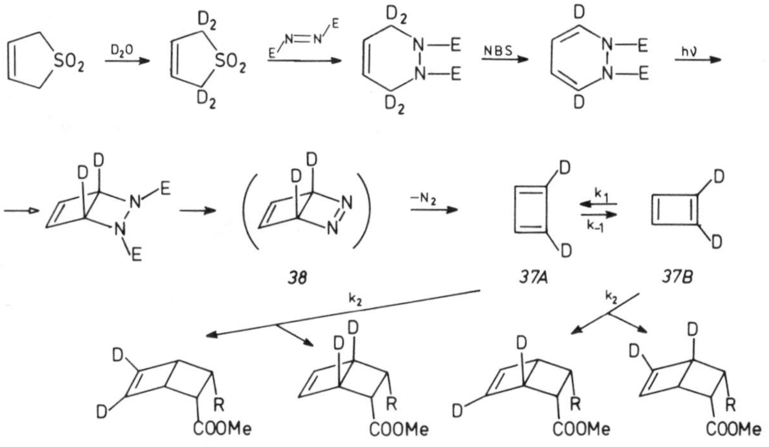

38   37A   37B

39   40   41   42

A. Substituted Derivatives of Cyclobutadiene

Practically simultaneously with Nenitzescu's attempts to obtain cyclobutadiene, Criegee investigated the chemistry of tetramethylcyclobutadiene.[39]

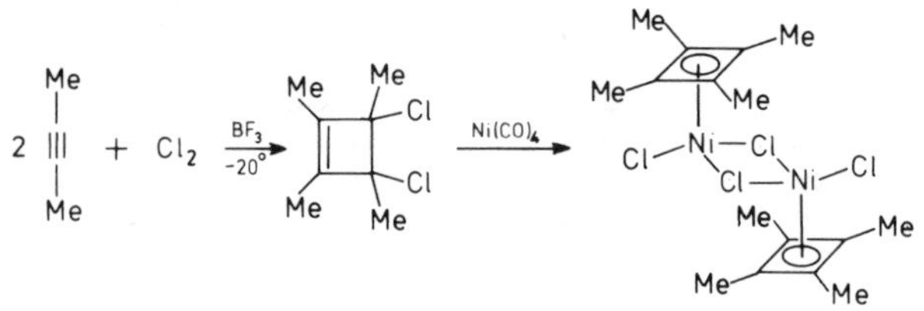

43   44

Criegee and co-workers[40] scaled up to kilogram amounts the procedure for obtaining 1,2-dichloro-1,2,3,4-tetramethylcyclobutene **43** from 2-butyne and chlorine,[41] raising the yield to 50%. Further treatment with nickel carbonyl gave a violet nickel complex of tetramethylcyclobutadiene **44** whose molecular structure was determined by X-ray crystallography.[42] The structure confirms the theoretical prediction of Longuet-Higgins and Orgel.[43] The dangling chlorine ligands in **44** may be displaced by solvent molecules (water, chloroform). Hydrogenation of the complex yielded tetramethylcyclobutane, but pyrolysis failed to yield free $(CMe)_4$, **45**. On refluxing **44** in water, the *anti*-dimer **46** (Schröder's dimer) was obtained.[44] Treatment of the complex with phenanthroline followed by pyrolysis afforded an isomer of Me₄CBD, namely a methylene-trimethylcyclobutene.[45] A different (*syn*) dimer (Maier's dimer) **47** was obtained when the dichloro-derivative was treated with lithium amalgam in ethyl ether at 20°C.[46] The structures of the two dimers were determined by ozonolysis; the *syn*-dimer gave a tetraketal **48**, while the *anti*-dimer gave a tetraketone **49**. The ''free'' unstable **45** was generated photochemically in a matrix at −196°C by Maier; on thawing, it yielded the *syn*-dimer **47**.[48]

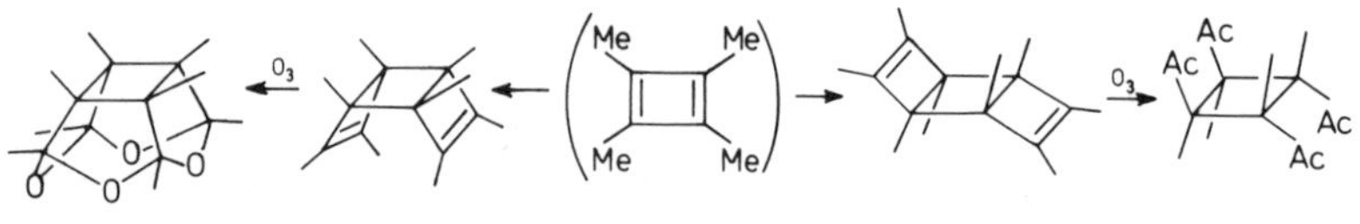

48   47   45   46   49

Cookson and co-workers[47] were able to trap elusive CBDs by dienophiles leading to Dewar-benzene derivatives. Photoirradiation of two different anhydrides **50** and **51** gave tetramethylcyclobutadiene **45** at −196°C but the electronic absorption spectra in the matrix were different. By contrast, the product formed from **52** via a third photolytic route gave the expected electronic absorption spectrum since in this case no charge transfer complex could result in the matrix between photolysis products.[49]

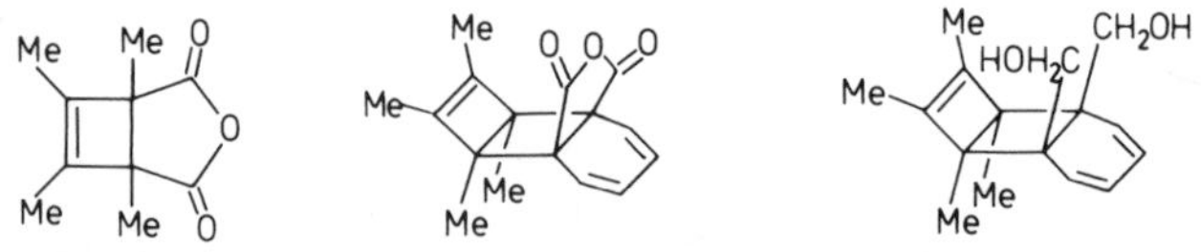

50   51   52

Perfluoromethylcyclobutadiene **53** (as an unstable species) was formed by photolysis of an ozonide **54** at −196°C or by pyrolysis of **55**.[50] Normally, **53** dimerized to a *syn*-dimer, but it could be trapped by ethyl azodicarboxylate.[51]

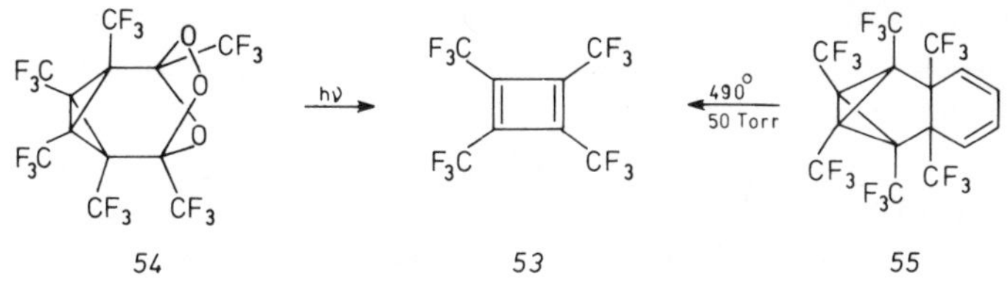

54   53   55

Similarly, tetrafluorocyclobutadiene **56** was formed photolytically, but dimerized to **57** and **58** in the absence of scavengers. It was trapped as Diels-Alder adduct **59** with furan.[52]

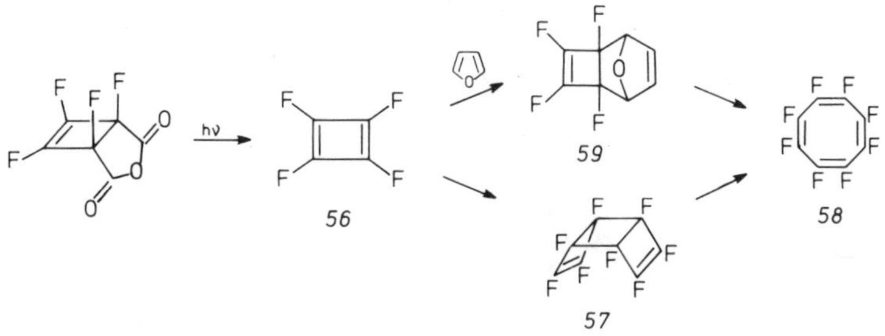

Analogous photoirradiation[53] of the tetrachlorocyclobutane-1,2-dicarboxylic anhydride at −196°C or at room temperature affords as isolable product octachloro-COT, **60**.

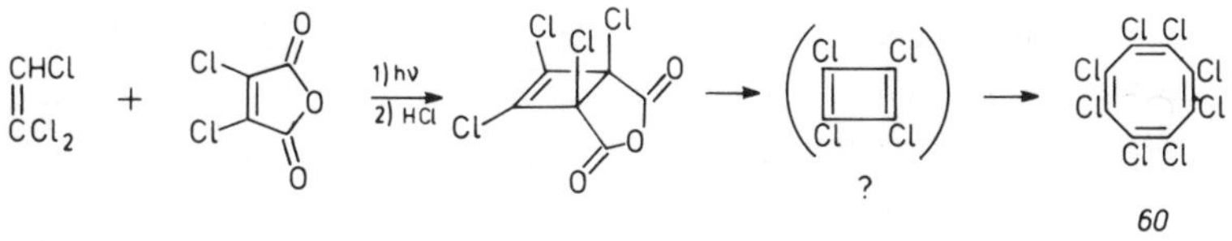

The first stable derivatives of CBD to be isolated, **61**, had electron-donating and electron-withdrawing groups (push-pull, or capto-dative stabilization);[54] since, however, in these cases a large contribution of dipolar limiting structures **61B** was expected, less perturbed CBD derivatives were investigated.

Starting from the idea that steric hindrance might curtail the [2 + 2]-cyclodimerization, Kimling and Krebs[55] prepared the crystalline tricyclic disulfide **62** whose molecular structure had a planar rectangular CBD ring with single (1.54 and 1.55 Å) and double bonds (1.41 and 1.38 Å) by X-ray crystallography. For the problem under examination this compound has the drawback that the two heterocyclic rings favor one of the two valence bond isomers, namely, that one which has the double bonds endocyclic in the heterocyclic rings.

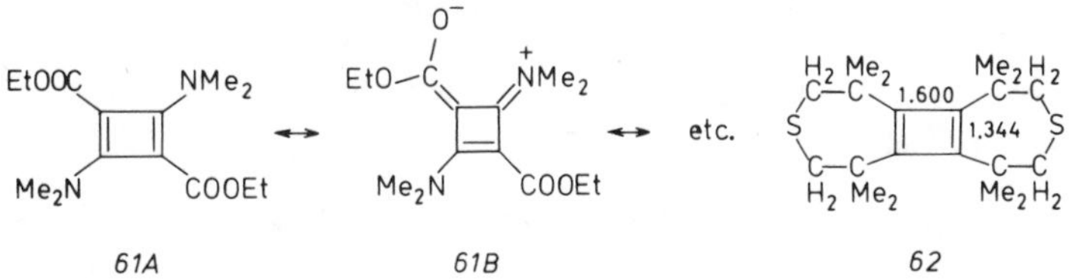

61A   61B   etc.   62

A smaller steric hindrance is present in tri-*t*-butylcyclobutadiene **63**[56,57] whose ^{1}H-NMR data are 2-tBu, $\delta = 1.14$; 1,3-tBu$_2$, $\delta = 1.05$; 4-H, $\delta = 5.35$ ppm. The last value is consistent with a paramagnetic ring current ($\Delta \delta = 1$ ppm).

Interestingly, this hydrocarbon is strongly basic,[58] comparable to alkoxide; electrophiles (e.g., ROO$^+$) add to the monosubstituted double bond, probably via a homocyclopropenium cationic intermediate **64**.

At room temperature, dimerization of **63** to the least hindered *syn*-dimer occurs; tri-*t*-butyl-CBD **63** may be trapped by dienophiles such as maleic anhydride as a *syn*-adduct. Low-temperature NMR spectra,[59] as well as photoelectron spectra of **63**[60] gave evidence for a singlet ground state with rectangular geometry with a bond-switching barrier of ~2.5 kcal/mol.

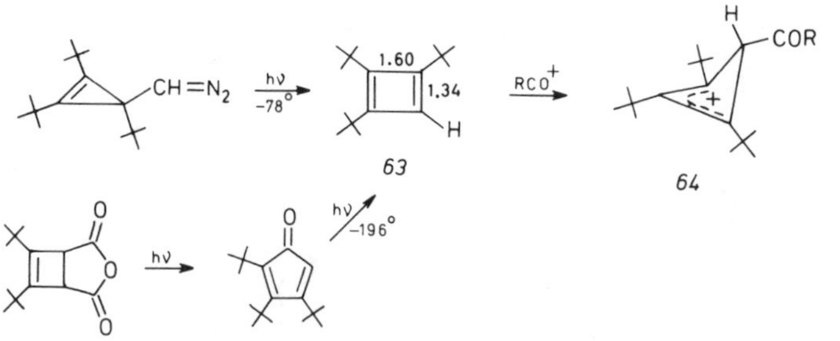

Masamune et al.[34] reacted at −110°C tri-*t*-butylcyclopropenium fluoroborate **65** with the anion of the diazo-ester **66** and photolyzed at −78°C the resulted cyclopropene **67** in oxygen-free pentane. Methyl tri-*t*-butylcyclobutadienecarboxylate **68** precipitated and was purified by sublimation at 50°C under reduced pressure; this tetrasubstituted CBD is thermally stable but reacts with oxygen. The ¹H- and ¹³C-NMR data indicate approximate twofold symmetry as requested by as a CBD structure and not a threefold symmetry in the case of a tetrahedranic structure. The rectangular planar geometry of **68** is confirmed by X-ray crystallography[61] which also indicates that the carbomethoxy and C=C systems are orthogonal owing to steric factors.

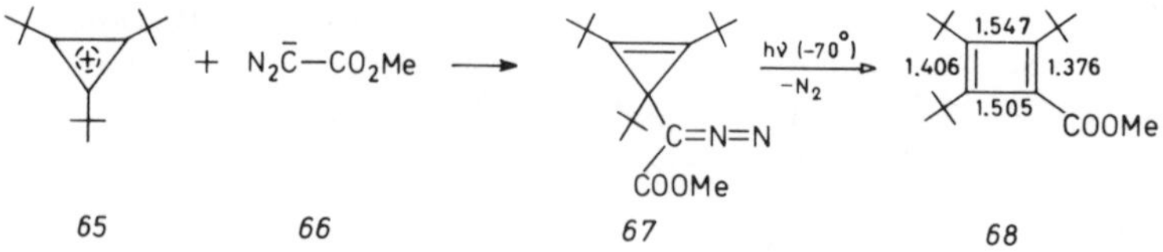

Both **63** and **68** react with dienophiles such as maleic anhydride yielding normal *endo* adducts. The adduct **69** of **68** with acetylenedicarboxylic esters yielded a Dewar-benzene (with the CBD ester group bonded to the bridgehead carbon) which does not ring-open to a benzene derivative because this would have three vicinal *t*-butyl groups in addition to three carbalkoxy groups. Tri-*t*-butylcyclobutadiene **63** reacts also as if it could easily adopt a triplet ground state: with CCl₄ an addition product **70** is formed at the singly substituted double bond.[62] Interestingly, the bond shift in tri-tBu-CBD (**63**) cannot be frozen even at −185°C (¹³C-NMR spectral evidence);[63] this fact is consistent with Carpenter's tunneling hypothesis mentioned above.

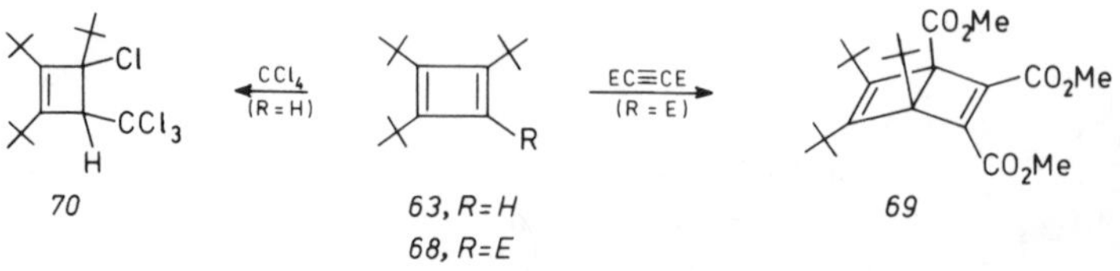

A striking difference exists between tri- and tetra-*t*-butyl-CBD, illustrated by their photo- and thermochemistry (**63, 71** to **76**).

The most spectacular substituted CBD derivative is the stable crystalline tetra-*t*-butyl-CBD (**74**) obtained by Maier,[64] via **71, 72,** and **73** and also discussed in Chapter 6 because its chemistry is intimately connected to that of its stable valence isomer, tetra-*t*-butyltetra-hedrane **73**. The geometry of **74** was obtained by low-temperature X-ray crystallography:[65] a planar ring with alternating double (1.441 Å) and single bonds (1.524 Å), and with four *t*-butyl groups in practically square arrangement. At room temperature the bond lengths are closer (1.464 and 1.483 Å),[66] yet this geometry has still alternating bonds, as shown also by photoelectron spectra.[67]

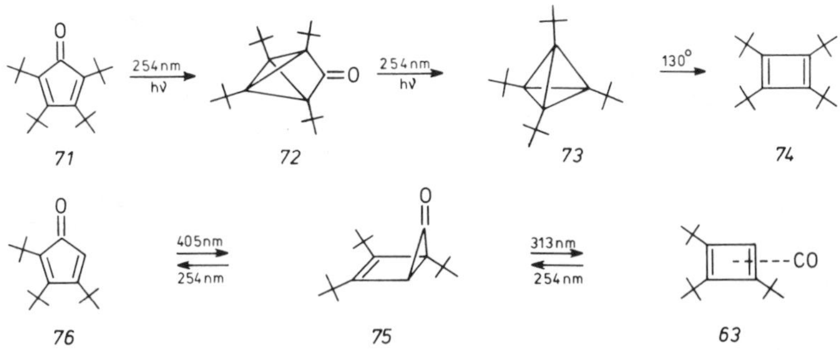

Reactions of t-butyl tri-t-butylcyclobutadiene carboxylate with tetracyanoethylene, acid chlorides, aldehydes, N-acylimines or diaromethane were reported by Regitz and co-workers.[324]

B. Metallic Complexes of CBD

These deserve a special discussion because they were prepared before CBD itself, because they are aromatic unlike CBD, and because they are the starting materials for obtaining CBD via thermal, photochemical, or redox reactions.

Malatesta et al.[68] discovered that palladium chloride dimerized diphenylacetylene **77** in ethanol to a tetraphenyl CBD-palladium complex. Soon afterwards, Blomquist and Maitlis[69] confirmed the structure of the product and found that the same reagents formed in aprotic solvents hexaphenylbenzene **80** and a salt-like complex of CBD, **81**.

By ligand transfer, complex **79** may be converted into complexes with other metals such as nickel, or (via metal carbonyls) iron, cobalt, etc.[70] By X-ray crystallography the Malatesta reaction was found to involve *endo*- and *exo*-ethoxy-tetraphenylcyclobutenyl complexes, **78** and **82**,[71] respectively.

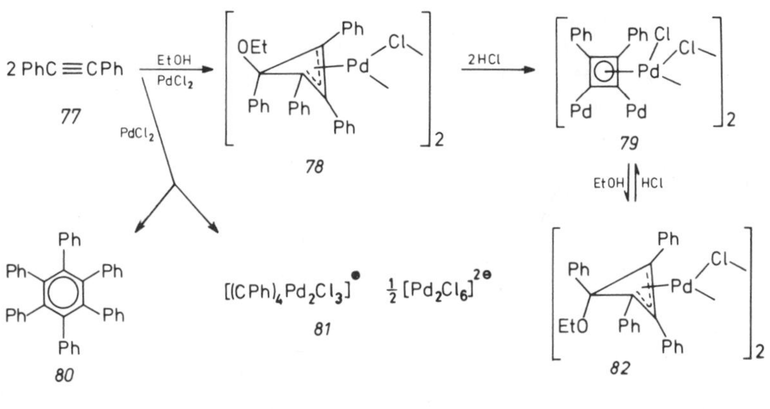

t-Butyl-methyl-acetylene **83** dimerizes with $PdCl_2$, affording σ-butadienylpalladium complexes (**84, 86**) which rearrange to salt-like (**85**), then to neutral complexes (**17**). Since the first reaction leads to the systems with the smallest steric interaction (**84, 86**) the resulting CBD has a larger strain (vicinal di-*t*-butyl-groups) than if the cyclization had proceeded directly, when 1,3-di-*t*-butyl-substituted systems would have resulted.

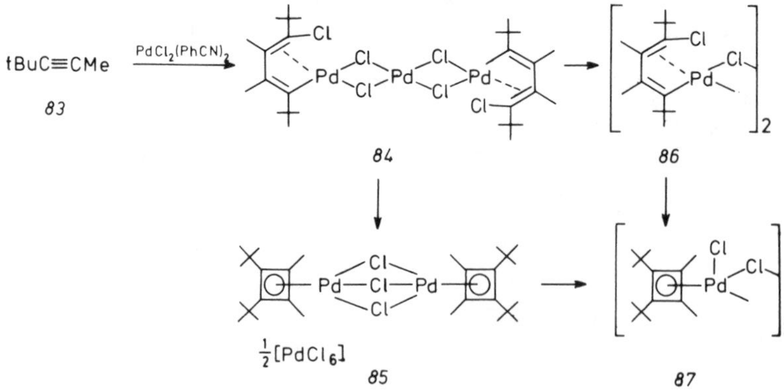

Cyclobutenyl- (**78, 82**) and butadienylpalladium complexes (**86**) are interconvertible via conrotatory electrocyclic reactions.[72-75]

Platinum(II) complexes of CBD (**90**) are analogous to those of palladium, but are more stable so that even dialkylacetylenes **88** lead to tetraalkylcyclobutadiene complexes. However, the inorganic starting materials require activation (either X = CO or X = $SnCl_3$ ligands are effective).[73-77]

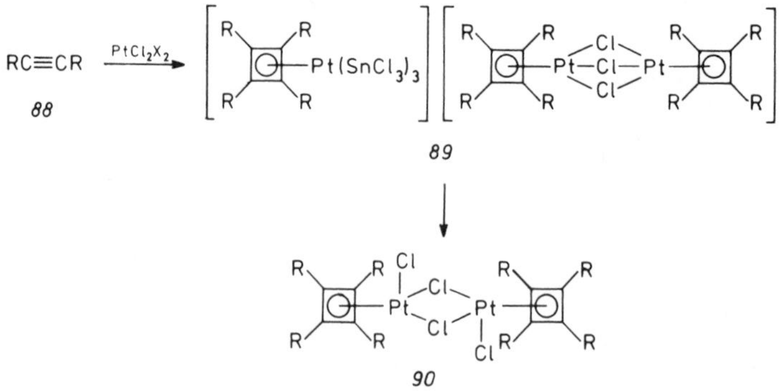

Nickel-CBD complexes were predicted to be stable by Longuet-Higgins and Orgel;[43] tetramethyl-CBD was indeed shown by Criegee to yield such complexes **44**.[44-46] The tetraphenyl-CBD-Ni complex **92** was prepared by Freedman from a tin-heterocycle **91**, but the insolubility of such complexes makes them difficult objects of study.[82]

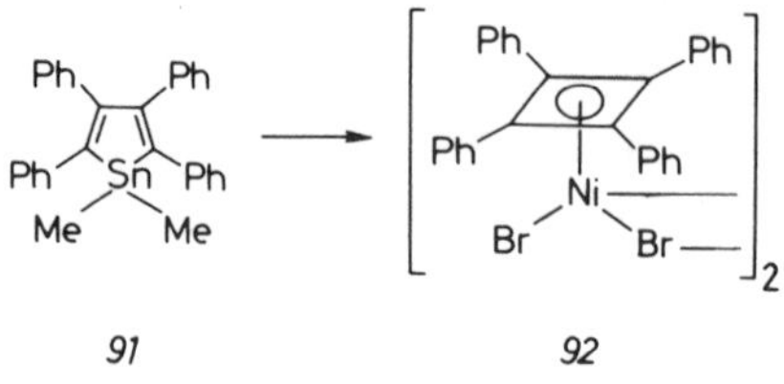

The unsubstituted CBD forms a stable iron tricarbonyl complex **18** which may be prepared from: (1) acetylene and iron pentacarbonyl, (2) *cis-* or *trans*-dichlorocyclobutene and diiron eneacarbonyl, and (3) by photolysis of α-pyrone in the presence of iron pentacarbonyl.[22-24]

It was established by mass spectrometry that photolysis of tricarbonyl-cyclobutadieneiron **18** forms free CBD.[30]

On photoirradiation, the mononuclear complex **18** leads[78] to a CO-bridged dinuclear complex **93**. In addition to such aromatic complexes (*vide infra*), CBD is able to form complexes in which only one double bond is involved, such as the dinuclear complex **94** and isomeric forms thereof.[79]

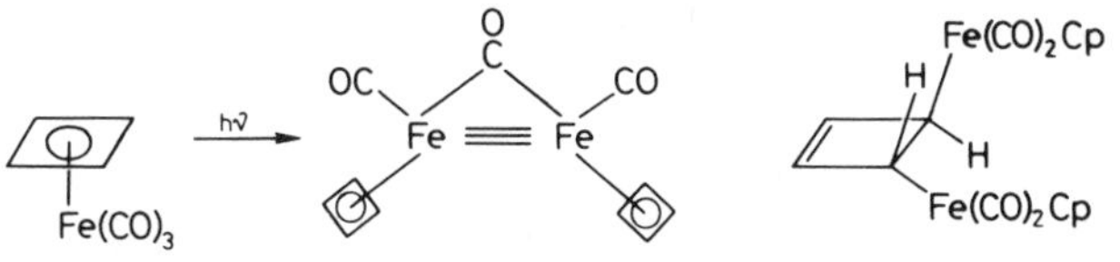

Hübel[80] was the first to obtain a tetraphenylcyclobutadieneiron carbonyl complex from diphenylacetylene and $Fe(CO)_5$; its X-ray crystallographic investigation established the structure.[81]

The iron-CBD complex **18** served as starting material for the first trapping of CBD via oxidation with Ce(IV) salts; the resulting CBD **19** dimerizes to **16** or may be trapped, e.g., with an acetylene derivative leading to a bicyclic Dewar-benzene adduct, **95** or to a tetracyclic product **96**.

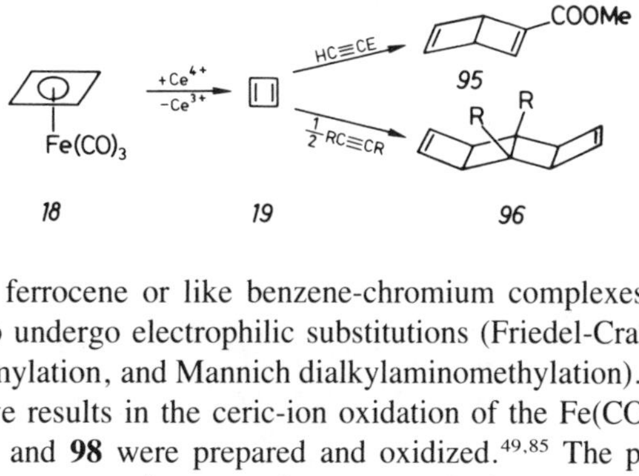

Interestingly, like ferrocene or like benzene-chromium complexes, tricarbonylcyclobutadieneiron is able to undergo electrophilic substitutions (Friedel-Crafts acylation, mercuration, Vilsmeyer formylation, and Mannich dialkylaminomethylation). In order to prove that a free CBD derivative results in the ceric-ion oxidation of the $Fe(CO)_3$ complex, optically active complexes **97** and **98** were prepared and oxidized.[49,85] The products were adducts with added dienophiles, or *syn*-dimers, and were racemic; hence, symmetrical intermediates must have been formed, for which free cyclobutadienes are the most logical candidates.

The explanation of these phenomena resides in the aromaticity of CBD-metal complexes. Various theoretical explanations for this aromaticity were advanced. The first was formulated by Longuet-Higgins and Orgel[43] and restated in terms of the "18-electron rule" by Mingos;[83]

in simple terms, the four π-electrons of the ring and the d-electrons of the metal fill completely the d-shell. Alternatively, Aihara[84] pointed out that a significant contribution to the stabilization energy is provided by Möbius-type conjugation involving singly twisted p-orbital sequences of any three-ring carbon atoms and one d-orbital of the metal (the nodal characteristic of d-orbital requires this twisting). NMR spectra of p-substituted phenylcyclobutadiene-Fe(CO)$_3$ complexes indicate electron donation via the phenyl π-system and electron donation via the σ-core.[86]

Experimental confirmation of the aromaticity was first provided by Criegee and co-workers[44-46] when they isolated the tetramethylcyclobutadiene complex [(CMe)$_4$NiCl$_2$]$_2$, and then by Pettit[22-24] with the (CH)$_4$Fe(CO)$_3$ complex: in both cases, X-ray crystallographic determinations showed equal C–C distances in the rings (1.46 Å in the unsubstituted CBD complex). The IR absorption spectrum of this (CH)$_4$-Fe(CO)$_3$ complex and its normal coordinate analysis are in agreement with a square planar geometry of the CBD ring.[87]

A finer detail concerns the hydrogens or groups attached to the carbon atoms in CBD complexes: microwave studies and X-ray crystallography show that hydrogens in (CH)$_4$Fe(CO)$_3$ point slightly away from the iron atom;[88] similar conclusions were obtained from X-ray crystallography of tetramethyl- and tetraphenylcyclobutadieneiron tricarbonyl complexes.

III. [6]ANNULENE OR BENZENE

[6]Annulene or benzene is too well known to be discussed here.

IV. [8]ANNULENE OR CYCLOOCTATETRAENE (COT)

Unsubstituted cyclooctatetraene (COT[89-91]) (**100**) was first prepared by Willstätter et al.[4] starting from the alkaloid pseudopelletierine (**99**) and using Hofmann eliminations from quarternary ammonium bases.

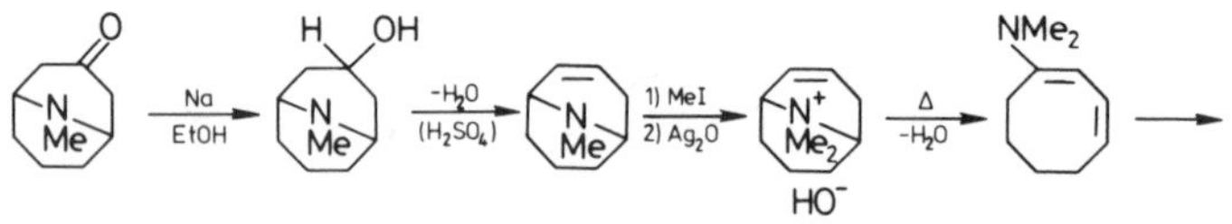

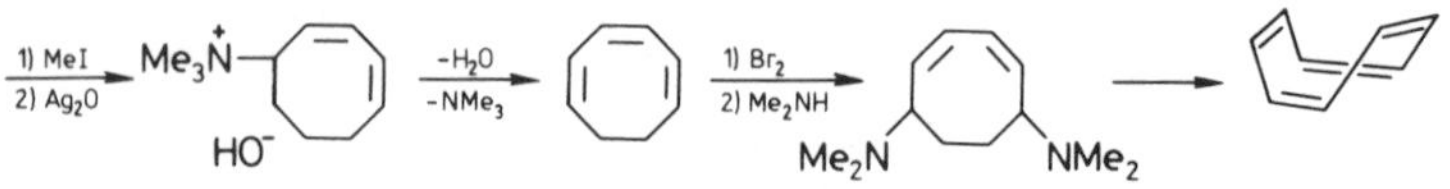

Although the overall yield was quite small, it was clearly demonstrated by the above authors that COT (**100**) was a yellow liquid showing typical olefinic behavior. Today the unsubstituted COT is obtained by Reppe's procedure[92] involving the tetramerization of acetylene at 70 to 120°C under pressure (15 atm) in the presence of Ni(II) complexes with XH = acetylacetone, cyanide, acetoacetic esters, or salicylaldehyde using tetrahydrofuran as solvent. Since handling compressed acetylene is not without danger, it is better to buy the product than to prepare it.

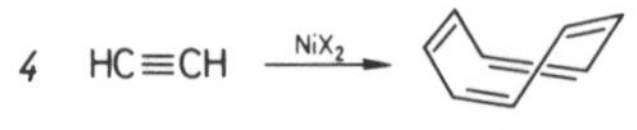

The ground state of COT is tub shaped with alternate single (1.476 Å) and double bonds (1.34 Å). The C–C–C bond angle is 126.1° and the C=C–H bond angle is 117.6°.

Dewar and Merz[309] investigated theoretically whether chair-cyclooctatetraene may exist.

Reppe showed, and Huisgen[93] substantiated by detailed kinetic investigations, that COT is in a dynamic equilibrium with its bicyclic valence tautomer, bicyclo[4.2.0]octa-2,4,7-triene (**103**). Vogel[94] prepared in pure state this valence isomer and found that it is less stable by about 7 kcal/mol than COT. On the other hand, photoirradiation of COT affords semibullvalene. All these valence isomerizations will be discussed in detail in Chapter 6, but the kinetics of the reversible valence tautomerization will be briefly mentioned in the next paragraphs. By ^{13}C and ^{2}H isotope labeling and NMR techniques, Anet,[95,96] then Schröder and Oth[97-101] in extensive investigations, have shown that COT and its derivatives undergo both ring inversion (RI) via a flattened bond-localized (polyolefinic) transition state **101** ($\Delta G^{\ddagger} = 9.5$ kcal/mol) and bond shift (BS) via a delocalized planar [8]annulenic antiaromatic transition state **102** ($\Delta G^{\ddagger} = 13.3$ kcal/mol). Schröder and Oth found similar barriers for alkoxy-COT derivatives. Evidently, the higher latter barrier reflects the destabilization energy associated with the antiaromaticity of a planar [8]annulene.

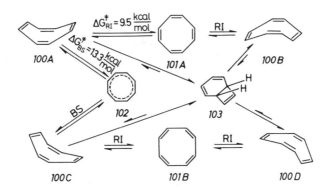

A. Substituted COT Derivatives

In addition to the "traditional" methods for the synthesis of unsubstituted COT, the detailed analysis of all isodynamical processes in which COT participates required the development of new synthetic methods for substituted COT derivatives. One of these is based on sulfones which may be converted into both bond-shift isomers of one and the same substituted COT derivative, in isomerically pure form. However, owing to the symmetry of the sulfones with one or two double bonds (**104**), this method is unsuitable for obtaining chiral COT derivatives. The bicyclo[4.2.0]octadiene valence isomer , **105** (resulted in the reaction) ring-opens under control of the cyclohexadiene ring for orbital symmetry reasons. Conversion into an urazole (**107, 110**, R = Ph) or into the iron tricarbonyl complex (**111, 112**) facilitates structural assignments by means of ^{1}H-NMR spectra. The tetramethyl-COT with contiguous methyl groups was obtained by Paquette et al.[102] as two shelf-stable bond shift isomers **106** and **108**; the isomer **108** with a twofold symmetry axis (but not the isomer **106** with a symmetry plane) is in equilibrium with its bicyclic valence isomer (**109**). From urazoles (**107, 110**) the COT derivatives may be regenerated by hydrolysis with hydroxides, then by treatment with manganese dioxide. The explanation of the stability of such bond-shift isomers resides in the steric hindrance of contiguous methyl groups to flattening of the molecule.

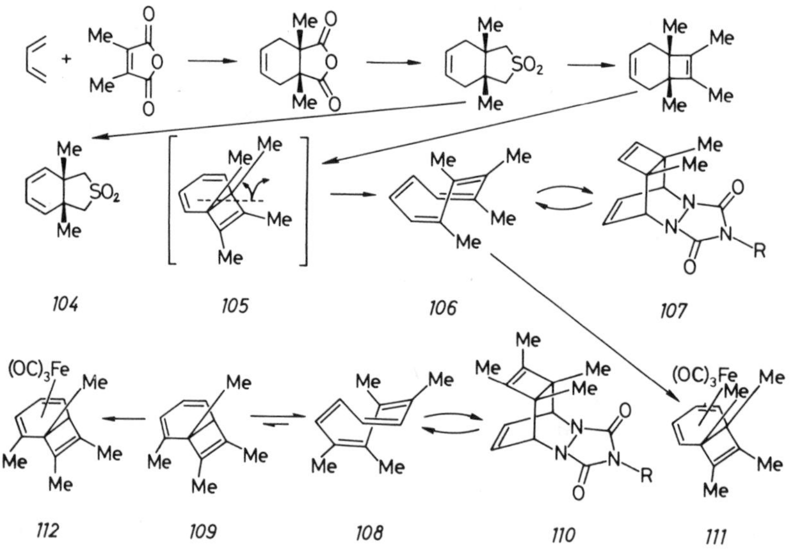

104   105   106   107

112   109   108   110   111

For obtaining enantiomerically pure COT derivatives, Paquette[103] developed a method via bromolactonization of monoamides of cyclohexene-4,5-dicarboxylic acids with optically active amines; X-ray crystal structure analysis allowed the determination of absolute configurations. In the case of *vic*-tetramethyl-COT, the bond-shift isomer **108** is chiral, while **106** is achiral; thus a direct convenient kinetic method is available by investigating the racemization of an enantiomer of **108** (in analyzing the kinetic data, one must take into account also the ring inversion). For obtaining larger amounts of chiral COT derivatives, an alternative method is their direct resolution using an urazole **107, 110** with R = *endo*-bornyl (**114**), formed by cycloaddition with (−)-*endo*-bornyltriazolinedione, followed by successive recrystallization or, preferably, HPLC resolution.[104] The absolute configuration was confirmed via X-ray analysis of urazoles and by ¹H-NMR in the presence of lanthanide shift reagents such as Eu(hfcam)₃, in the case of 1,3-di-*t*-butyl-COT (**113**).

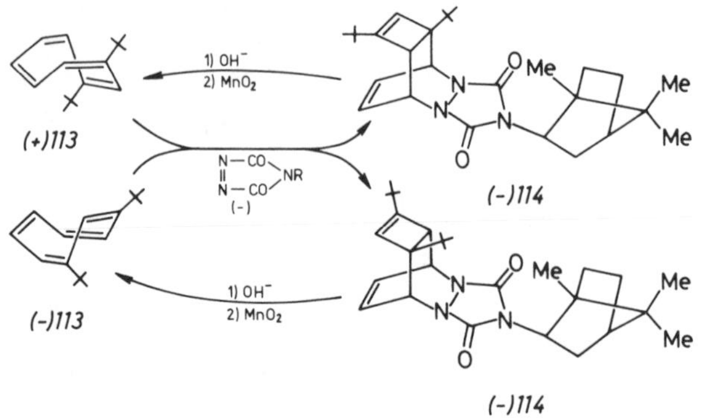

(+)113   (−)114

(−)113   (−)114

Finally, a photochemical decarbonylation method was developed starting from Nenitzescu's *cis*-1,2-dichlorocyclobutene and based on cycloaddition with substituted cyclopentadienones (**115**) or *ortho*-benzoquinones (**119, 122**). Photolytic decarbonylation (loss of one CO molecule in the former case, and of two CO molecules in the latter) affords COT derivatives, exemplified by the synthesis of 1,3- (**121**) and 1,4-di-*t*-butyl-COT (**125**);[105] 1,4-dimethyl-2,3-diphenyl-COT (**116**) was prepared likewise.[106] Interestingly, in all these cases the bicyclic valence isomers are more stable than the COT derivative.

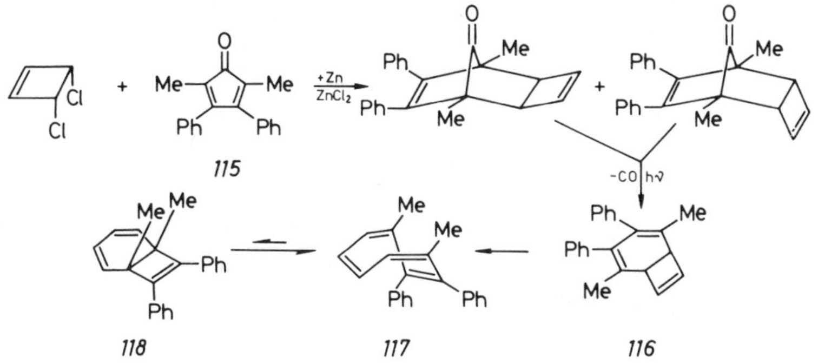

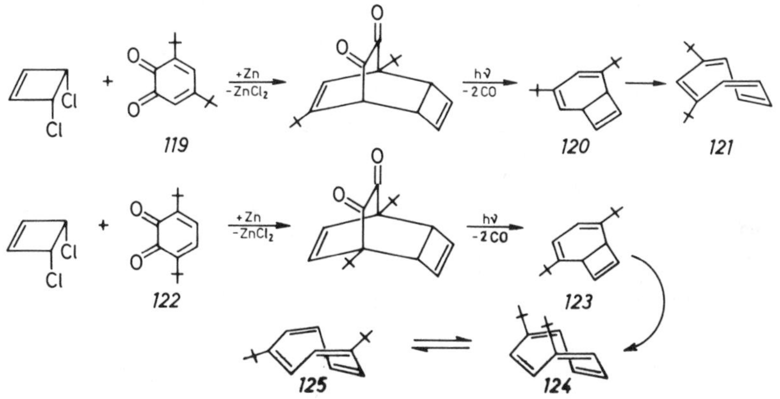

1,2,3-Trimethyl-COT affords enantiomeric forms (**126, 127**) which racemize by a first-order process.[103] The presence of three or four adjacent methyl groups destabilizes the planar forms relative to COT: the localized form is destabilized more strongly than the delocalized one. Therefore, as seen in Table 1, the free energies of activation become practically equal for the two vicinal tetramethyl-derivatives of COT, **106** and **108**: $\Delta G^{\ddagger}_{RI}$ = 31.8; $\Delta G^{\ddagger}_{BS}$ = 32 kcal/mol, much higher values than for the unsubstituted COT.[103] With other substitution patterns, $\Delta G^{\ddagger}_{RI}$ is smaller than $\Delta G^{\ddagger}_{BS}$: for 1,3-di-*t*-butyl-COT (chiral, with two chemically different *t*-butyl groups) $\Delta G^{\ddagger}_{BS}$ = 23.6 and $\Delta G^{\ddagger}_{RI}$ = 21.6 kcal/mol at 25°C; the isomeric 1,4- and 1,6-di-*t*-butyl-COT interconvert by bond shift more easily; in this case the two compounds are nonidentical isomers; one of them is chiral, the other is *meso*, but racemization is quite fast; the sterically congested bond-shift isomer is more stable by 0.44 kcal/mol in CDCl$_3$ owing to London attraction forces between nonbonded *t*-butyl groups.[105] 1,2,3-Trimethyl-COT (**126,127**) has higher activation barriers: $\Delta G^{\ddagger}_{RI}$ = 24.4 and $\Delta G^{\ddagger}_{BS}$ = 26.4 kcal/mol. Finally, 1,4-Me$_2$-2,3-Ph$_2$-COT (**128**) and 1,6-Me$_2$-7,8-Ph$_2$-COT (**117**) are bond-shift isomers which coexist in equilibrium with high concentrations of bicyclo[4.2.0]octatrienic valence isomers **118**, **131**; the absolute configuration of the optically active **117** was determined by X-ray crystal structure analysis; the racemization rates and the BS isomerization rates allow the determination of RI activation parameters indicated in Table 1.[107] Interestingly, 7,8-di-*t*-butylbicyclo[4.2.0] octatriene **132** shows no tendency to isomerize to the 1,2- or 1.8-di-*t*-butyl-COT; in this case the customarily less stable bicyclic valence isomer becomes favored thermodynamically owing to steric factors.[108]

Table 1

ACTIVATION PARAMETERS FOR BOND SHIFTING (BS) IN STERICALLY CONGESTED, AND FOR RING INVERSION (RI) IN CHIRAL COT DERIVATIVES

Compound	Process	E_a (kcal/mol)	lnA	$\Delta H^{\ddagger}_{25°C}$ (kcal/mol)	$\Delta S^{\ddagger}_{25°C}$ e.u.	$\Delta G^{\ddagger}_{25°C}$ (kcal/mol)	$\Delta H^{\ddagger}_{RI} - \Delta H^{\ddagger}_{BS}$ (kcal/mol)	Ref.
100 	BS	—	—	—	—	13.3	−3.8	95,96
	RI	—	—	—	—	9.5		
124 125 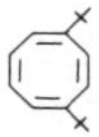	BS	—	—	14.8	−26.3	22.6	—	113
129 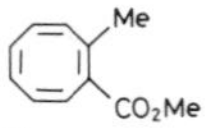	BS	—	—	—	—	18.8	—	95,96
130 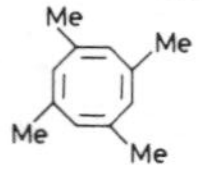	BS	—	—	—	—	22.5	—	112
121 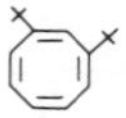	BS	23.3	29.0	22.7	−2.9	23.6	−3.4	105
	RI	19.9	26.7	19.3	−7.5	21.6		
126 127 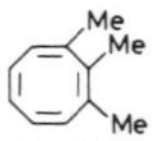	BS	23.5	24.4	22.9	−12.0	26.5	(−0.5)	111
	RI	23.0	26.6	22.4	−7.7	24.4		
106 108 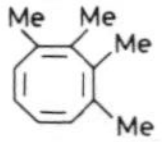	BS	28.7	24.4	28.1	−13.3	32.0	(+0.4)	103
	RI	29.1	24.9	28.5	−11.0	31.8		
117 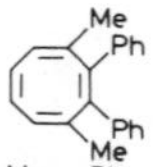	BS	35.4	33.1	33.3	+5.2	33.3	—	107
128 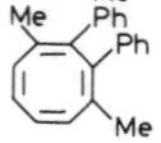	BS	34.8	32.4	32.4	+3.9	33.1	−5.4	107
	RI	27.6	26.1	27.0	−8.8	29.7		

126   *127*

(+)128   *(-)128*   *131*

(±)117   *(±)117*   *118*

The thermally allowed valence-isomerization of COT (**100**) and its bicyclic valence tautomer (**103**) converts the COT system devoid of π-conjugation into a genuine 1,3-diene system in **103** which is able to undergo Diels-Alder reactions; therefore, many reactions of COT under conditions of valence-tautomerization are actually reactions of the bicyclic valence isomer, which will be discussed in Chapter 6. However, the catalytic hydrogenation of COT to cyclooctane and many other reactions prove that at equilibrium the monocyclic system prevails.

The bicyclic valence isomer **103** of COT was obtained in pure state by Vogel and co-workers;[114] they measured the activation parameters of the valence isomerization. For the unsubstituted COT, Huisgen et al.[93] determined an equilibrium concentration of 0.01% bicyclic valence isomer **103** at 100°C (for ring closure $\Delta H^{\ddagger} = 28.1$ kcal/mol, $\Delta S^{\ddagger} = +1$ e.u.; for ring-opening $E_a = 18.7$ kcal/mol); however, the equilibrium between **128** and **131** favors slightly the latter bicyclic system.[107]

Paquette's group was able to obtain a direct measure (for substituted COT derivatives) of the energies separating the tub-shaped localized COT, the planar-localized COT, and the planar-delocalized (antiaromatic) COT. The conclusion is that angle strain, van der Waals repulsions, and related phenomena account for 85% of the energy gap between the tub-shaped ground state and the planar-delocalized state. Only 15% of this energy gap is due to resonance destabilization (antiaromaticity); hence, the nonplanarity of COT is minimally connected with its antiaromaticity: the experimental values of 2 to 4 kcal/mol (destabilization due to antiaromaticity) are so low that the σ-network would prefer to be planar and delocalized.

The barrier to bond shift (the transition state is the planar antiaromatic **102**) is due to HOMO-LUMO orbital crossing, making the interconversion a forbidden one. The barrier to ring inversion (the transition state is the planar localized **101**) in compounds which do not have high steric congestion is lower. The difference $\Delta\Delta H^{\ddagger} = \Delta H^{\ddagger}_{RI} - \Delta H^{\ddagger}_{BS}$ (a negative number in Table 1) may be equated to the destabilization associated with antiaromaticity. It may be noted that the experimentally determined values for $\Delta\Delta H^{\ddagger}$ with COT (**100**), 1,3-di-*t*-butyl-COT (**121**) and 1,4-dimethyl-2,3-diphenyl-COT (**128**) are in the range from -3.4 to -5.4 kcal/mol, respectively, in good agreement with the Dewar resonance energy (-4 kcal/mol) calculated by Dewar and Gleicher.[115] It will be observed that though this difference is practically constant, the individual values increase tremendously from about 10 kcal/mol

Table 2
**KINETIC PARAMETERS (IN KCAL/MOL) FOR RING-
OPENING OF BICYCLO[4.2.0]HEXATRIENES TO COT
DERIVATIVES**

Compound	E_a	$\Delta G^{\ddagger}_{25°C}$	Ref.
103 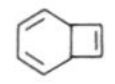	18.7	—	114
120 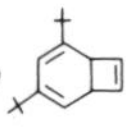	28.4	23.2	105
123 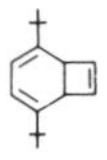	31.1	32.9	91
118 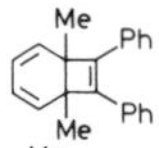	22.2	19.7	107
131 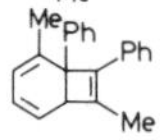	23.7	21.1	107
132 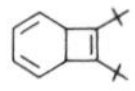	No ring opening, this is the thermodyn- amically favored valence isomer		108

for COT, to about 20 kcal/mol for **121**, and to about 30 kcal/mol for **128**, in parallel with increasing steric strain in the planar systems (cf. Table 1).

The values of the activation energy of the reversible valence isomerization (e.g., **100** ⇌ **103**) increase in the same order (Table 2) but the range is less wide; in this case the *t*-butyl-substituted systems have the highest barriers, unlike the previous case presented in Table 1.

Among substituted COT derivatives, the octafluoro-COT constitutes an exception because its geometry was reported to be planar.[116]

An interesting observation is that in 1,2-disubstituted COT-derivatives, the isomer with a longer (single) bond between the substituents usually predominates at equilibrium. An extreme case is provided by 1,2-dicarbomethoxycyclooctatetraene (**A**, R = CO_2Me), which in the temperature range -40 to $+200°C$, exists exclusively as one valence bond isomer **A** that presents no bond exchange.[109] The corresponding diacid does not afford a cyclic anhydride.[109]

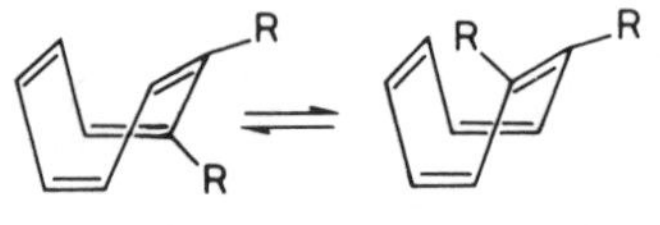

A   *B*

1,2-Dimethyl-COT rearranges pyrolytically (400 to 600°C) to the 1,4-dimethyl derivative, hence the unsubstituted COT must automerize similarly via a tetracyclic valence isomer **C**.[101]

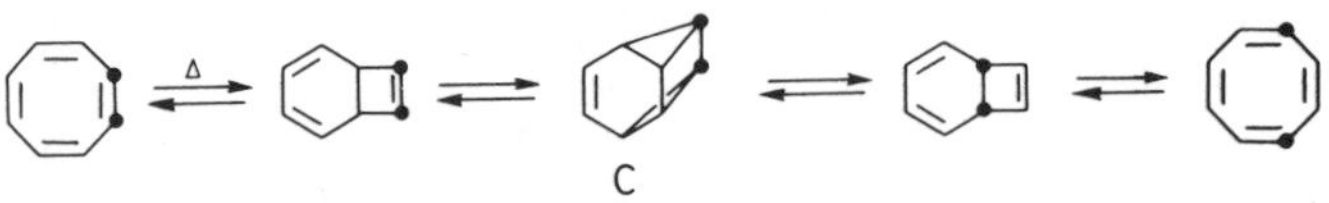

C

Among the metallic complexes of COT, the most outstanding one, uranocene, **133**, was obtained by Streitwieser[117] following an ingenious reasoning for f-shell complexes by analogy with ferrocene and other *d*-shell metal complexes. Analogous complexes were formed with other actinides such as protactinium.[118] In these complexes the eight-membered rings are flat, indicating aromaticity.

Rösch and Streitwieser[314] reported the molecular structure of di-π-[8]-annulene-uranium(IV), i.e., of uranocene, and its Th and Ce analogs.

In the COT-iron tricarbonyl complex **134**[119] only four electrons coordinate with the metal,[120] hence the complex undergoes cycloadditions more readily than COT; with tetracyanoethene a 1,3-cycloaddition takes place yielding the complex **135** and the Fe(CO)$_3$ adduct **136**.[121] The X-ray structure of complex **135**[122] has established clearly its unusual geometry.

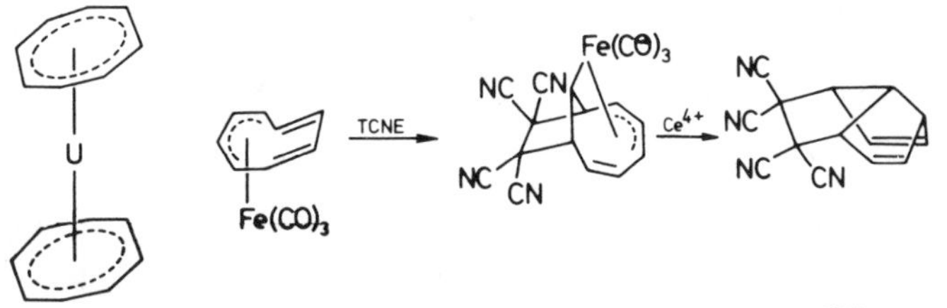

Protonation of **135** at low temperature yields an eight-membered ring allylic cation **137**[123] which at room temperature contracts forming a [5.1.0]bicyclic cation **138**.[124]

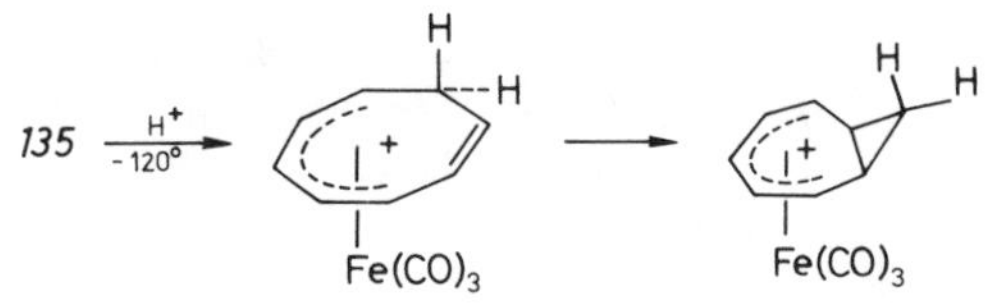

Other metallic complexes of COT with Mo(CO)$_3$[125] or ruthenium and osmium carbonyls[126] are also known.

Low-temperature ^{13}C-NMR spectroscopic investigations of the metal 1,5-shift in Os(η^6-cyclooctatetraene) η^4-1,5-cyclooctadiene) showed[325] that the COT ligand presents four separate fluxional processes with $\Delta G^\ddagger$ = 8.9, 11.5, 16.5, and 19.5 kcal/mol.

V. [10]ANNULENE OR CYCLODECAPENTAENE

Unlike the previously discussed smaller annulenes, [10]annulene is able to exist in several different geometries and has a substantial steric strain in each of these geometries. Since according to Hückel's rule, [10]annulene is the next aromatic annulene following benzene, a considerable effort was invested in trying to synthesize and characterize cyclodecapentaene. The topic was reviewed not only in the general monographs on annulenes or aromaticity cited in the introduction, but also in reviews dedicated to [10]annulene.[127-130]

The early history may be found in the above reviews, including unsuccessful attempts to obtain [10]annulene by dehydrogenation of cyclodecane[131] analogously to the formation of benzene from cyclohexane or by gradual introduction of unsaturation in ten-membered ring systems.[132,133]

Mislow[134] discussed in 1952 the stabilities of various annulenes on the basis of normal bond lengths and angles and of van der Waals radii. Owing to the 1,6-interaction in the di-*trans* isomer of cyclodecapentaene (**139**) and to similar interactions in higher annulenes, he presumed that the smallest annulene capable of achieving planarity, hence aromatic stabilization, would be [30]annulene. Experiments have shown, however, that with slight deviations from planarity, smaller [*n*]annulenes (*n* ≥ 14) are stable and aromatic. The Woodward-Hoffmann rules link three stereoisomers of cyclodecapentaene (**139, 141, 143**) with two stereoisomers of 9,10-dihydronaphthalene (**140, 142**) via thermally or photochemically allowed valence isomerization.

139	*140*	*141*	*142*	*143*
UN: TCTCC		TCCCC		CCCCC
di-trans	trans	mono-trans	cis	all-cis
2DC: 00111001		00111111		11111111

The first cyclodecapentaene derivative **145** was probably obtained as an unstable reaction intermediate by Avram et al.,[135] although the structure of 2,6-dicarbomethoxy-9,10-dihydronaphthalene **146** was disputed.[136] Doering and Rosenthal[137] provided evidence that the pyrolysis of the related Nenitzescu's hydrocarbon (i.e., **144** with H instead of E = COOMe group) did lead to 9,10-dihydronaphthalene. Van Tamelen and Burkoth[138] synthesized *trans*-9,10-dihydronaphthalene starting from the known 1,6-naphthalenediol (**147**)[139] via reduction with LiAlH$_4$, reaction with 48% aqueous HBr, then with *N*-bromosuccinimide (see Chapter 6). The mixture of epimeric tetrabromides **149** was dehalogenated with lithium amalgam, affording a hydrocarbon which is oxidizable by air to naphthalene. For this hydrocarbon the probable structure is **140**, but various stereoisomeric cyclodecapentaene structures are also possible. On UV irradiation with 253.7-nm photons, this hydrocarbon gave at room temperature a reaction mixture from which *cis*-9,10-dihydronaphthalene **142** and its photolysis products (bicyclo[4.2.2]deca-2,3,7,9-tetraene (**152**), bullvalene and naphthalene) were separated. On low-temperature irradiation (− 190°C) in ether-isopentane-ethanol (5:5:2) a complex mixture resulted, whose UV spectrum was recorded. On brief thawing and recooling to − 190°C, a new UV band at 247 nm due to **142** appeared. The thermally labile progenitor of **142** was hydrogenated by diimide at − 78°C to cyclodecane in 40% yield and identified thereby to be a cyclodecapentaene stereoisomer.

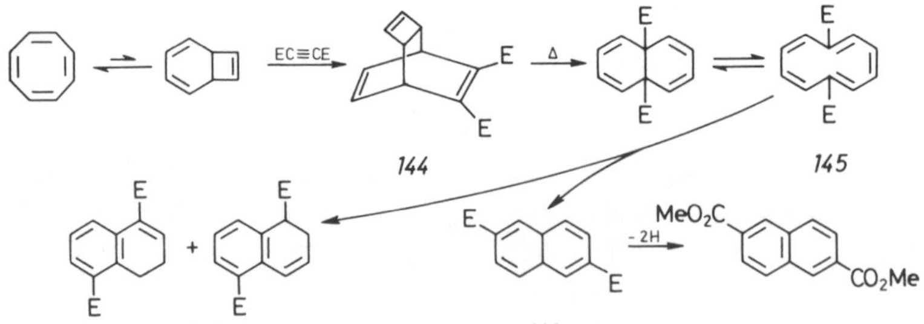

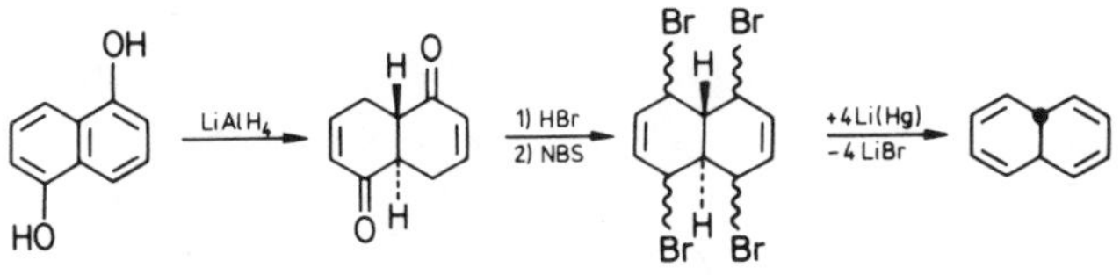

147   148   149   140

Van Tamelen and Greeley[140] announced that on UV irradiation at low concentrations (6 · 10^{-4} *M*) of *trans*-9,10-dihydronaphthalene at 77 K in a matrix, a cyclodecapentaene was trapped in 14 to 40% yield; it was not certain whether this was the *all-cis* or the *di-trans* stereoisomer;[141] in view of solution data reported below, the latter stereoisomer is unlikely.

Masamune and co-workers[142] synthesized bicyclo[6.2.0]deca-2,4,6,9-tetraene **151** by UV irradiation (pyrex-filter) of the tosylhydrazone anion **150**. Hydrocarbon **151** is thermally unstable and rearranges to *trans*-9,10-dihydronaphthalene **140** ($\Delta H^{\ddagger} = \Delta G^{\ddagger} = 25$ kcal/mol) via an unstable mono-*trans*-cyclodecapentaene **141**.

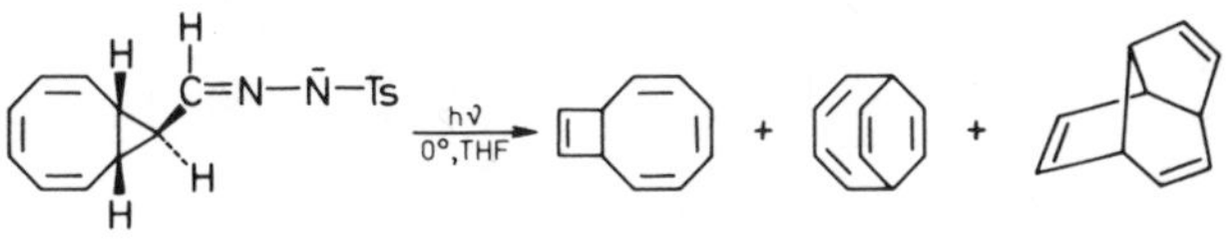

150   151   152   153

UV irradiation (253.7 nm) of **151** or of *cis*- (**142**) or *trans*-9,10-dihydronaphthalene (**140**) at −110°C led to a photostationary state consisting of the two 9,10-dihydronaphthalenes **140,142** (*cis*, 8%; *trans*, 19%) and two other (CH)$_{10}$ valence isomers, bullvalene (7%) and 65% tetracyclo[4.4.0.0^{2,10}.0^{5,7}]deca-3,8-diene (**154**) which rearranges easily to **152**. Using knowledge from the photolysis of C_9H_{10} hydrocarbons,[129] the UV irradiation was carried out at −50 to −70°C in order to obtain evidence for the cyclodecapentaene intermediates. By chromatography on alumina the *all-cis* (**143**) and mono-*trans*-cyclodecapentaene (**141**) stereoisomers were isolated in crystalline form (at −60°C!), and characterized by ¹H- and ¹³C-NMR, UV spectra, and thermolysis to *cis*- and *trans*-9,10-dihydronaphthalene, respectively.[143]

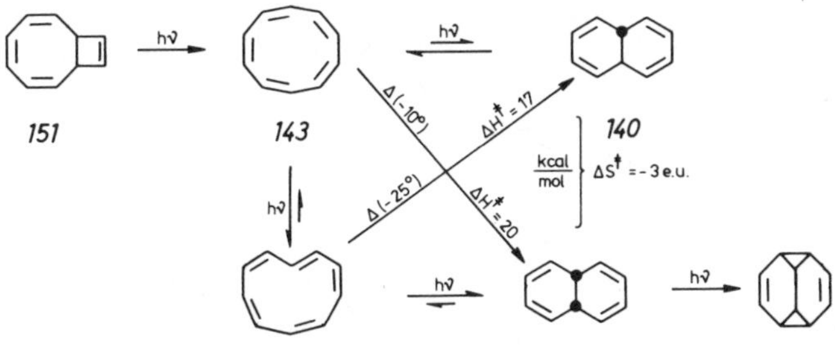

151   143   140

141   142   154

The *all-cis* (CCCCC) isomer **143** is nonplanar with a tub-shaped portion attached to a near-planar diene fragment and undergoes a facile conformational change by pseudorotation as in the twist-boat cyclohexane conformers; both the ¹H-NMR ($\delta = 5.47$) and ¹³C-NMR spectra ($\delta = 130.4$ ppm) are sharp singlets at temperatures as low as −160°C.

The mono-*trans* (TCCCC) isomer **141** presents at $-100°C$ a noise-decoupled ^{13}C-NMR spectrum with five peaks which coalesce to one singlet at $-40°C$. The interpretation of these conformational and $\sigma - \pi$ bond changes involves stable puckered conformers, e.g., **141A** and less stable, nearly planar, transition states (**155**). The most interesting conclusion from the NMR spectra (two multiplets in the ^1H-NMR spectrum at $-100°C$ which coalesce to one singlet, $\delta = 5.86$ at $-40°C$) is the presence of a minimal, if any, diagmagnetic ring current owing to the nonplanarity of the molecule, which in turn is due to the repulsion between inner hydrogens. Therefore, this compound, like COT, does not present tropicity.

In conclusion, from the eight possible geometrical arrangements of *cis/trans* double bonds in cyclodecapentaene, only three are structurally possible for steric reasons; the *all-cis* and the mono-*trans* isomers were isolated in crystalline state, but the evidence for the 1,3-*di-trans* stereoisomer is still inconclusive.[129,143]

This stereoisomer **139** in turn may exist in three different conformations, unlike the other ones. An *ab initio* MO study[144] of stereoisomeric [10]annulenes confirmed Masamune's conclusion: the lowest-energy isomer is TCCCC (**141**); the nonplanar *all-cis* isomer **143** lies 2 kcal/mol higher in energy; the TCTCC (symmetries C_2 and C_s) isomers **139A**, **139B** lie 3 and 11 kcal/mol, respectively, higher than TCCCC, so that the former isomer **139A** (C_2) might be a reasonable target for observation. By contrast, the planar *all-cis*-isomers with alternating (D_{5h}) and equal (D_{10h}) bond lengths have energies of 7.7 and 7.2 kcal/mol, respectively, higher than the nonplanar congener.

139 A	139 B	139 C
UN: TCTCC (C_2)	TCTCC (C_S)	TCTCC (C_1)

Vogel's stable bridged [10]annulene derivatives, which present a marked diamagnetic ring current are reviewed separately in Chapter 5.

VI. LARGER [*n*]ANNULENES ($n \geqslant 12$)

A. Synthesis via Oxidative Coupling of Terminal Acetylenes

Sondheimer[145] adapted Glaser's method[146] for forming a new C–C bond between acetylenic

CH groups with copper(II) chloride and oxygen for intermolecular coupling and intramolecular ring closure. Alternatively, the reaction may be carried out under Eglinton conditions,[147] i.e., with copper(II) acetate in pyridine. The products are polyenes which, after separation and purification, may be isomerized by potassium *t*-butoxide to give dehydroannulenes. Subsequent selective hydrogenation with palladium catalysts leads to annulenes. Thus the syntheses of many annulenes and dehydroannulenes were performed by Sondheimer's group. Two examples follow. Preparation of [16]annulene **157** and 1,2,9,10-tetradehydro-[16]-annulene **156** in Glaser conditions:

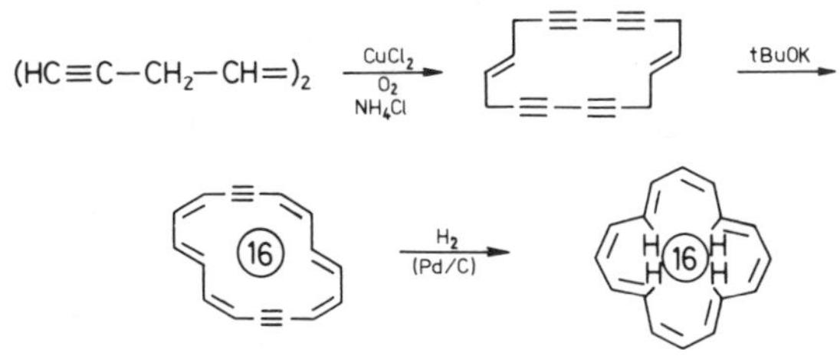

156   **157**

Preparation of [18]annulene **159** and its hexadehydro-derivative **158** under Eglinton conditions:[148]

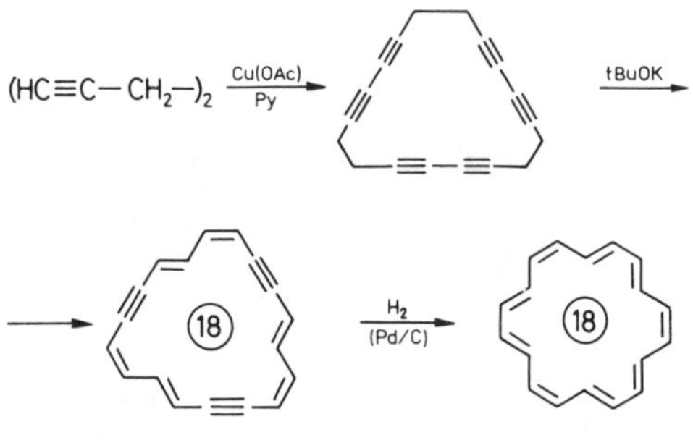

158   **159**

B. Photoirradiation of Polycyclic Valence Tautomers[149]

For less stable annulenes which have to be obtained at low temperatures and/or in solid matrices, the photochemical ring opening of a stable polycyclic isomer is particularly suitable because it obviates the necessity of other reagents or of solution chemistry; an example was seen in van Tamelen's and Masamune's preparation of cyclodecapentaene stereoisomers. For larger annulenes, two examples are presented.

The preparation of [12]annulene **161** was performed either from the tricyclic adduct **160** of cyclobutadiene with COT, or from *cis-* or *trans-*bicyclo[6.4.0]dodecapentaene (**162, 163**) by photoirradiation at low temperature ($-100°C$).[150] [10]Annulene **157** was prepared from Schröder's [2 + 2]-*cis*-dimer of cyclooctatetraene **164** in a clean reaction.[151]

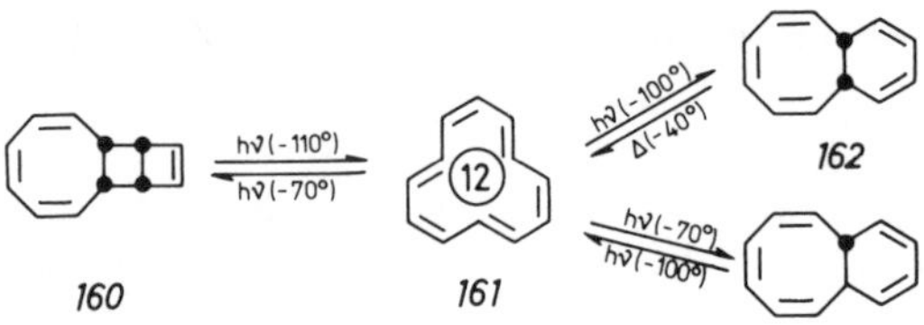

By comparing the yields and the number of steps, it is clear that, when applicable from suitable starting materials as in the last reaction, this method requires less labor than Sondheimer's method.

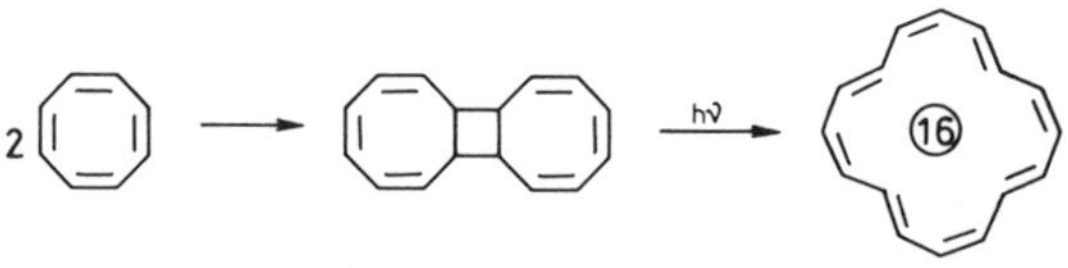

C. [12]Annulene

[12]Annulene obtained via photoirradiation[150] has a temperature-dependent ^{1}H-NMR spectrum: at $-170°C$ two bands at $\delta = 7.83$ (3H) and 5.88 ppm (9H) are due to inner and outer protons in the C_3 structure, respectively; on heating to $-80°C$ the spectrum presents two bands at $\delta = 6.88$ (6H) and 5.97 ppm (6H) due to the fluctuating geometry of the molecule interchanging the specified hydrogens of the three *trans* double bonds between inner and outer positions (**161A** $\rightleftarrows$ **161B**); the nonspecified hydrogens of the *cis* double bonds are not affected by this conformational change. The chemical shifts indicate that [12] annulene is slightly paratropic.

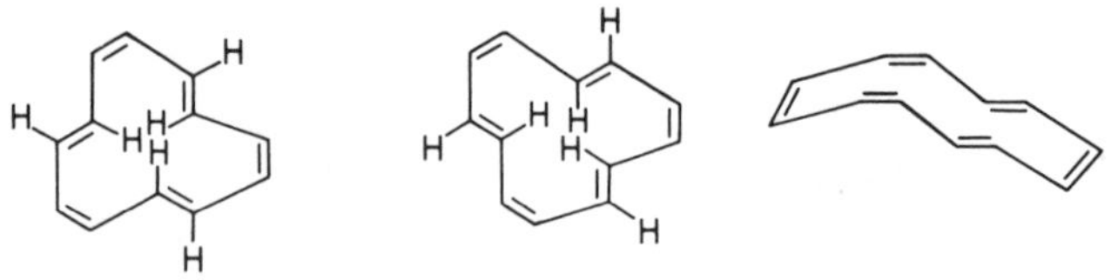

On heating to $-40°C$ [12]annulene undergoes ring closure to *cis*-bicyclo [6.4.0]dodeca-2,4,6,9,11-pentaene, **162**; the *trans*-isomer of the above bicyclic system (**163**) is obtained on photoirradiation at $-70°C$, probably via the nonisolable *di-trans*-isomer **165** of [12] annulene.

D. [14]Annulene

[14]Annulene was obtained by Sondheimer et al.[152,153] starting from the acyclic tetradeca-4,10-diene-1,7,13-triyne via oxidative coupling of terminal acetylene carbons. The crystalline form was shown by X-ray crystallography[154] to have a nonplanar centrosymmetric configuration with four "inner" hydrogens and with ring bond lengths in the range 1.35 to 1.47 Å (a wider range than for [18]annulene).

In solution the compound exists as two stereoisomers in a ratio of 11.5:1 at $-10°C$; the major component **166** gives rise at $40°C$ to one band at $\delta = 5.58$ ppm, which misled initially but which on cooling to $-30°C$ becomes flattened and at $-60°C$ is split into two bands at δ = 0 (4H) and 7.6 (10H) ppm; these data indicate a diatropic aromatic ring current. On rapid dissolution of the crystals, only the above ^{1}H-NMR peaks may be seen. The minor stereoisomeric (**167**) which appears in solution after about 30 min gives rise at 40°C to one band at δ = 6.07 ppm which decoalesces at much lower temperatures than the major isomer, hence is more mobile conformationally.[155] The difference between the two stereoisomers was first ascribed to conformational isomerism between two tri-*cis* conformations due to a different pattern of mutual overlap between the inner hydrogens (**166** and **167A**). However, determination of the ^{1}H-NMR spectrum at very low temperatures showed that the minor component is a configurational stereoisomer, **167B**, with a tetra-*cis*-configuration.

167A   166   167B

168

A 1,8-dimethyl-[14]annulene and its dihydro-octalenic valence isomer were prepared by Vogel et al.[156] Semiempirical MO studies[157] were employed for studying equilibria of six stereoisomeric [14]annulenes and also of several [16]- and [18]annulene stereoisomers.

The reaction of [14]annulene with triamminetricarbonylchromium(0) led to the formation of a *trans*-dihydro-octalenic bis-chromium complex **168**, whose structure was established by X-ray crystallography.[158] Interestingly, in solution this complex decomposes slowly at room temperature, regenerating the monocyclic [14]annulene rather than its bicyclic valence isomer.

E. [16] Annulene

[16] Annulene **157** (prepared as a yellow oil both via oxidative coupling[159] or via photolysis[151]) is a crystalline stable compound which was extensively studied and presents a temperature-dependent NMR spectrum. At 30°C a time-averaged ^{1}H-NMR singlet at δ = 6.7 ppm appears, but at −120° two bands appear at δ = 10.4 (4H) and 5.4 ppm (12H); these data indicate strong paratropicity, in agreement with the negative diamagnetic susceptibility exaltation (Λ = −5). The X-ray crystallographic data reveal that [16]annulene **157** has alternating single and double bonds in a geometry with a small deviation from planarity. The four inner protons minimize steric intereactions by alternating above and below the mean molecular plane.

In solution the predominant isomer **157** resembles that existing in crystalline state, but in addition a minor stereoisomer **169** appears. The minor configurational stereoisomer (**169**), the only one with all angles equal to 120°, has five inner hydrogens instead of four as in the two other stereoisomers. It will be observed that both for [14]- and [16]annulene the major configurational isomer in solution is the exclusive stereoisomer in the crystal and is centrosymmetric; in the former case this isomer has all angles equal to 120°, in the latter case different angles. A third stereoisomer (**170**) plays an important part in the thermolytic and photolytic reactions of [16]annulene which lead to *cis*-transoid-*cis* and to *trans*-transoid-*trans* tricyclo[10.4.0.0^{4,9}]hexadecahexaene, respectively.

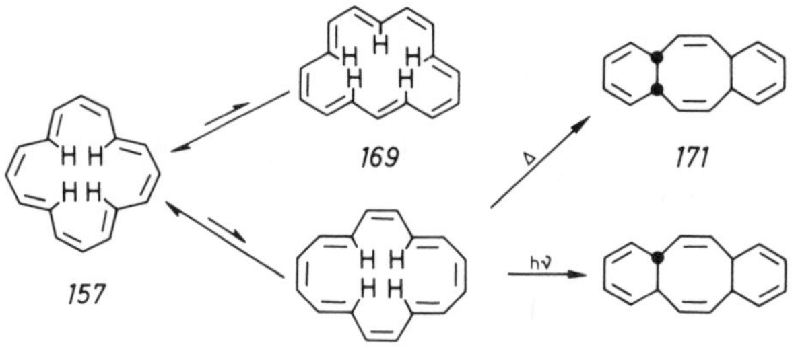

F. [18]Annulene

[18]Annulene is, undoubtedly, one of the most remarkable Hückel annulenes. Its preparation was included in *Organic Synthesis*,[148] together with its ^{13}C-NMR spectrum.

It was the first macrocyclic annulene to be obtained[160] and illustrated brilliantly Hückel's prediction, both in so far as its chemical properties are concerned and with respect to its stability and physical properties.

[18]Annulene forms brick-red crystals which react by substitution with acetylating, formylating, brominating, Vilsmeier, and nitrating reagents under certain conditions. However, it decomposes slowly in solution at room temperature, especially in the presence of air or light; it adds hydrogen to yield cyclooctadecane; it undergoes cycloadditions with maleic anhydride and may add halogens.

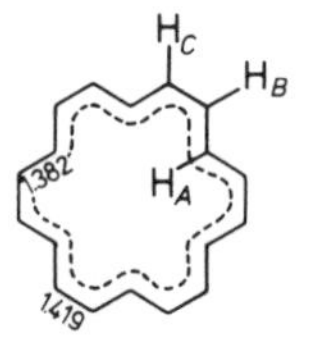

159

X-Ray crystallographic analysis revealed that [18]annulene is nearly planar.[161] Its six *cisoid* bonds are slightly longer than the 12 remaining ones, providing evidence that the structure has a center of symmetry and that there is no bond length alternation. This interesting inequality of bond lengths arises in the σ skeleton where the σ-conjugation favors the *trans* rather than the *cis*-arrangement.[162]

The red color is due to a very low-lying singlet excited state (λ_{max} = 768 nm, lg ε = 1.15 m^2 · mol^{-1}). More about the electronic absorption spectra will be said in the paragraph with conclusions.

The marked diatropic ring current is evident from the low-temperature ($-70°$C) ^{1}H-NMR spectrum presenting two signals at δ = 9.28 (12H, multiplet) and -2.99 ppm (6H, triplet). On heating, these signals coalesce and at 110°C a sharp singlet appears at the weighted average field, δ = 5.45 ppm. The strong shielding of inner protons and deshielding of outer protons is convincing evidence for the aromatic character. Similarly, at $-70°$C two broadband decoupled signals appear in the ^{13}C-NMR spectrum at δ = 128 (12C) and 121 (6C) while at 60°C one peak at δ = 126 ppm appears. The fluxional geometry involves inversion between three equivalent structures (invertomers).[163]

The stabilization energy, measured from the heat of combustion, was determined to be 100 ± 6 kcal/mol.[164] A more recent determination of the delocalization energy (DE)[165] based on the thermochemical decomposition (in DMF, at 130°C) to benzene and dihydrobenzocyclooctene **171C** is 37 ± 6 kcal/mol, a value which represents about 1/3 of the DE per π-electron from that of benzene.

This splitting of benzene probably occurs via valence isomerization to a nonisolable tetracyclic compound **171A**.[166] The second reaction intermediate, probably **171B**, affords [12] annulene **161** photochemically.

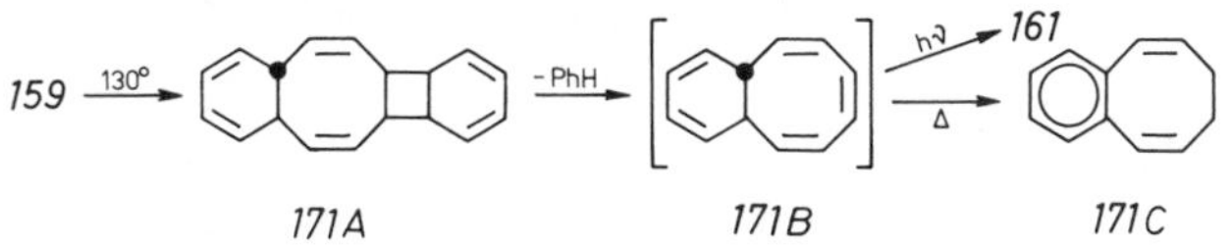

A lower value (around 20 kcal/mol) had resulted from the enthalpy of activation of the conformational exchange making all protons magnetically equivalent above 50°C.[167]

Various substituted [18]annulenes have been prepared. It was predicted, and the experiment confirmed this prediction, that 1,2-disubstituted [18]annulenes will exist in one conformation with both substituents directed outwardly; 1-chloro-2-fluoro- as well as 1,2-difluoro-[18]annulenes (**176**) were prepared via COT-dimers **172, 173**, addition of tetrahaloethenes leading to **174**, dehalogenation to **175**, and photoirradiation; their ^{1}H-NMR spectra are temperature independent as predicted.[168]

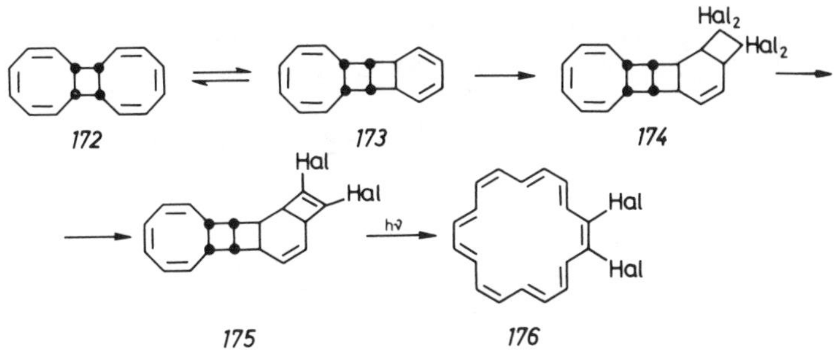

The symmetrical 1,2,7,8,13,14-hexamethyl-[18]annulene is less stable than the parent compound.[169] The mononitro-[18]annulene (**177**) has the nitro group fixed in an outer position and so are the five associated hydrogen atoms (denoted by dots). In this case, from the three invertomers **177A** to **C**, only two (**177A** and **177B**) are able to exist, for obvious steric reasons; the ^{1}H-NMR spectrum is still temperature dependent.[170] Some of the derivatives of [18]annulenes were prepared by electrophilic substitution,[17] others by subsequent reactions (Br → CH$_3$, CHO → CO$_2$Me), and still others by partial hydrogenation of acetylenic precursors. This is the case of the above hexamethyl derivative and of the hexadeutero derivative.[172] The ratio of inner to outer proton ratio in the ^{1}H-NMR spectrum of the latter product is 1:2 and not 1:3 as expected for a "frozen" conformation.

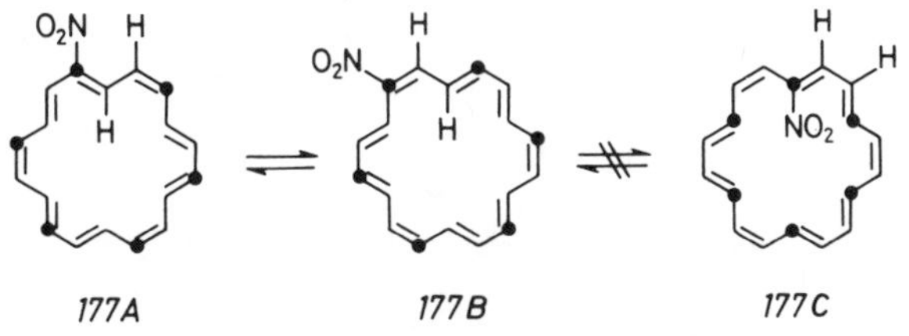

Although bridged annulenes are discussed separately in Chapter 5, it may be mentioned here that various systems with a periphery of [18]annulene are known: UV irradiation of the cyclophane **179** yields[173] the bridged [18]annulene **180** whose four inner protons give ^{1}H-NMR signals at $\delta = -2.58$ and -2.86 ppm, indicating strong diatropicity. An even stronger diamagnetic ring current is present in the hexahydrocoronene **181** whose six inner protons give rise to multiplets centred at $\delta = -6.44$, -6.82, and -7.88 ppm.[173] The difference in diatropicity may be due to the fact that **180** has a reentrant "bay region", whereas **181** has exactly the perimeter of [18]annulene. Sexipyridines[174] which yield strong complexes with Na^+ ions have a hetero [18]annulene circuit.

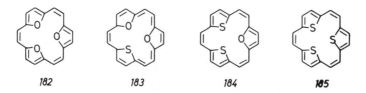

Several heteroatom-bridged [18]annulenes with no inner protons were prepared, derived from furan and thiophene heterocycles. The outer protons of the trioxide **182** resonate at around 8.7 ppm indicating diatropic deshielding;[175] however, when the heteroatoms are progressively replaced by sulfur atoms, the diatropicity decreases in the monosulfide bis-oxide **183**, and vanishes in the bis-sulfide monoxide **184** and trisulfide **185** which are atropic;[176] this effect is due probably both to the larger volume of the sulfur atoms causing deviations from planarity of the peripheric π-electron system, and to the stronger local aromaticity in the thiophene rings than in furan leading to lack of through-conjugation.

An even more interesting planar system with two perpendicular symmetry axes is the bis-furan **186** formed in 3% yield by spontaneous dehydrogenation in a Wittig reaction.[177] The solid green product, m.p. 235°C, presents, in agreement with its structure (**186**) five types of carbons and four types of hydrogens in the NMR spectra. The strong diatropicity is manifest in the high-field position of the inner H_a signals ($\delta = -5.89$) relative to the outer H_b and H_c signals (10.00 and 9.20 ppm); the furanic protons resonate at 9.13 ppm and are singlets. The coupling constants of the other hydrogens are $J_{ab} = 13.5$ Hz, $J_{bc} = 10.8$ Hz. The difference between inner and outer proton chemical shifts is higher than in the unsubstituted [18]annulene (12.13 ppm) or in tetra-*t*-butyl-bisdehydro[18]annulene (14.90 ppm), and this led the authors to suppose that the bonds may be completely equalized, as in benzene, according to the two equivalent Kekulé structures **186A** and **B**. By contrast, the isomeric bridged [18]annulene **187** is less planar and less diatropic.[178]

G. [20]Annulene

[20]Annulene, red-brown needles, has at $-105°C$ a ^{1}H-NMR spectrum with two broad multiplets at $\delta = 13.9$ to 10.9 and $\delta = 6.6$ to 4.1 ppm which at $25°C$ are replaced by a singlet at δ 7.18 ppm. This indicates paratropicity, flexible geometry, and probably the existence of several configurational isomers of which **188** appears to be the most likely.[179]

188

H. [22]Annulene

[22]Annulene obtained via the oxidative method is a dark purple compound which cannot be conserved owing to decomposition.[180] Its ^{1}H-NMR spectrum at $-90°$ has two multiplets at $\delta = 9.65$ to 9.3 and at -0.4 to -1.2 ppm which coalesce on heating at $65°C$ to give a singlet at $\delta = 5.65$ ppm. The large separation between the chemical shifts at low temperature indicates diatropicity due to ring current in a probable conformation **189**, whereas the high-temperature coalescence shows conformational changes as in [18]annulene.

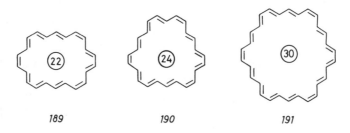

189   *190*   *191*

I. [24]Annulene

[24]Annulene, a deep purple crystalline compound, was obtained from the Eglington tetramerization of hexa-1,5-diyne followed by prototropic rearrangement with tBuOH and partial hydrogenation.[181] Its ^{1}H-NMR spectrum at low temperature ($-80°C$) has two multiplets at $\delta = 12.9$ to $11.2(9H)$ and 4.73 ppm $(15H)$ indicating paratropicity. At $30°C$ a singlet at $\delta = 7.25$ is observed. From the possible stereoisomers present in the mixture, **190** is the most probable.

J. [30]Annulene

[30]Annulene (**191**) is the largest annulene yet prepared.[182] Its spectra were not recorded.

K. Conclusions for Macrocyclic Annulenes

The preparation and properties of macrocyclic ($n \geqslant 12$) [n]annulenes shows clearly that, in the absence of steric destabilizing interactions between inner hydrogens, ($4n + 2$)-membered annulenes are paratropic and aromatic, whereas ($4m$)-membered annulenes are paratropic, usually nonplanar systems.

A comparison of electronic absorption spectra is provided by Table 3 which reveals that the ($4m + 2$)-membered annulenes absorb lower energy photons than ($4m$-annulenes; this is due to the fact that the π-electron system of the latter nonplanar compounds has less extensive conjugation. The high intensity of absorptions is a little higher in the ($4m + 2$)-membered rings.

Table 3

PHYSICAL DATA (MAIN UV-VIS ABSORPTION MAXIMA, FREE ENERGIES OF ACTIVATION FOR AUTOMERIZATION, AND RING CURRENTS) OF [n]ANNULENES

n	12	14	16	18	20	22	24	—	—	30	
λ_{max}	—	317	285	349	323	400	363	—	—	432	nm
lg ϵ_{max}	—	4.48	4.86	5.48	5.16	5.15	5.30	—	—	5.15	
$\Delta G^{\ddagger}$	5.5	10.5	8.6	14.3	9.8	12.8	11.0	—	—		kcal/mol
Tropicity	*para*	*dia*	*para*	*dia*	*para*	*dia*	*para*	—	—	—	

All macrocyclic annulenes have temperature-dependent ^{1}H- and ^{13}C-NMR spectra, indicating fluctuating geometry. The free energy of activation $\Delta G^{\ddagger}$ (c.f. Table 3) is higher for the $(4m + 2)$-membered systems which show diatropicity than for the $(4m)$-membered systems: a diagram $\Delta G^{\ddagger}$ vs. n shows a clear alternation, with the $\Delta G^{\ddagger}$ values for $[4m + 2]$ annulenes increasing steadily with m.

VII. GENERAL CONCLUSIONS FOR ANNULENES

The whole physicochemical behavior of all known annulenes confirms the brilliant predictions of Hückel and Dewar: when $n = 4m + 2$ (where m is an integer $m = 1,2,...$) is the absence of adverse steric interactions which prevent the attainment of planarity, the $[n]$ annulenes are aromatic, presenting to a larger or smaller extent the phenomena associated with aromatic character and discussed in Chapter 2; when $n = 4m$, with the same proviso as before, the $[n]$annulenes are antiaromatic. The experimental manifestations of antiaromatic character are more diverse than those of aromaticity: planar antiaromatic systems exhibit a paramagnetic ring current and display in most cases alternating bond lengths leading to automerizations via bond shifts, as for cyclobutadiene. More often, [4m]annulenes adopt nonplanar conformations which again may give rise to interesting conformational automerizations, e.g., ring inversion in the case of COT (which in this case is superimposed on bond-shift processes).

There exist two limitations for Hückel's rule. The first is of electronic nature and is apparent when considering the variation of the Dewar resonance energy (DRE) vs. the ring size, n: for sufficiently large n values the distinction between aromatic and antiaromatic systems vanishes as far as energies are concerned; because experimental data for $n > 24$ are scarce, it is not yet possible to say for certain which is the critical n value, or whether various properties associated with aromatic/antiaromatic character (such as bons alternation, dia/paratropicity, etc.) will have different critical n values.

The second limitation has a steric origin and leads to interesting stereoelectronic consequences. Among the lower $[n]$annulenes with $n = 4, 6,$ and 8, angle strain is nil for [6] annulene and significant for the two antiaromatic annulenes; [4]annulene adopts a planar rectangular geometry in its ground state, associated with antiaromatic destabilization and high reactivity (both as diene and as dienophile) so that in solution [4]annulene is unable to subsist. By contrast, [8]annulene adopts a tub form which allows the substance to exist and to exhibit typical nonaromatic behavior. Larger $[n]$annulenes may adopt several conformations; the higher n, the larger the number of possible conformations. This phenomenon has two consequences: (1) fluxional behavior, encountered both for Hückel and anti-Hückel annulenes and (2) the possibility of valence isomerization from certain conformations according to the Woodward-Hofmann rules. Examples of the latter process are [8]annulene which valence-tautomerizes reversibly with a low activation barrier to bicyclo [4.2.0] octatriene, *all-cis*-[10]annulene which affords *cis*-9,10-dihydronaphthalene at temperatures above − 10°C, *all-trans*-[10]annulene which leads to *trans*-9,10-dihydronaphthalene above − 25°C (the photochemical interconversions are opposite, in agreement with orbital symmetry control), [12]annulene which yields thermally above − 40°C the *cis*-bicyclo [6.4.0]dodecapentaene (the photochemical reaction affords the corresponding *trans*-bicyclic valence isomer), [16] annulene which on heating above 20°C gives the *di-cis*-tricyclo [10.4.0] hexadecahexaene, whereas photoirradiation yields the corresponding *di-trans* valence isomer, and [18]annulene which valence-isomerizes at 90°C to a tetracyclic valence isomer which then splits off benzene leaving dihydro-benzocyclooctene. A further consequence of steric factors, due to the presence of colliding inner hydrogens, is the preparation of stable systems when these hydrogens are either replaced by bridging groups, as will be shown in Chapter 5, or removed by introducing triple bonds, as will be described in the next paragraph. In the latter case, triple bonds introduce supplementary angle constraints (sp-hybridized atoms normally form rectilinear chains with angles of 180°, which in this case have to be accommodated in rings).

VIII. DEHYDROANNULENES

A. Introduction

Dehydroannulenes are cyclic continuously conjugated systems having double and triple bonds (in some cases, cumulated double bonds, too). The delocalized molecular orbitals of the planar dehydroannulenes involve, for each carbon atom, the π atomic orbitals of the sp^2-hybridized olefinic carbons, and one of the corresponding π atomic orbitals of the sp-hybridized acetylenic or cumulenic carbons.

The first stable dehydroannulenes, like the first macrocyclic annulenes were prepared by Sondheimer's group; the former were actually intermediates in the synthesis of the latter. According to the IUPAC Nomenclature Rule C-41.2, the even number of hydrogens which have to be removed from the annulene to afford the specified dehydroannulene is indicated as a prefix (di, tetra, etc.); sometimes the prefix di is omitted, but this will not be done here. The position of these removed hydrogens is indicated by numbers when necessary. Alternatively, the location and position of triple bonds resulting by removal of $2k$ hydrogens is denoted by specifying the number k (which may be odd or even) as a prefix as in the following examples: 1,8-bisdehydro [14] annulene or 1,7,13-trisdehydro[18]annulene. Confusion between these two alternative nomenclature systems is unlikely because of the ending -kis in the latter system. Monodehydro in the latter system means didehydro in IUPAC nomenclature.

We shall not discuss here the chemistry of benzyne (1,2-didehydrobenzene) which can be viewed as a dehydro[6]annulene. Reviews on this topic are available;[183-185] despite its instability, it can be readily prepared and undergoes many cycloadditions. Its chemistry was developed by Roberts, Huisgen, and Wittig. Dehydro[8]annulenes are known with fused benzo rings, as will be shown in Chapter 7. Masamune and co-workers[186] investigated dehydro[10]annulenes: the lithium salt of 1,2-diethynylcyclohexene (**192**) was obtained by reaction with *n*-butyllithium; the cyclized product **193**, resulted in 10% yield, was converted with trityl fluoroborate into cation **194**, whose hydration yielded the hydroxybenzoate **195**; successive conversion into **196** and **197**, followed by treatment with sodium methoxide led to a mixture of two isomeric hydrocarbons: 1,2,3,4-tetrahydroanthracene **200**, and **199**. For the formation of the latter compound, identified by X-ray crystallography, a Cope rearrangement of resonance structure **198B** seems likely. Thus, the 1,2,5,6-tetradehydro[10]-annulene **198** appears to have at most the thermodynamic stability of benzene.

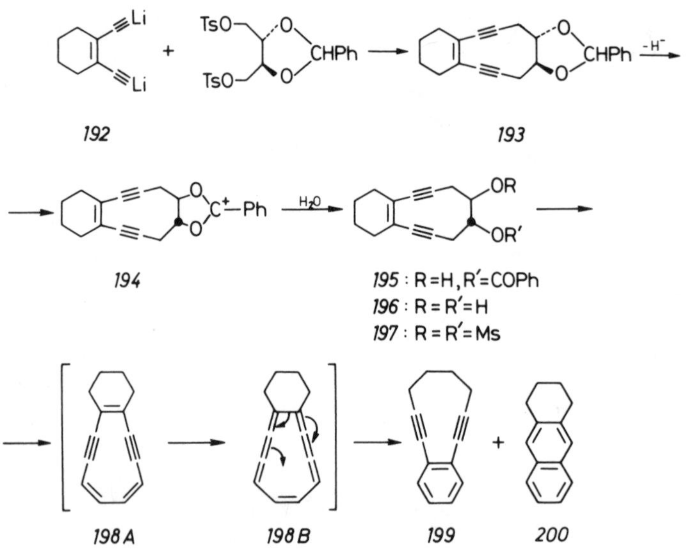

Li
Li
TsO
TsO
O
O
CHPh
-H⁻
192
193
C⁺-Ph
H₂O
OR
OR'
194
195 : R=H, R'=COPh
196 : R = R' = H
197 : R = R' = Ms
+
198A
198B
199
200

Another interesting system would be the 1,2,6,7-tetradehydro[10] annulene **201C** because it would posess two equivalent Kekulé structures (**201A** and **B**). Its synthesis was attempted by Sondheimer et al.[8] unsuccessfully.

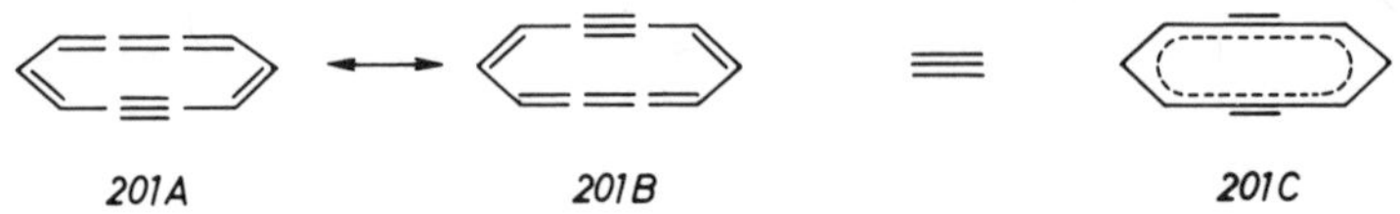

201A
201B
201C

B. Syntheses

Oxidative coupling (Glaser[146] or Eglinton[147] conditions) of acetylenes or polyacetylenes with two terminal methinic groups leads to the formation of polymers, among which cyclic polymers exist (Sondheimer's method[145]). Probably both acetylenic methinic groups are associated with the copper atom, explaining thereby the formation of cyclic products. These products are then separated and converted into dehydroannulenes by base-catalyzed prototropic isomerization (potassium *t*-butoxide in benzene/*t*-butanol. Thus didehydro[14]annulenes (**205,206**) 1,2,7,8- (**208**) and 1,2,8,9-tetradehydro[14] annulene (**207**) were prepared by intramolecular oxidative coupling of $C_{14}H_{14}$ (**202**).[152,153,187]

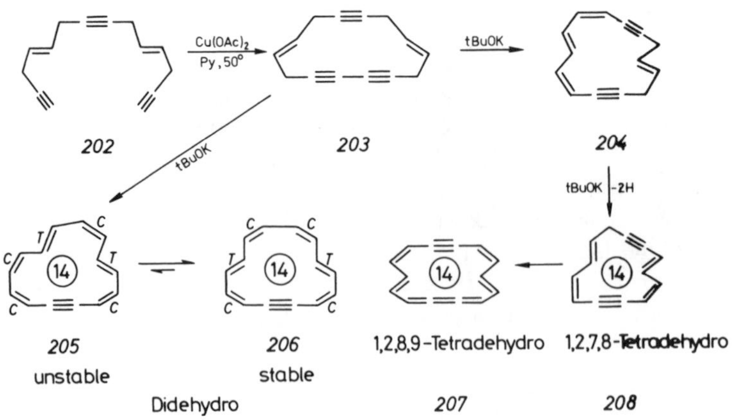

Cu(OAc)₂
Py, 50°
tBuOK
tBuOK
tBuOK
-2H
202
203
204
14
C
T
205
unstable
206
stable
Didehydro
1,2,8,9-Tetradehydro
1,2,7,8-Tetradehydro
207
208

On the other hand, tri- and tetradehydro[18]annulenes **211** to **213** were prepared by intermolecular coupling of 1,5-hexadiyne (**209**, C_6H_6),[188] a reaction which gave a cyclic trimer **210** in 6% yield along with cyclic di-, tetra-, penta-, hexamers as well as to larger polymers.

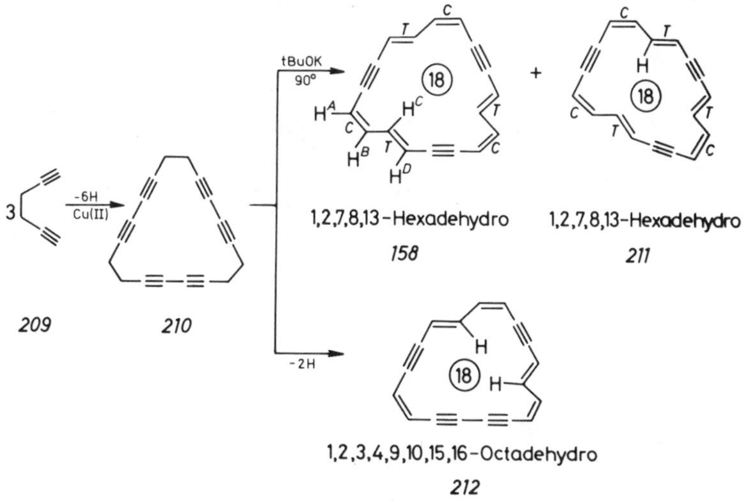

Nakagawa[189,190] devised another method for substituted dehydroannulenes which gives higher yields. The reactions involve Favorsky-type condensations between lithium acetylides and conjugated *cis*-enones (or aldehydes, or acetals, e.g., **216**) followed by Eglinton oxidative coupling and dehydration of diols in the presence of stannous chloride and hydrochloric acid.[191,192] The yields are remarkably high (60 to 80%).

219	R	R′
a	Me	Me
b	Me	Ph
c	Ph	Ph
d	An	An
e	tBu	Ph
f	tBu	tBu

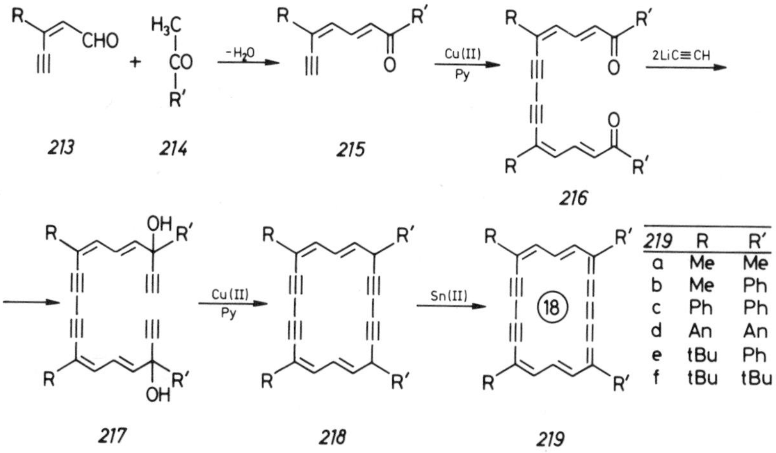

With R and R′ being Me and/or especially Ph groups, the products had unsatisfactory solubilities for NMR spectroscopy; however, *t*-butyl groups imparted good solubility and gave high yields of stable crystalline products with much less electronic perturbation than with aryl substituents (as indicated by electronic absorption spectra). When the synthesis of the above product with R = PH and R′ = tBu was repeated by reverting the substituents (R′ = Ph and R = tBu), the same product was obtained, proving thereby the complete delocalization, i.e., the equivalence of the polyacetylene and cumulene moieties.[193]

The above synthesis was generalized both in "length and width" of the ring system, i.e., by varying the number of conjugated double and triple bonds, or the nature of substituents for obtaining **223**[154] and **227**.[195]

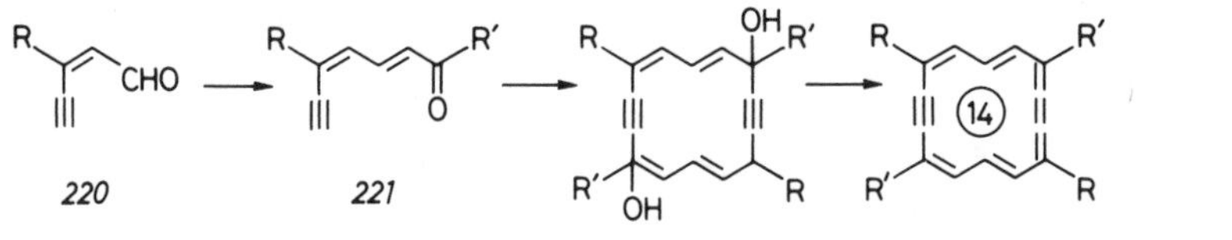

220	**221**

222

223 a: R = R'= tBu
b: R = R'= Ph
c: R = Ph, R'= tBu

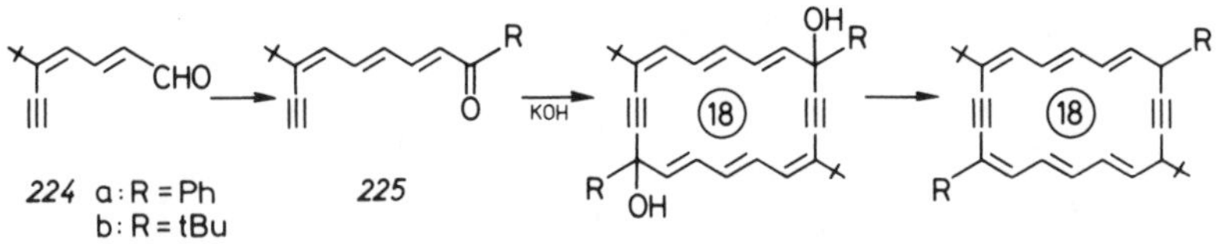

224 a: R = Ph
b: R = tBu

225

226

227

Similarly were obtained **228**,[196] **229**,[197] and **230**.[198]

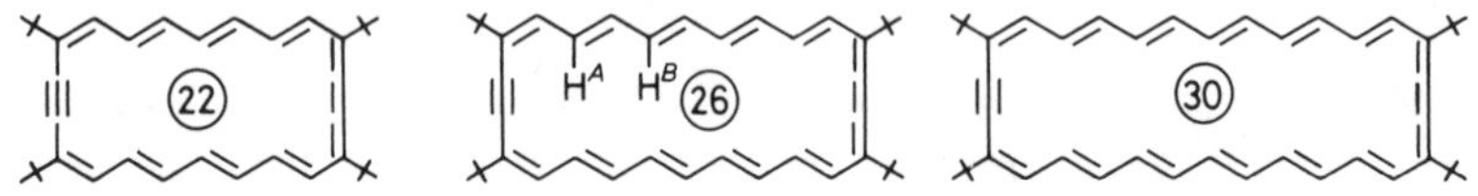

228	**229**	**230**

Analogously to the above [4*m* + 2]-membered systems, [4*m*]-membered substituted compounds **231** to **233** were synthesized by Nakagawa's group.

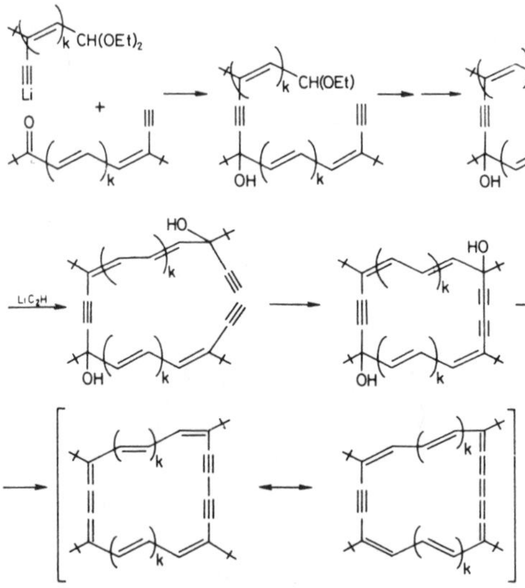

231 (k = 1)   Trisdehydro [16] annulene[228]
232 (k = 2)   Trisdehydro [20] annulene[229]
233 (k = 3)   Trisdehydro [24] annulene[230]

C. 12-Membered Rings

On oxidative coupling of 1,5-hexadiyne under Glaser conditions (the trimerization under Eglinton conditions leads to **158** and **159** with 18-membered rings) followed by rearrangement of the cyclic dimer **234** with potassium *t*-butoxide, Sondheimer et al. obtained[199] biphenylene as the main product (7% yield) along with 1,2,5,6-tetradehydro[12]annulene **235** (1.5% yield) and 1,2,5,6,9,10-hexadehydro[12]annulene **236** (0.6%). The latter product was obtained in higher yield by Untch et al.[200]

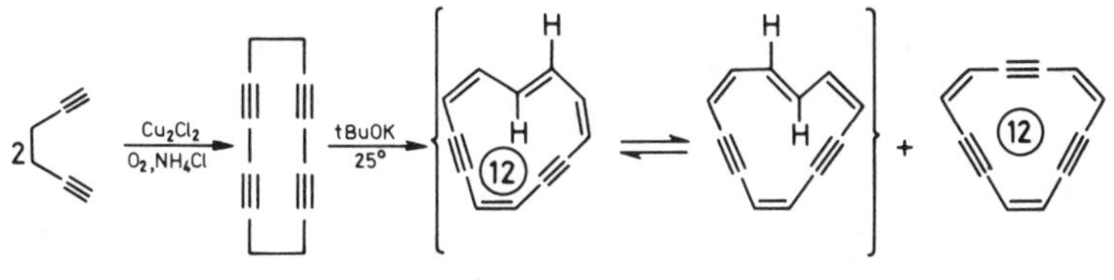

234   235 A   235 B   236

Both last systems have ^{1}H-NMR spectra which indicate paratropicity: the inner proton of the former compound **235** and its *transoid* proton appear around $\delta = 11$ ppm as a quartet, while the outer protons appear at $\delta = 4$ to 5 ppm. The calculated chemical shift of the inner proton is about $\delta = 17.9$ ppm. The ^{1}H-NMR spectrum appears to represent an average, even though it remains unchanged to $-130°C$, because 9-bromo-1,5-bisdehydro[12]annulene where there is no exchange of *trans* double bond protons has the inner proton at $\delta = 16.4$ ppm. The ^{1}H-NMR spectrum of the latter compound **236** is a singlet at $\delta = 4.42$ ppm (in CCl_4); the structure is interesting because it has D_3-symmetry, with two nonequivalent Kekulé structures: one with three cumulenic systems, the other with three triple and three double bonds; there is no inner hydrogen. This compound had been initially believed to be an isomer of bisdehydro[12]annulene.

D. 14-Membered Rings

Some of the earliest and most interesting data were obtained with dehydro[14]annulenes, which according to Hückel's rule are expected to be aromatic. The most thoroughly studied member from this series is the 1,2,8,9-tetradehydro[14]annulene; the unsubstituted system was prepared by Sondheimer et al.,[199] while tetrasubstituted derivatives were prepared by Nakagawa and co-workers;[195] both groups obtained consistent results.

Since two equivalent Kekulé structures may be written for **207**, **219**, **223**, and **227** to **230**, one expects complete electronic delocalization, which may be formulated as **207C**.

207 A   207 B

Since this system has both inner and outer hydrogens, one can apply the criterion of ^{1}H-NMR ring current in $CDCl_3$. Indeed, the inner protons H^C appear as a triplet at $\delta = -5.48$ ppm, the outer protons H^B as a doublet ($\delta = 8.54$), and H^A as a quartet at 8.54 ppm. The difference δ (outer-inner) amounts to 14 ppm (see also the conclusions of this section). As will be seen, in the tetrasubstituted compounds this difference is a little lower. Interestingly, the ^{1}H-NMR spectrum of **207** is not temperature dependent, and likewise the ^{1}H-NMR spectra of **223a** to **c**: $\delta = -4.39$ and 9.42 for the red **223a**, -3.42 and 9.88 for **223b**, and -2.56 and 9.94 ppm for the violet **223c**. Interestingly, the ^{13}C-NMR spectrum of **223a** presents only one signal for the sp-hybridized carbons at 116.7 ppm, indicating complete electronic delocalization. The electronic absorption main maximum of **207** at 310 nm has a very high intensity (lg $\epsilon = 5.3$). The X-ray crystallographic structure determination indicates that the molecule is centrosymmetric,[201] and that C(sp^2) to C(sp^2) bonds have bond lengths similar to those of benzene without bond alternation.

Its chemical behavior is consistent with an appreciable aromatic stabilization: it is stable in solid state and in solution, but at 200°C it rearranges to benz[*a*]azulene. Several electrophilic substitutions were performed: nitration with cupric nitrate in acetic anhydride, sulfonation with oleum, acetylation with acetic anhydride, and boron fluoride. In all cases the substitution occurs at H^A (the position adjacent to the acetylene/cumulene moiety).[202]

Nakagawa and co-workers[203] prepared various diphenyl-di-*t*-butyl-derivatives of **223c** corresponding to *o*-, *m*-, and *p*-terphenyl but with the central phenylene replaced by a di-*t*-butyl-tetradehydro[14]annulenylidene ring.

Measurements of fluorescence excitation spectra of **223a** and of the *para* isomer **223b**, and of the polarized reflection spectrum of a single crystal of **219f** showed that in all these systems the longest-wavelength (^{1}L$_b$) and the second electronic transitions (^{1}L$_a$) have the directions of polarization showed by the arrows in formula **207C**, in agreement with theoretical calculations.[190]

Acetic anhydride and zinc chloride cause *ipso* substitution of one *t*-butyl group of **223a** at 140°C for 0.5 min; stepwise synthesis confirmed the structure of the acetyl derivative[204] which was strongly diatropic. Also by stepwise synthesis, a related compound with carboxyl instead of acetyl groups was prepared.[205] Its pK$_a'$ value was determined spectrophotometrically to be 5.92 at 18°C; benzoic acid and *p*-*t*-butylbenzoic acids have pK$_a'$ values of 5.55 and 5.64, respectively, whereas allenecarboxylic acid has pK$_a'$ 3.96, and 3-pentynoic acid has pK$_a'$ 3.60. The fact that the value for the dehydro[14]annulenecarboxylic acid is much closer to that of benzoic acid is in agreement with extensive delocalization.

A different isomer is 1,2,7,8-tetradehydro[14]annulene (**208**). It is obtained on mild treatment of **203** with potassium *t*-butoxide in *t*-butyl alcohol and benzene at 10°C for 1 min; further treatment of **208** with the same basic reagent for 10 min causes isomerization to the thermodynamically stable 1,8-bisdehydro-isomer, **207**. The ^{1}H-NMR spectrum of **208** has the inner protons at $\delta = -0.7$ as a double doublet, and the outer protons as a complex multiplet at $\delta = 7.3$ to 8.6, indicating strong diatropicity.

Whereas the difference between **207** and **208** was constitutional isomerism (due to which **208** has nonequivalent diacetylenic and dicumulenic Kekulé structures), Sondheimer and co-workers obtained two stereoisomers (**205** and **206**) of didehydro[14]annulene.[145,152,153] The stable isomer was obtained in 20% yield, the unstable isomer in 10% yield (their separation was effected by chromatography on AgNO$_3$-coated alumina or silica gel), and 1,8-bis-dehydro[14]annulene **207** (resulted simultaneously in the tBuOK-caused rearrangement of **203** and **204**) in 10% yield. The unstable isomer rearranged to the stable one after conservation in ether for 24 hr without protection from diffuse daylight, or after photo-irradiation in boiling benzene for 30 min. Both isomers are dehydrogenated to phenanthrene in 50% yield on refluxation in dimethylsulfoxide or dimethylformamide.

The 60-MHz ^{1}H-NMR spectra of the two isomers are fairly similar: the stable isomer presents a double doublet centered at $\delta = -0.7$, due to two inner protons, in addition to a multiplet at $\delta = 7.3$ to 8.6 ppm indicating the presence of two *trans* and four *cis* double bonds, in addition to the triple bond. The unstable isomer has the inner protons at $\delta = -0.6$ ppm and the two coupling constants are such that at 60 MHz the double doublet appears as a 1:2:1 triplet. Initially it was believed that the two compounds were conformational isomers, but in his last review published in *Accounts of Chemical Research*[145]

Sondheimer revised this idea in favor of configurational isomers; the assignment of formulas **205** and **206** is provisional and might be reverted; X-ray crystallographic investigation of the stable isomer was inconclusive. On partial catalytic hydrogenation, the stable isomer gave [14]annulene in 30% yield but the unstable isomer gave no [14]annulene. Both isomers (apparently involving conversion to the stable one) gave electrophilic nitration to the same mononitro-didehydro[14]annulene with copper(II) nitrate in Ac_2O. They also afforded the same mixtures of two or three isomeric monoacetyl derivatives (with Ac_2O + BF_3) and methyl sulfonates (on treatment with oleum, then with silver nitrate, and finally with methyl iodide).

The most highly unsaturated compound in this series is the red 1,5,9-trisdehydro[14]-annulene (**237**) obtained in 1% yield starting from 1,6-heptadiyn-4-ol methanesulfonate; oxidative coupling led to a cyclic dimer which on treatment with potassium hydroxide in methanol-dimethylsulfoxide at 20°C underwent elimination of 2 mol of methanesulfonic acid; the major products were bi- and tricyclic compounds, and they were separated chromatographically.[206] The inner proton of **237** appears as a triplet at $\delta = -4.96$, and the seven outer protons occur as a complex signal at $\delta = 8.08$ to 9.47 ppm. Again, this indicates very strong diatropicity.

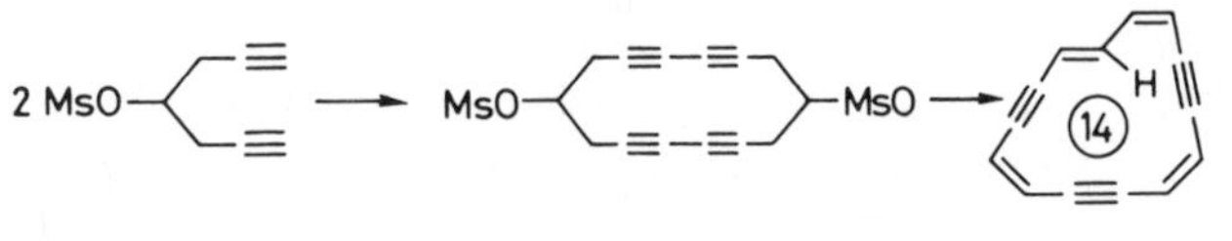

237

E. 16-Membered Rings

Sondheimer's group[159] obtained from *trans*-4-octene-1,7-diyne on Glaser oxidation a cyclic dimer which gave on brief treatment with tBuOK + tBuOH (at 40°C for 1 min) 1,9-bisdehydro[16]annulene (**156**). The same compound was obtained directly under Eglinton conditions. Its ¹H-NMR spectrum in perdeuteroacetone at −80°C displays the two inner hydrogens H^D at $\delta = 9.80$, the outer protons H^B, H^C, H^E at $\delta = 6.08$, and the outer H^A and H^F type protons at $\delta = 5.45$ ppm. At room temperature the peaks due to the *trans* double bond protons H^C and H^D coalesce giving rise to a new peak at $\delta = 7.88$ ppm. These data indicate marked paratropicity.

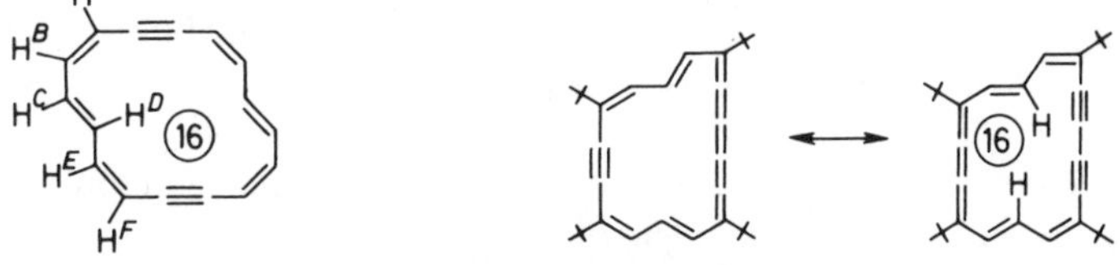

156   *231 A*   *231 B*

Nakagawa's group[207] obtained tetra-*t*-butyl-substituted trisdehydro[16]annulene **231**, whose ¹H-NMR spectrum (CDCl₃) is temperature independent: inner protons at $\delta = 17.10$, outer protons at 4.17 ppm, $J = 11$, and 15 Hz. The ¹³C-NMR chemical shifts of sp-hybridized carbon atoms are 86.6, 90.5, and 153.3 ppm, i.e., two signals in the acetylenic region (70 to 90 ppm) and one in the cumulenic region (150 ppm). All these data are consistent with an alternating bond structure **231B**.

F. 18-Membered Rings

We shall discuss the two isomeric trisdehydro[18]annulenes **158** and **211**, and the tetrakisdehydro[18]annulene (**212**), obtained by Sondheimer's group, as well as the tetra-*t*-butyl-bis- and -tetrakisdehydro [18] annulenes **227** and **219** obtained by Nakagawa and co-workers.

1,7,13-Trisdehydro[18]annulene (**158**), light-brown plates,[160] presents a ¹H-NMR spectrum with the double doublet due to the three inner protons H^C at high field (δ = 1.74 ppm) and the peaks of the outer protons at low field (H^A, 7.02; H^C, 7.56; H^B, 8.10 ppm), indicating that the compound is diatropic. The spectrum is temperature independent from −60 to 150°C, presumably because rotation of the *trans* double bond would lead to a nonequivalent conformer of higher energy than **158**. The electrophilic nitration of **158** leads to a 3-nitroderivative, whose ¹N-NMR spectrum is in agreement with expectations.

The configurational isomer **211** is formed from **210** in lower yield (0.5%) than **158** (1.5%).[208] The inner protons give rise to two quartets: one (1H) centered at δ = 1.58, the other (2H) at δ = 1.73 ppm; the nine outer protons form a complex multiplet at δ = 6.9 to 8.3 ppm.

Tetrakisdehydro[18]annulene **212** was formed in very low yield (0.1%) since the reaction involves the prototropic isomerization of **210** plus loss of two hydrogens.[208] The ¹H-NMR spectrum of 212 has two quartets, corresponding each to one inner proton, centered at 2.34 and 2.20 ppm, respectively; the outer protons form a complex band at δ = 6.6 to 8.2 ppm.

Both **211** and **212** were not detected initially in the reaction mixture because conventional column chromatography on alumina or silica gel fails to achieve separation; however, on coating these adsorbents with silver nitrate, satisfactory chromatographic separation could be made.[208]

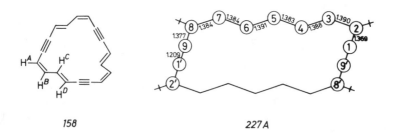

158   *227A*

We come now to the tetra-*t*-butyl-substituted systems. An X-ray analysis[209] of the bis-dehydro[18]annulene **227** gave the bond lengths indicated on formula **227A**. The distance between the inner hydrogens is 1.91 Å; CCC bond angles of sp-hybridized bonds are 178.9 to 179.5°, and of sp²-hybridized bonds are 120.3 to 126.1°. The molecule is nearly planar, and the polyene parts involving seven carbon atoms each show almost no bond alternation (average bond length 1.387 Å). The formal acetylenic and cumulenic linkages are found to be equivalent, in agreement with a highly delocalized electronic structure and marked diatropicity for which the ¹H-NMR data give evidence: inner protons at δ = −3.42, outer protons at δ = 9.87 ppm.

The ¹³C-NMR spectrum presents at 36°C in $CDCl_3$ only one signal for both acetylenic and cumulenic carbons at 115.7 ppm; this value is intermediate between the corresponding ranges, indicating complete electronic delocalization.

The tetrakisdehydro[18]annulenes **219a** to **f** with R = R′ have outer protons at δ = 9.66 (R = R′ = Me), 9.98 (R = R′ = *t*-Bu), 10.04 (R = R′ = *p*-anisyl), or 10.31 (R = R′ = Ph).[194] When R ≠ R′, the outer protons appear as two distinct multiplets in the same range. The inner protons give rise to one multiplet at δ = − 3.00 (R = R′ = *p*-anisyl), −3.19 (R = R′ = Ph), −4.92 (R = R′ = *t*-Bu), −5.24 ppm (R = R′ = Me),

−4.20 for **219b**, and −3.90 for **219e**. Compounds **219a** and **219b** gave temperature-independent ¹H-NMR spectra indicating a high conformational stability of the skeleton. It can be seen that also in this case the system is highly diatropic, indicating aromaticity and delocalization as required by Hückel's rule.

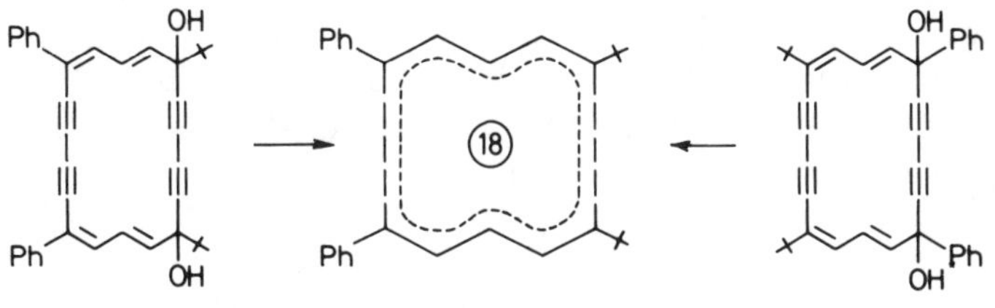

238   219e   239

By independent syntheses, two isomeric glycols **238** and **239** were converted into the same compound, **219e**, as attested by the same decomposition point (about 190°C), IR, electronic, ¹H-NMR, and argon laser Raman spectra.[193]

G. 20-Membered Rings

1,11-Bisdehydro[20]annulene (**240**) and a monodehydro[20]annulene were synthesized by Sondheimer and Gaoni.[210] In the ¹H-NMR spectrum of **240** at room temperature, four four-proton bands are apparent at 5.09, 5.68, 7.45, and 8.40 ppm, assigned to H^C, H^D, H^B, and H^A, respectively, on the basis of decoupling experiments. On cooling to −60°C the last two bands become invisible but the first two bands remain essentially unchanged. At −80°C two new low-field bands appear at δ = 11.60 (2H, $H^{C'}$) and 10.45 ppm (2H, H^D), along with a new band at 5.60 ppm (H^C, $H^{D'}$). This temperature-dependent ¹H-NMR spectrum indicates that at room temperature the H^A, $H^{A'}$, H^B, $H^{B'}$ protons of the *cis* double bonds are not affected, while the protons H^C, $H^{C'}$, H^D, and $H^{D'}$ of the *trans* double bonds undergo rapid exchange and become averaged. The fact that inner protons appear at high field and outer protons at low field, as in 1,9-bisdehydro[16]annulene, indicates paratropicity.

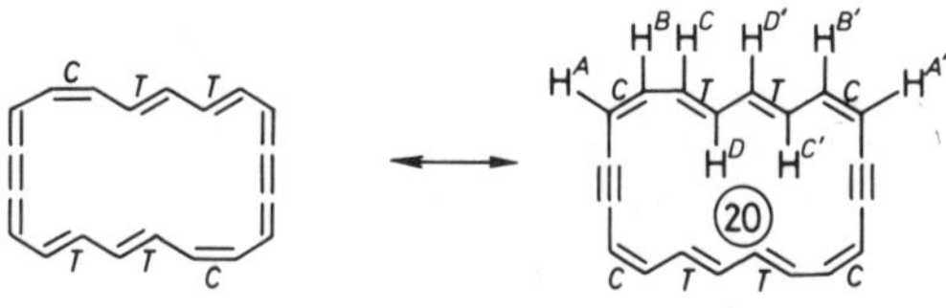

240 B   240 A

Tetra-*t*-butyltrisdehydro[20]annulene (**232**) was investigated by Nakagawa et al.[211] The inner proton resonates at 13.78 ppm, and the outer protons at 4.39 ppm at room temperature. In the ¹³C-NMR spectrum at −20°C acetylenic (δ = 85.2, 86.5 ppm) and cumulenic peaks appear (δ = 148.3 ppm) exactly as in the case of the corresponding[16]annulene, substantiating the view that the structure probably has alternating bond lengths.

No data have been reported for the NMR spectra of monodehydro[20]annulene.

H. 22-Membered Rings

Monodehydro[22]annulene was prepared by McQuilkin and Sondheimer.[212] Its ¹H-NMR spectrum presents signals of inner protons at δ = 0.70 to 3.95 ppm, and of outer protons at δ = 6.25 to 8.45 ppm indicating some diatropicity. Nakagawa's tetra-*t*-butyl-bisdehydro

[22]annulene **228** was investigated by X-ray crystallography and electronic and NMR spectroscopy. The molecular structure (**228A**) presents[213] an interesting difference from that of the analogous bisdehydro[18]annulene (**227**): for **228** the linkage C^{10}–C^{11}–C^{12}–C^{13} shows a more pronounced cumulenic character than the linkage C^2–C^1–C^{22}–C^{21}, which is closer to an acetylenic one. At the same time, the bonds C^2–C^3 and C^{21}–C^{20}, adjacent to the more acetylenic linkage, are shorter than the bonds C^{10}–C^9 and C^{13}–C^{14} which are adjacent to the more cumulenic linkage, as expected from the resonance formula **228**. These data indicate an increased bond alternation in **228** compared to **227**. Distances between inner hydrogens in **228** are larger than in **227**, namely, 2.14 to 2.25 Å. CCC-Bond angles for sp-hybridized carbons are 176.5 to 179.5° and for sp^2-hybridized carbons 119.0 to 127.8°.

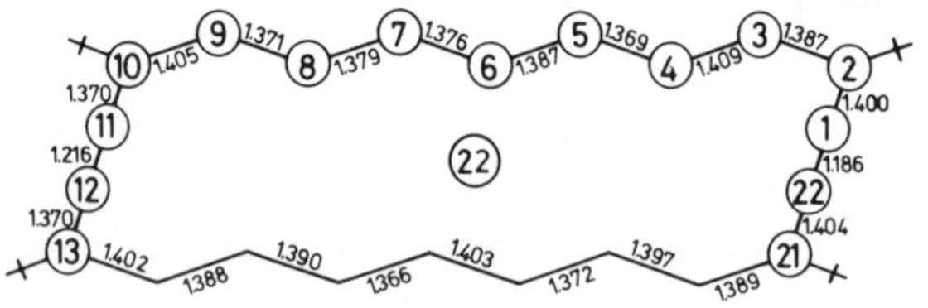

228 A

Tetrakisdehydro[22]annulenes with four phenyl or *t*-butyl substituents (**241a,b**) were synthesized by Nakagawa and co-workers.[198,214] Though they are unstable, the ¹H-NMR spectra show stronger diatropicity than the unsubstituted system: with R = Ph, the inner H^B signal appears at −2.80 ppm (triplet), and the outer H^A and H^B protons at 10.72 (doublet) and 10.79 (triplet), respectively; the coupling constant J is 14 Hz. With R = tBu, the corresponding values are −3.70, 9.91, and 10.40 ppm, respectively; the multiplets are the same but the J value is 13 Hz.

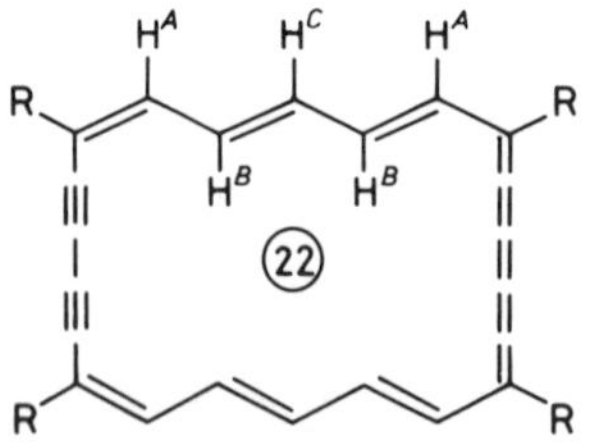

241 a : R=Ph
     b : R=tBu

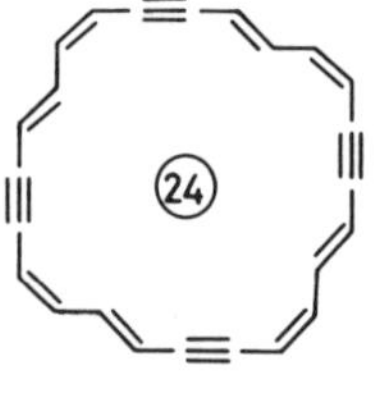

242

I. 24-Membered Rings

Sondheimer and co-workers prepared 1,7,13,19-tetrakisdehydro[24]annulene (**242**). Its ¹H-NMR spectrum[152] at room temperature or at 150°C exhibits an averaged multiplet (a 4-proton double doublet at δ = 8.40) due to inner protons and a 12-proton complex band at δ = 4.98 to 5.60 ppm. At −80°C the inner proton band appears at 11.2 to 12.9 ppm, and the outer proton band at δ = 4.73 ppm, showing the presence of a paratropic system.[215]

The tetra-*t*-butyl-trisdehydro[24]annulene **233** has a ¹H-NMR spectrum[216] with inner protons at 11.80 and outer protons at 4.70 ppm, indicating much stronger paratropicity than in **242**.

J. 26-Membered Rings

Sondheimer's group[217] obtained a monodehydro- (**243**) and a trisdehydro[26]annulene (**247A**). They were found to be atropic and weakly diatropic, respectively. Apparently, the

Volume I

fact that **244** was found to be atropic is due to the presence of three acetylenic bonds in one resonance structure, making it energetically preferred to a delocalized system with two nonequivalent resonance structures (the cumulenic one is apparently of higher energy). The geometries **243** and **244** are the most probable ones and are consistent with the ratio of inner/outer protons in the ^{1}H-NMR spectra.

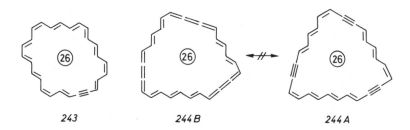

243   244 B   244 A

Nakagawa's[197] tetra-*t*-butyl-bisdehydro[26]annulene **229** shows much more intense diatropicity with inner protons H^A at $\delta = 1.95$ and H^B at 1.82 ppm, and outer protons at $\delta = 8.23$ ppm.

K. 28-Membered Rings
28-Membered rings have not been reported so far in the dehydroannulene series.

L. 30-Membered Rings
Sondheimer and co-workers[160] obtained in very low yield two dehydro[30]annulenes, but they showed no discrete inner and outer protons, even when cooled to $-60°C$, exhibiting a broad resonance in the $\delta = 5.5$ to 7.5 ppm range. Apparently, in this case, the aromatic resonance energy is too small to warrant delocalization leading to a cumulenic Kekulé structure, so that the compound remains with the localized acetylenic-polyenic nonequivalent Kekulé structure.

However, the ^{1}H-NMR data of Nakagawa's[218] unstable black-violet tetra-*t*-butylbisdehydro [30]annulene **230** (inner protons at $\delta = 3.50$, outer protons at $\delta = 7.50$ ppm at $-60°C$ in CDCl$_3$), shows that this system, which (unlike the foregoing one) has two equivalent Kekulé structures, is still able to sustain an induced diamagnetic current.

M. Conclusions
Dehydroannulenes are even more strongly diatropic in the $(4m + 2)$-membered series or paratropic in the $(4m)$-membered series than the annulenes. With the introduction of one or several triple bonds the following effects appear: (1) conformations become more rigid owing to angle strain but in many $(4n)$-membered systems, *trans* double bond protons exchange their sites fairly rapidly; (2) the number of inner hydrogens decreases and may attain zero, e.g., in **236**; (3) the chemical stability increases because there is usually less chance for valence isomerization by intramolecular cycloaddition; (4) resonance structures may or may not be equivalent, even if the geometry is ignored, because cumulenic linkages are possible, e.g., nonequivalent for **158**, **231** to **233**, **237**, **240**, and **244** and equivalent for **207**, **219**, **223**, and **227** to **230**.

As seen in Tables 3 and 4, it was found that normally the $(4n + 2)$-series presents strong diatropicity (stronger than related annulenes) and a temperature-independent ^{1}H-NMR spectrum; when equivalent Kekulé structures can be written, complete electronic delocalization is observed up to bisdehydro[22]annulene, but with the corresponding bisdehydro[26] annulene some bond localization is apparent from X-ray crystallographic data. The $(4m$-series presents marked paratropicity, as seen in Tables 3 and 4. The ^{13}C-NMR data[190] of

Table 4
¹N-NMR DATA OF TETRA-*t*-BUTYL-BIS- AND TRIS-
DEHYDRO[*n*]ANNULENES (CDC1₃, 36°C) FOR INNER (δ_i)
AND OUTER PROTONS (δ_o, PPM)

($4m + 2$)-Membered series				($4m$)-Membered series			
n	Inner δ_i	Outer δ_o	$\delta_o - \delta_i$	*n*	Inner δ_i	Outer δ_o	$\delta_o - \delta_i$
14	−4.44	9.32	13.76				
				16	17.10	4.17	−12.93
18	−3.42	9.87	13.29				
				20	13.78	4.39	−9.39
22	−0.83	9.16	9.99				
				24	11.80	4.70	−7.10
26	1.95	8.23	6.28				
30	3.50	7.50	4.00				
	Diatropic				Paratropic		

trisdehydro [4*m*] annulenes **231** and **232** exhibit three C(sp) peaks whereas those of bisdehydro [4*m* + 2]annulenes **223a** and **227** have one C(sp) peak, indicating electronic localization (bond alternance) in the former case, and delocalization in the latter case.

Another conclusion which can be drawn from Table 4 is that, in agreement with theoretical predictions,[219] the magnitude of the dia- or paratropicity, expressed by [$\delta_o - \delta_i$] decreases with increasing ring size.[189,190] The chemical shift due to diamagnetic ring currents in aromatic compounds is approximately proportional to ISR^{-3}, where I is the intensity of the ring current, S is the ring area, and R is the distance of the protons from the center of the molecule. One can assume R to be independent of the ring size for outer protons, and one derives the conclusion that [$\delta_o - \delta_i$]/S will decrease monotonously for [4*m* + 2]annulenes with increasing ring size.

It was predicted theoretically[115,219] that planar [4*m* + 2]annulen up to [22]annulene will be aromatic, but that [26]annulene and larger systems will no longer be aromatic. The experimental data indicate that for dehydroannulenes, although some bond alternance sets in for [26]annulene, as predicted, the distinction between diatropic [4*m* + 2]annulenes and paratropic [4*m*]annulenes is still valid up to [30]annulene, provided that two equivalent Kekulé structures may be written.

The most important conclusion is that the study of dehydroannulenes by Sondheimer, Nakagawa, Schröder, Oth, and co-workers validated all theoretical conclusions based on the Hückel rule for annulenes.

IX. SYSTEMS RELATED TO ANNULENES

Three areas related with the chemistry of annulenes will be mentioned briefly in the remaining sections of this chapter: monocyclic ions derived from annulenes, bi- and poly-cyclic annuleno-annulenones, and zero-atom-bridged annulenes. These areas will be presented without many details because, although they are highly pertinent to the problem of aromaticity and to the general framework of annulene chemistry, their special features (e.g., stability of Hückel-type annulenium ions) are restricted to the parent structures so that no valence-isomers have been isolated.

A. Aromatic and Antiaromatic Singly Charged Ions

The first impetus for new nonbenzenoid aromatics in the post-Hückel years was represented by Dewar's prediction[220] that tropone and tropolone might present aromatic character, explaining thereby the reactions of several natural tropolonic products such as thujaplicin, colchicine, stipitatic acid, etc. whose structure had baffled chemists until then. Soon after this revival of Hückel's prediction that the tropylium cation ought to be aromatic, Doering and Knox[221] prepared tropylium salts and noted that Merling[222] had already performed the same reaction; in the absence of a theoretical support, however, it had been impossible to guess that bromocycloheptatriene would prefer energetically to exist as tropylium bromide, for the same reason that cyclopentadiene and its benzoderivatives were recognized to have an unusual acidity: both the tropylium cation and the cyclopentadienide anion are possessed of aromatic character. Vice versa, early attempts to prepare cyclopentadienylium cations or cycloheptatrienide anions failed, confirming that they would be antiaromatic. It was recently predicted theoretically by Stohrer and Hoffmann[224] and then confirmed experimentally by Masamune et al.[225] that one can obtain a $(CH)_5^+$ cation and that its structure is a tetragonal pyramid rather than a planar regular pentagon. Further theoretical work substantiated the pyramidal structure of $(CH)_5^+$ using a variety of theoretical procedures,[226] including a graph-theoretical analysis[227] similar to that explaining the bonding in boranes ("tridimensional aromaticity").[228] Analogously, the $(CH)_6^{2+}$ dication is a pentagonal pyramid.[229] Pyramidal cations were recently reviewed.[326]

The simplest cation which according to Hückel's rule should possess aromatic character is the three-membered 2 π-electron cyclopropenium cation (**248**, R=H): triphenylcyclopropenium salts were prepared by Breslow in 1957,[230] and the unsubstituted system was obtained 10 years later by two independent syntheses.[231,232] The latter hexachloroantimonate is stable at room temperature but water or moisture cause its decomposition. The stability of substituted salts **248** is higher as shown by increasing pK_{R+} values: R=H, −7.4; R=Ph, 3.1; R=n-Pr, 7.2 (see References 6, 7, and 233).

Tropylium bromide **249** (see Reference 234) has a pK_{R+} value of 4.75[221] and can be most readily prepared by treating cycloheptatriene with trityl salts in acetonitrile or with phosphorus pentachloride in CCl_4.[235] Volpin and Kursanov[236] demonstrated the equivalence of all seven carbon atoms using ^{14}C-labeling. X-Ray crystallographic analysis revealed the planarity and, of course, ^{1}H- and ^{13}C-NMR spectra of **248** and **249** indicate the equivalence of all CR or CH groups at all temperatures.

A one-carbon bridged undecapentaenium cation **250** was obtained as tetrafluoroborate. Its ^{1}H-NMR data indicate diatropicity, and the X-ray crystallographic analysis confirms the delocalized 10 π-electron periphery rather than a benzohomotropylium structure.[237]

The most stable carbenium ion in this series is the two-carbon bridged [15]annulenyl cation **251**. The tetrafluoroborate has a $pK_{R+} = 8.4$.[238]

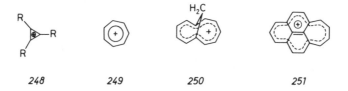

248	*249*	*250*	*251*

In addition to cyclopentadienide, other annulenide anions were obtained by deprotonation: bicyclo[6.1.0]nonatrienes **252** (R=Cl, MeO) yielded on treatment with potassium in THF at −80°C the rapidly topomerizing mono-*trans* cyclononatetraenide anion **254**, which at room temperature slowly isomerizes to the *all-cis* isomer **253**.[239] ^{1}H-NMR spectra of **254** at −40°C show the outer protons at low field (δ = 7.5 to 6.4 ppm) and the inner proton at δ = −3.52 ppm. The *all-cis* isomer **253** has a singlet ^{1}H-NMR peak.

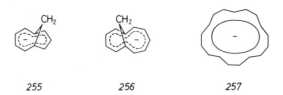

The 1,6-methanocyclononatetraenyl anion **255** is diatropic as shown by ^{1}H-NMR, whereas the 12 π-anion **256** is strongly paratropic with the bridge methylene protons at very low field (δ = 10.31 and 14.19 ppm).[240]

The [17]annulenyl anion **257** presents ^{1}H-NMR peaks for inner protons at δ = −7.97 and for outer protons at 8.2 and 9.5 ppm, indicating strong diatropicity[240] due to the 18 π-electron system.

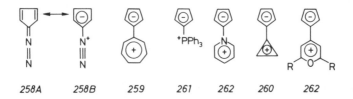

Many other aromatic singly or multiply charged ions have been obtained during the last 20 years. Their structure may be described by resonance formulas which are constitutional (molecular) graphs consisting of vertices with degree three and as many vertices of degree two as the electrical charge. Such graphs were discussed in Chapter 3 as being derived from general cubic graphs by deleting the loops and their pendant edges.

Closely related to these aromatic ions are the corresponding carbonyl derivatives and their conjugated acids, e.g., tropone is related to tropylium and its conjugated acid, hydroxytropylium. Vice versa, cyclopentadienide is related to fulvene. Although the aromaticities of the unsubstituted tropone and fulvene are low or nonexistent (see Chapter 9), the ions derived from them do have aromatic character. Diazocyclopentadiene **258**[241] owes its unusual stability in part to aromaticity, similarly to other aromatic zwitterions, such as sesquifulvalene **259**, calicene **260**, the phosphonium salt **261**, and the pyridinium (**269**) or pyrylium derivatives (**262**). In all but one of these cases one can write limiting structures with single or double exocyclic bonds such as **258A** and **258B**; the exception is **262**, where the exocyclic bond can be only single; in all these cases, physicochemical data indicate considerable contribution of dipolar limiting structures.

Truly antiaromatic systems (approximately planar) are difficult to synthesize. One example, presented above, is **256**. The cycloheptatrienide anion[242] and its substituted derivatives have either triplet or unsymmetrical singlet ground states and are unstable. Destabilized carbocations were reviewed by Tidwell;[242a] among them, the antiaromatic ones reveal extremely strong destabilization manifested in pK_{R+} values or in kinetic barriers.

The ESR spectra of tetrasubstituted [4]annulene radical-cations were discussed by Court-neidge, et al.[315] and Chan et al.[316] Wilhelm et al.[317] investigated the pentalene radical-anion. Stevenson et al.[318] reported the thermal generation of the [16]annulene radical-anion from the [8] annulene radical-anion.

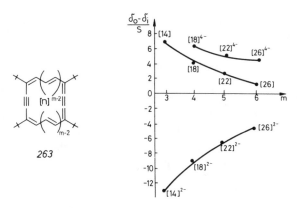

FIGURE 1.   Ring currents (evaluated by relative differences between chemical shifts of outer and inner protons) of [*m*]annulenes (**263**), their dianions and tetraanions.

Tricyclic diatropic systems with double bond localization in a planar 14 π-electron perimeter are cyclopent[*f*]azulenide (**i**)[319] and the isomeric cyclopent[*e*]azulenide (**ii**).[320]

The area of aromatic ions and of their congeners was reviewed in several monographs.[6,7] Since the present book is not centered on aromaticity but on annulenes, derivatives, and their valence isomers, we shall discuss annulenones, annulenediones, annulenium ions, and their valence isomers in Chapter 9. In the following paragraph, brief mention will be made of multiply charged ions derived from annulenes.

B. Multiply Charged Ions

Highly reduced and highly oxidized annulenes form di- or tetraanions and dications, respectively, according to the pioneering work of Katz,[243] Oth,[244] Schröder,[244] Sondheimer,[244-246] Garratt,[245,246] Boekelheide,[247] Paquette,[248] Müllen,[249] and Nakagawa[249] for polyanions, and of Olah and co-workers,[250a] or of Pagni,[250b] for dications. Together with the corresponding neutral system, one may thus compare for one and the same ring size the ¹H- and ¹³C-NMR spectra of two Hückel and two anti-Hückel π-electron systems. It is known that the ring size influences the ring current, according to the London-McWeeny theory, by virtue of the ring area, S; Haddon[251] showed that the reduced ring current (RC) is linearly related to S and, for [4*m* + 2]annulenes, with the Dewar resonance energy (DRE) by the simple relationship:

$$RC = S(3\pi^{-2})DRE$$

The diamagnetic ring currents may be viewed as analogous to nondissipative currents in superconductors. Of course, in polyanions or in dications, correction factors must be introduced for charge density and for the effect of counterions, but the overall data are in excellent agreement with theoretical expectations according to the Hückel rule. Figure 1 presents the reduced RC effects (expressed as difference between ¹H chemical shifts of outer and inner protons divided by S) in Nakagawa's tetra-*t*-butyl substituted bisdehydro [4*m* + 2]annulenes **263**, then tetranions, and dianions with *m* = 3, 4, 5, and 6.[249] It is evident that the neutral systems and even more so the tetraanions support diamagnetic induced RCs, whereas the dianions support paramagnetic currents; in all these cases, the magnitude of reduced ring currents decreases with ring size.

Vice versa, not only the ring currents but also the stabilities of [4*m*]annulene dianions indicate their aromaticity. An example is the dianion of [12]annulene,[252] obtained from the annulene electrochemically or with alkali metals at $-80°C$ and in tetrahydrofuran. The ^{1}H-NMR spectrum shows three inner protons at $\delta = -4.60$ and outer protons at $\delta = 6.98$ (6H) and 6.23 ppm (3H), indicating thereby a strong diatropicity. Unlike the parent annulene which rearranges at $-50°C$ to *cis*-bicyclo[6.4.0]dodeca-2,4,6,9,11-pentaene, the dianion is stable at $+30°C$, but at $+60°C$ it disappears probably by reaction with the impurities generated during the photolytic production of [12]annulene. The most interesting feature of the ^{1}H-NMR spectrum is its invariance in the temperature range between -90 and $+30°C$ indicating the absence of any isodynamic process ($\Delta G^{\ddagger} > 18$ kcal/mol). One should recall that [12]annulene which has the same geometry as its dianion equilibrates its hydrogens via a conformational change (ring inversion) with $\Delta G^{\ddagger} = 6.6$ kcal/mol, whereas [14] annulene which is isoelectronic with the [12]annulene dianion presents a dynamical proton exchange with $\Delta G^{\ddagger} = 1.2$ kcal/mol (both values calculated for 0°C).

The [18]annulenyl dianion[252a] is a 20 π-electron paratropic system with inner protons at $\delta = 29$ ppm and the outer protons at $\delta = -1.13$ ppm.

Since dianions or dications of annulenes may be written as charged bis-homoannulenes, these ions will be mentioned in Chapter 9 dedicated to homoaromatic systems. The aromatic COT dianion depicted as **264** is easily formed from COT and alkali metals, and on carbonation affords 2,5,7-cyclooctatriene-1,4-dicarboxylic acid **265**.[89]

264A   *264B*   *265*

Among the aromatic annulenic dications, we shall mention briefly the dications of substituted cyclobutadienes or COTs and of [16]annulene. While the dications of unsubstituted [4]- and [8]annulene have not yet been obtained, those of substituted derivatives show with certainty their aromaticity: 1,2-diphenyl-, 1,2-difluoro-3,4-diphenyl-, tetramethyl-, and tetraphenylcyclobutadiene dications were prepared at $-78°C$ from the corresponding 1,2-dihalocyclobutenones in magic acid or $ClSO_2F + SbF_5$.[253] The tetramethyl-derivative presents one ^{1}H-NMR singlet at $\delta = 3.68$ ppm and two ^{13}C-NMR peaks at $\delta = 18.8$ (CH$_3$) and 209.7 (ring carbons). According to the Spiesecke-Schneider relationships (which correlate electrical charge with ^{1}H or ^{13}C chemical shifts)[254] a calculated value $\delta = 208.8$ ppm is obtained for two π-electrons delocalized over a four-carbon periphery, in good agreement with the experiment.

1,4-Dimethyl- and 1,3,5,7-tetramethyl-[8]annulene dications were obtained in the same solvent (SbF_5/SO_2ClF at $-78°C$) and their NMR spectra (^{1}H and ^{13}C) indicate practical planarity despite the angle strain; the latter dication, **266**, R=H, rearranged at $-20°C$ to a diallylic dication **267**.[255]

266   267   268

Octamethyl-[8]annulene or its valence isomers (Me$_8$-tricyclo[4.2.0.0^{3,5}]octadiene and Me$_8$-semibullvalene) form diallylic cations (267, R=Me) in magic acid, but the tricyclic bullvalenic system yields a different dication, **268**, in SbF$_5$-SO$_2$ClF, with homocyclopropenyl moieties.[256]

The [16]annulene dication was obtained by Schröder et al.[257] from [16]annulene in FSO$_3$H/SO$_2$/CD$_2$Cl$_2$ at $-80°$C along with polymers, as indicated by ^{1}H-NMR and ^{13}C-NMR spectra. The two-electron oxidation apparently involves a radical-cation intermediate, rather than protonation followed by evolution of molecular hydrogen.

Müllen et al.[321] described paratropic tetra-dehydro[14]annulene dianions; Mortensen and Heinze[322] studied electrochemically generated acepleiadylene tetraanions and showed that unlike earlier claims, pyrene gives only a dianion.

X. ANNULENO-ANNULENES

Just as napththalene may be considered to be a benzobenzene (benzoannulenes and their derivatives are described in Chapter 7 in Volume II), one can annelate annulenes with other rings than benzenoid ones: the products of these annelations are called annuleno-annulenes. The annelation may be effected (1) at two vicinal carbon atoms as in naphthalene or (2) at nonvicinal carbon atoms. An example for type (1) annelation is Sondheimer and Cresp's double tetradehydro[14]annulene **271**. Type 2 annelation is illustrated by Nakagawa's compounds **273**[259] with [18]annulene and **275**[260] with [14]annulene units. The peripheries of **271**, **273**, and **275** are dehydro[26]-, -[26]-, and [22]annulenes, respectively, i.e., Hückel-type [4m + 2]annulenes. The red-brown crystalline **271** exhibits ^{1}H-NMR spectra indicating stronger diatropicity than of **269** but weaker than **270a** to **c**. The dark green crystals of **273** give ^{1}H-NMR inner proton signals at δ = -2.85 ppm and outer proton peaks at 10.64 and 10.06 ppm, indicative of strong diatropicity. The corresponding values for **275** are very similar. On considering the electronic absorption spectra, the X-ray crystallographic analysis of **275** (which reveals alternating bond lengths along the periphery as indicated on formula **275**) and the difference between chemical shifts of inner and outer protons, the electronic structures of **273** and **275** are better described as two isolated 14π- or 18π-electron systems rather than peripherally delocalized [26]- and [22]annulenes, respectively. Similarly, naphthalene is better described as two condensed benzenoid rings than as a zero-atom bridged [10]annulene. Indeed, for naphthalene as for **271**, **273**, and **275** one may write three Kekulé structures each (two of which match one another; for **271**, **273**, and **275** we have shown therefore only two Kekulé structures each); structures **A** predominate over those (**B**) with peripheral delocalization. In the series of Sondheimer's tetrakisdehydro-bicyclo[12.12.2]-(**271**), [14.12.2]-(**276**), and [16.12.2]-annuleno-annulenes (**277**), ^{1}H-NMR methyl peaks indicate that the [14]annulene rings are all diatropic, the more so as the other ring is larger; in **276**, the inner proton resonance of the 14-membered ring indicates that the 16-membered ring makes a paratropic contribution, whereas in **271** and **277** the 14- and 18-membered annelating rings have diatropic contributions.

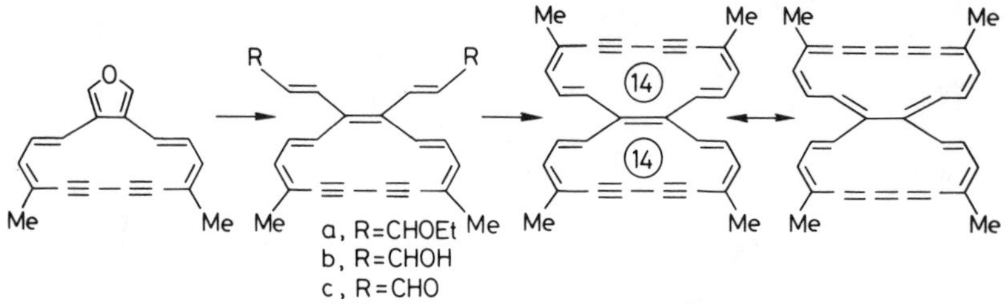

269   *270*

a, R=CHOEt
b, R=CHOH
c, R=CHO

271A   *271B*

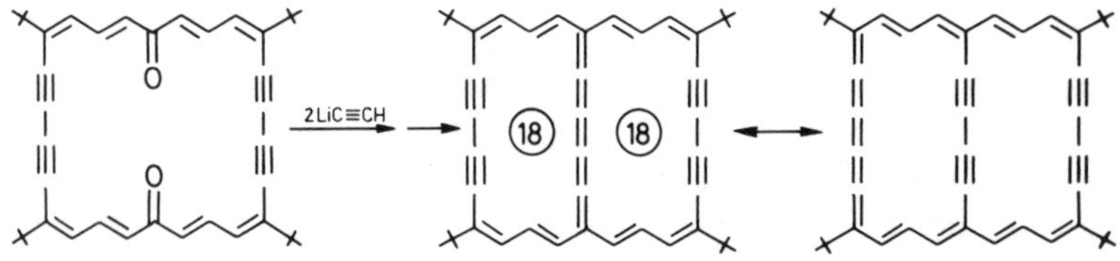

272   *273A*   *273B*

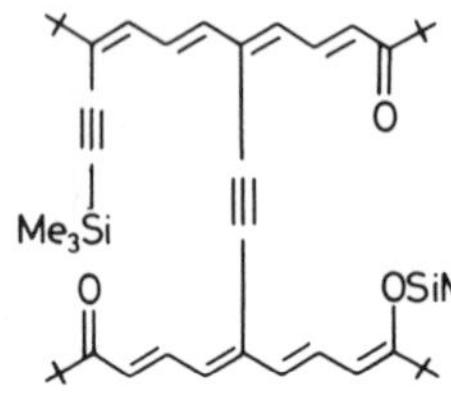

274 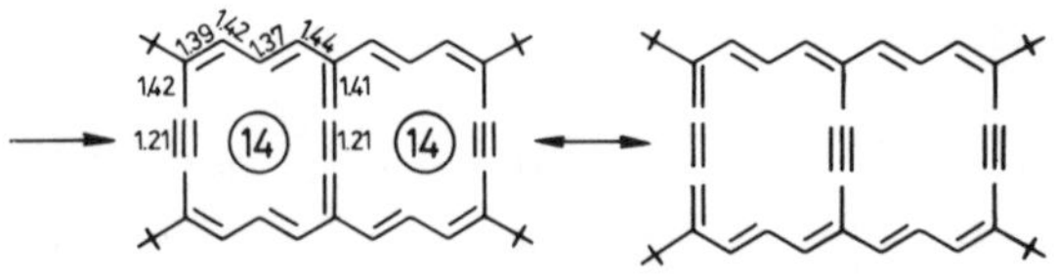   *275A*   *275B*

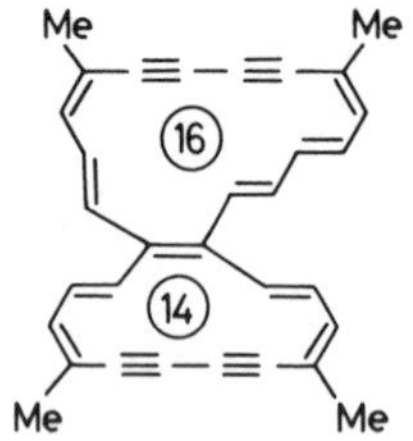

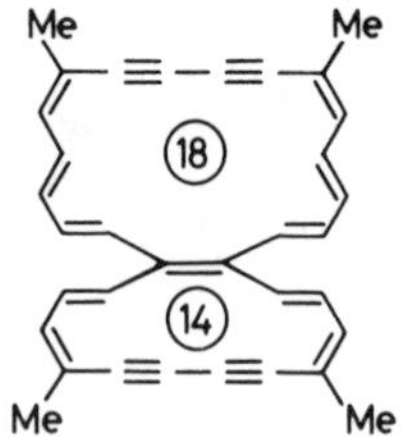

276   *277*

Nakagawa and co-workers also prepared various series of annuleno-annulenes for comparison purposes, in addition to **271** and **273**. The literature prior to 1979 is summarized in Nakagawa's review.[262] Structures and references are indicated in formulas **278** to **280** with two (4m + 2)-membered rings, **281** to **287** with one (4m + 2)-membered and one 4n-membered ring, either symmetrically or unsymmetrically condensed (leading to three isomeric systems, **283, 285,** and **286**), and finally, **288** which combines a Nakagawa-type dehydroannulene with a Sondheimer-type one, both having (4m + 2)-membered rings.

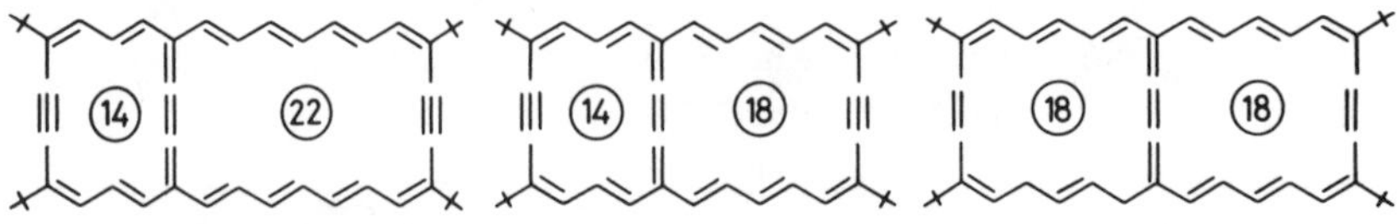

278[263a]   *279*[263b]   *280*[264]

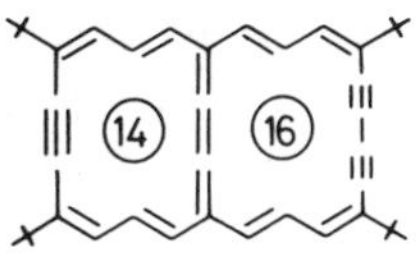

The conclusion for these investigations based on electronic absorption, ¹H-NMR, and ¹³C-NMR spectra, is the "independent behavior" of the two moieties, each acting as diatropic if $(4m + 2)$-membered and paratropic if $(4m)$-membered. The limiting structures with the central butatriene bonds seem to predominate, explaining thereby the prevalence of local delocalization over peripheral delocalization. The marked suppression of diatropicity in a $(4m + 2)$-membered moiety is due to the tendency to preserve the inherent paratropicity of $4m$-membered moieties when these are condensed. However, the contribution of the peripheral resonance cannot be ignored.

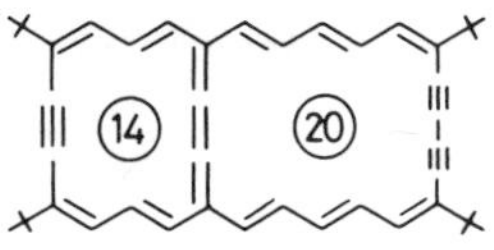

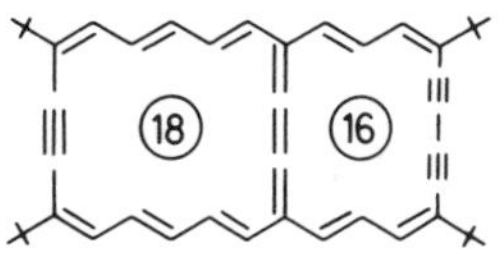

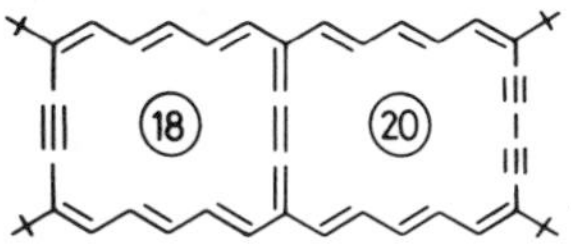

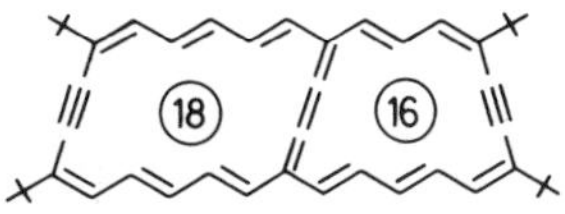

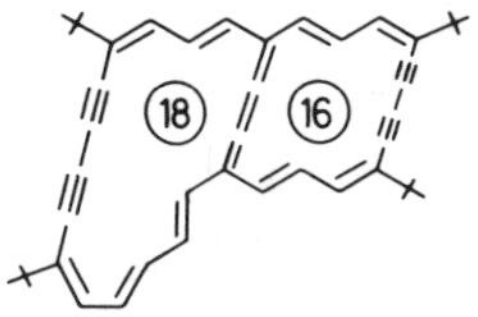

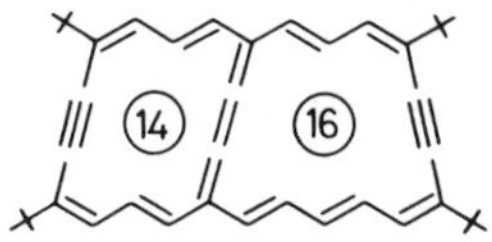

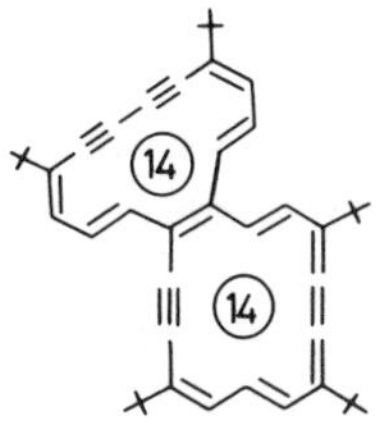

On comparing the tris- (**285**), tetrakis- (**283**), and pentakisdehydro[16]annuleno[18]annulene (**286**), the latter system appears to be conformationally mobile unlike the other two.

The interesting system **288** shows strong diatropicity in the Nakagawa-type *t*-butyl-substituted ring, and suppression of diatropicity in the Sondheimer-type methyl-substituted ring; it had been observed that (1) the extent of suppression of diatropicity of an annulene moiety annelated with benzenoids or annulenes is proportional to the resonance energy of that benzenoid or annulene system, and that (2) the less diatropic annulene ring suffers more suppression of annelation than a more strongly diatropic annulene moiety.

Theoretical treatment of such annuleno-annulenes substantiates the above conclusions.[272,273]

XI. ANNULENES WITH ZERO-ATOM BRIDGES

In Chapter 2 it was mentioned briefly that although the Hückel rule was devised strictly for monocyclic systems, the literature mentions that *cata*- or *peri*-condensed benzenoid polycyclic systems may be considered as having bridged annulene perimeters. While this is true for bridges formed from saturated chains such as CH_2 or CH_2CH_2 groups (see Chapter 5 for details), it was shown in Chapter 2 that nonbenzenoid *cata*-condensed or *peri*-condensed systems with zero-atom bridges show many exceptions to the Hückel rule. We shall discuss briefly a few bicyclic continuously conjugated condensed systems and ions derived therefrom: C_6H_4 (**289** and **290**), C_8H_6 (**291** to **293**), $C_{10}H_{17}$ (**294** to **296**), $C_{12}H_{10}$ (**297, 298**), and $C_{14}H_{12}$ (**299**).

Butalene (1,4-dehydrobenzene, **289**) was generated and trapped by Breslow et al.[279] Despite its having the aromatic benzene periphery, the MOs and the ring strain preclude any stability; the same reasons, even more strongly, apply to 1,3-dehydrobenzene.

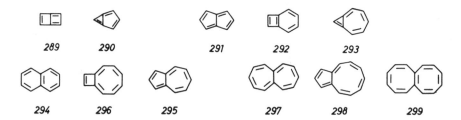

| 289 | 290 | 291 | 292 | 293 |
| 294 | 296 | 295 | 297 | 298 | 299 |

Pentalene **291** and heptalene **297** were erroneously predicted by simple HMO calculations to possess a sizable resonance energy. Only when more sophisticated methods were employed, or when Dewar resonance energies were calculated, was the fallacy of the HMO method revealed.

Another "false friend" had been the fact that two Kekulé structures may be written for pentalene and heptalene, just as in the case of the aromatic benzene or azulene. By means of the superposition procedure, it is easy to demonstrate, however, that the two Kekulé structures of pentalene and heptalene are of opposite sign, whereas the two Kekulé structures of benzene and azulene have the same sign. All Kekulé structures of any benzenoid polycyclic aromatic hydrocarbon (PAH) also have the same sign. The signs of the three valence-bond (limiting) structures of benzocyclobutene were discussed in Chapter 2.

Molecules such as pentalene and heptalene with 4*m*-membered periphery have a very low stability attributable to antiaromaticity. We shall not discuss the pentalene valence-isomer **293**, which has a considerable angle strain; another valence-isomer, benzo[4]annulene (benzocyclobutene), like all other benzoannulenes, is discussed in detail in Chapter 7.

Hafner discovered the first high-yield synthesis of azulene which does not involve dehydrogenations, by treating pyrylium salts[276] with sodium cyclopentadienide. Subsequently he explored the chemistry of azulene,[277] pentalene, heptalene, and of heterocyclic congeners, in pioneering investigations involving ingenious synthetic methods.[278] 1-Methylpentalene[279] and 1,3-dimethylpentalene[280] can be trapped at $-196°C$ in solid matrices and are extremely unstable. However, benzopentalene and hexaphenylpentalene,[280] as well as 1,3-bis-(dimethylamino)pentalene[281] are isolable crystalline compounds which, however, react readily with various reagents.

The latter compound is electron rich, and X-ray crystallographic analysis indicates a dipolar structure (**300**);[282] even more electron rich is the pentalene dianion[283] with a periphery having 10 π-electrons. In this respect one should also mention the tricyclic *sym*-(**301**)[278,284] and *asym*-indacenes (**302**)[285] which also are known as stable dianions with 14 π-electrons. The *sym*-indacenes **301** with R = R' = NMe$_2$ group are stable towards air,[282] but those with one R = NMe$_2$ group and an alkyl or aryl R' group are sensitive towards air.[286] The tetra-*t*-butyl-*sym*-indacene[284] gives one tBu ^{1}H-NMR peak and two olefinic signals; it is reactive towards air, acids, and light. Two valence isomers of azulene were recently prepared: the tetracyclic azulvalene and the tricyclic Dewar-azulene;[286a] Both isomerize to azulene on heating.

| 300 | 301 | 302 | 303 |

In analogy to the vinylogation of cyclobutadiene affording benzene derivatives by [2 + 2]-cycloaddition followed by opening of the four-membered ring,[287] Hafner developed a vinylogation[284,288] converting pentalene into azulene derivatives and the latter (**304**) into either heptalene or unstable, rearrangement-prone cyclopentacyclononene systems (**298**, **307**). Vice versa, 1-cyanoazulene **309** can be converted via cycloaddition with ynamines followed by Alder-Rickert cleavage into the stable 5-cyano-1-diethylamino-2-methylpentalene **311**.[284]

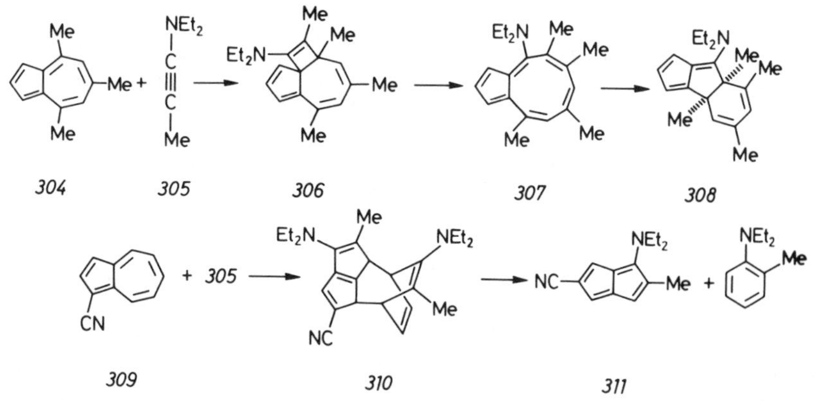

Pentalene dications have not yet been obtained, but dibenzopentalenes (**312**, R = H or Me) yield in SbF_5/SO_2ClF at $-78°C$ the corresponding dications.[289]

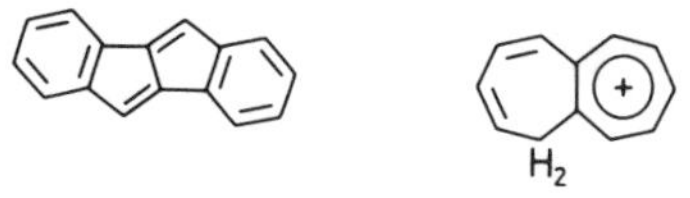

Schleyer and co-workers calculated (MNDO) and determined the structure of the 10π-electron dilithium pentalenide (X-ray diffraction of the crystal containing 2 mol of dimethoxyethane which solvate the two lithium atoms situated on opposite faces of the two five-membered rings); the pentalenide "dianion" is planar and has no bond alternation, whereas dilithium napthalenide (which has a similar structure with the two lithium atoms situated on opposite faces of each of the two six-membered rings) is nonplanar because it is antiaromatic (12π-electrons).[327]

Heptalene, **297**,[290] a liquid which was obtained at $-78°C$ and which polymerizes on warming or in the presence of oxygen, forms a monocation **313** in 96% H_2SO_4.[291] Further heptalene syntheses were described by Vogel and Paquette,[292] in addition to Hafner's vinylogation of azulene.[284] Though naphthalene **294**, and bicyclo[6.2.0]decapentaene **296** (see Chapter 2) will not be discussed here, it should be mentioned that some of their derivatives form stable dications in magic acid at $-78°C$; naphthalene yields the radical monocation, but tetra- and octamethylnaphthalene form dications;[293] 1,2-dimethylbenzocyclobutadiene dications were obtained from the 1,2-diol.[294]

Suglhara et al.[323] reported the synthesis and isomerizations of cyano-Dewar-heptalene leading both thermally and photochemically to cyanoheptalene. Double bond isomers of substituted heptalenes were shown to be interconverted on photoirradiation or on heating above 80° without loss of optical purity; according to the substitution pattern, the double bond shift may be degenerate.[328]

Octalene **299** was synthesized in 1977 by Vogel et al. via a Bamford-Stevens reaction of bistosylhydrazine **315d** and dehalogenation of **317**, with NaI in acetone, accompanied by valence isomerization of the resulting tetracyclic compound.[295]

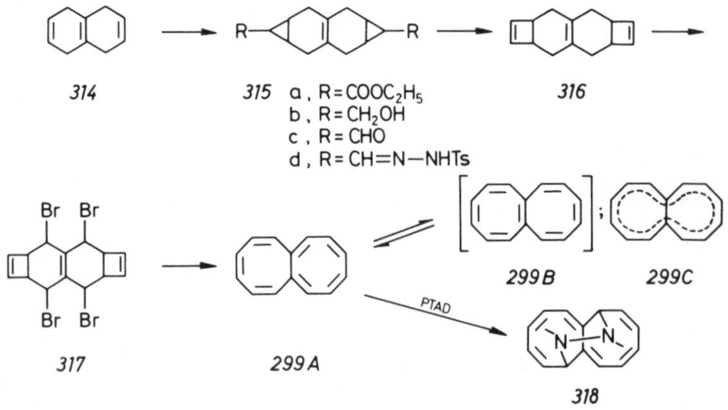

Octalene **299**, an air-sensitive yellow liquid (bp = 50 to 52°C/1 Torr) shows ¹H-NMR peaks at δ = 5.65 (m; 10 H) and 6.30 (d; J = 12 Hz) whereas in the ¹³C-NMR spectrum the signals appear at δ = 142.62 ppm; 136.34; 131.46; 131.43; 129.01; 127.59; 126.90 ppm. The ¹³C-NMR data are in agreement only with the structure **299A** (and not with **299B** or with the delocalized structure **299C** which should give a four-line spectrum).

The X-ray structural analysis of the adduct **318** of octalene with *N*-phenyltriazolinedione (PTAD) confirmed the structure of octalene.

Careful analysis of the temperature-dependent ¹³C-NMR spectrum of **299** showed that at −150°C a 14-line spectrum (C₁ symmetry) is obtained, in agreement with a fixed chiral conformation **319**. At higher temperatures independent ring inversions of cyclooctatriene (rate k₁) and cyclooctatetraene (k₂) rings occur (isodynamic processes). From the spectrum at −115°C the values (k₁ + k₂) = 300 sec⁻¹ and ΔG‡ (k₁) = 8 kcal · mol⁻¹ were estimated.[296] It has also been shown that the inversion of cyclooctatetraene ring (k₂) or the synchronous inversion of both types of rings in **319** (k₃) converts one enantiomeric form into another (R ⇌ S) whereas the single inversion of cyclooctatriene ring (k₁) has no effect on stereochemistry.[296]

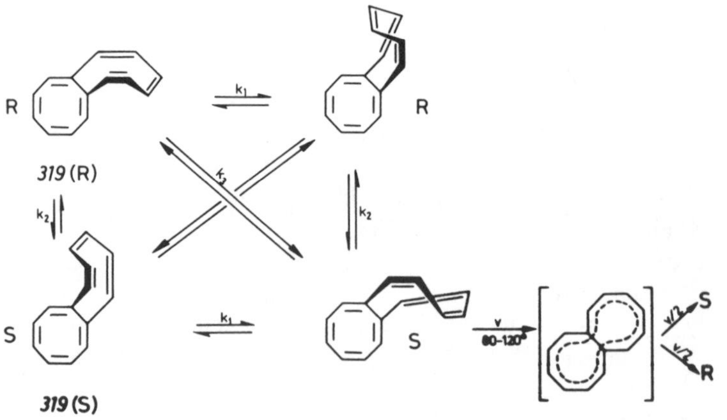

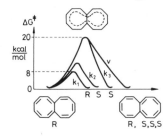

FIGURE 2.   Activation free energies in automerizations of octalene (**299**) or (**319**).

The analysis of the spectrum at 80 to 120°C allows the assessment of a π-bond migration (v) in octalene having a rate constant of ∼200 sec^{-1} and $\Delta G_v^{\ddagger} \cong 20$ kcal/mol. The transition state for π-bond migration was considered to be **299C** for which an approximate $\Delta H_{stab}; \cong$ 15 kcal · mol^{-1} (the energy gain from the bond delocalization) was estimated.[296] This value, sensibly lower than for [14]annulene, could be ascribed to the destabilization effect of the central bond.

The energy profiles describing the automerization of one octalene structure into another from the seven possible isomeric structures are presented in Figure 2.

From thermochemical data it was estimated that the isomer **299B** should be less stable than **299A** by 5 to 6 kcal/mol.[296] The above data are in agreement with previous results of calculations.[297,298]

Octalene forms[299] in THF and DMF stable diamagnetic di- and tetraanions. The former (**320**) has localized 10 π-electron systems; the latter (**320A**) is a delocalized 18 π-electron aromatic compound, explaining thereby the stability at room temperature despite the high Coulomb repulsions.

Recently a very interesting conversion of octalene into [14]annulene derivatives was announced:[300] the octalene dianion **320** was converted into 1,8-dimethyl[14]annulene (**322**) on treatment with dimethyl sulfate (most probably through the intermediacy of **321** which is valence tautomerized to **322**)

Similar to [14]annulene, the dimethyl derivative **322** shows a temperature-dependent ^{1}H-NMR spectrum due to conformational interconversions **322a** ⇌ **322b** ⇌ **322c** (a rotation of the *trans*-double bonds around the neighboring single bonds exchanging the inner and outer protons). In solid state only **322a** occurs. The carbon skeleton of **322** is considerably deviated from coplanarity (torsional angles up to 20°); however, the carbon-carbon bond lengths are typically aromatic (1.364 to 1.407 Å).

322a   322b   322c

A quadrupolar pentacyclic system (bis-calicene) with a 16π-electron periphery was recently reported to be stable and to possess equalized bond lengths.[301]

REFERENCES

1. **Sondheimer, F. and Wolovsky, R.,** *J. Am. Chem. Soc.,* 84, 260, 1962.
2. **Thiele, J.,** *Liebigs Ann. Chem.,* 306, 87, 1899; 308, 333, 1899; 314, 298, 1900; 319, 121, 1901.
3. **Willstätter, R. and Schmaedel, W., von,** *Ber. Dtsch. Chem. Ges.,* 39, 1992, 1905.
4. **Willstätter, R. and Waser, R.,** *Ber. Dtsch. Chem. Ges.,* 44, 3423, 1911; **Willstätter, R. and Heidelberger, M.,** *Ber. Dtsch. Chem. Ges.,* 46, 517, 1913.
5. **Allinger, N. and Yuh, Y. H.,** *Pure. Appl. Chem.,* 55, 191, 1983.
6. **Garrat, P. J.,** in *Comprehensive Organic Chemistry,* Vol. 1, Barton, D. and Ollis, W. D., Eds., Pergamon Press, Oxford, 1979, 361; *Aromaticity,* McGraw-Hill, New York, 1971.
7. **Lloyd, D. M. G., Ed.,** *Carbocyclic Non-Benzenoid Aromatic Compounds,* Elsevier, Amsterdam, 1966.
8. **Sondheimer, F.,** *Pure Appl. Chem.,* 7, 363, 1963; *Proc. R. Soc.,* A297, 173, 1967; *Acc. Chem. Res.,* 5, 81, 1972; **Sondheimer, F. et al.,** in *Aromaticity,* Chem. Soc. Spec. Publ. 21, London, 1967, 75.
9. The Open University, *Aromaticity* (Units 17 to 19), The Open University Press, Milton Keynes, 1977.
10. **Oth, J. F. M. and Gilles, J. -M.,** *Tetrahed. Lett.,* p. 6259, 1968; **Oth, J. F. M., Gilles, J. -M., and Anthoine, G.,** *Tetrahedron Lett.,* p. 6265, 1968.
11. **Balaban, A. T.,** *Tetrahedron,* 27, 6115, 1971.
12. **Gordon, M. and Davison, W. H. T.,** *J. Chem. Phys.,* 20, 428, 1952.
13. **Cava, M. P. and Mitchell, M. J., Eds.,** *Cyclobutadiene and Related Compounds,* Academic Press, New York, 1967.
14. **Maier, G.,** *Angew. Chem. Int. Ed. Engl.,* 13, 425, 1974.
15. **Masamune, S.,** *Tetrahedron,* 36, 343, 1980; *Pure Appl. Chem.,* 44, 861, 1975; **Bally, T. and Masamune, S.,** *Tetrahedron,* 36, 343, 1980.
16. **Kekulé, A., von,** *Liebigs Ann. Chem.,* 162, 77, 1872.
17. **Perkin, W. H., Jr.,** *J. Chem. Soc.,* 65, 967, 1894.
18. **Willstätter, R. and Schmädel, W., von,** *Ber. Dtsch. Chem. Ges.,* 38, 1892, 1905.
19. **Buchman, E. R., Schatter, M. J., and Reimes, A. O.,** *J. Am. Chem. Soc.,* 64, 2701, 1942.
20. **Nenitzescu, C. D., Avram, M., Marica, E., Maxim, M., and Dinu, D.,** *Studii Cercet. Chim. Acad. R. P. Romania,* 7, 481, 1959; **Nenitzescu, C. D.,** *Angew. Chem.,* 73, 300, 1961.
21. **Avram, M., Marica, E., and Nenitzescu, C. D.,** *Chem. Ber.,* 92, 1088, 1959; **Avram, M. et al.,** *Tetrahed. Lett.,* 1611, 1963; p. 21, 1961; *Tetrahedron,* 19, 187, 1963; *Chem. Ber.,* 97, 372, 1974.
22. **Emerson, G. F., Watts, L., and Pettit, R.,** *J. Am. Chem. Soc.,* 87, 131, 1965.
23. **Fitzpatrick, J. D., Watts, L., Emerson, G. F., and Pettit, R.,** *J. Am. Chem. Soc.,* 87, 3254, 1965.
24. **Watts, L., Fitzpatrick, J. D., and Pettit, R.,** *J. Am. Chem. Soc.,* 87, 3253, 1965.
25. **Corey, E. J. and Streith, J.,** *J. Am. Chem. Soc.,* 86, 950, 1964.
26. **Lin, C. Y. and Krantz, A.,** *Chem. Commun.,* p. 1111, 1972.
27. **Chapman, O. L., McIntosh, C. L., and Pacansky, J.,** *J. Am. Chem. Soc.,* 95, 1337, 1973.
28. **Maier, G., Hartan, H. G., and Sayrac, T.,** *Angew. Chem. Int. Ed. Engl.,* 15, 226, 1976.
29. **Hedaya, E., Miller, P. D., McNeil, D. W., D'Angelo, P. F., and Schissel, P.,** *J. Am. Chem. Soc.,* 91, 1875, 1969.

30. **Tyerman, W. J. R., Kato, M., Kebarle, P., Masamune, S., Strausz, O. P., and Gunning, H. E.,** *Chem. Commun.,* p. 497, 1967.
31. **Masamune, S., Nakamura, N., and Spadaro, J.,** *J. Am. Chem. Soc.,* 97, 918, 1975.
32. **Lage, H. W., Reisenauer, H. P., and Maier, G.,** *Tetrahed. Lett.,* 23, 3983, 1982.
33. **Rebek, J., Jr. and Gavina, F.,** *J. Am. Chem. Soc.,* 97, 3453, 1975.
34. **Coulson, C. A.,** *Chem. Soc. Spec. Publ.,* 12, 85, 1958.
35a. **Hehre, J. W. and Pople, J. A.,** *J. Am. Chem. Soc.,* 97, 6941, 1975; **Borden, W. T., Davidson, E. R., and Hart, D.,** *J. Am. Chem. Soc.,* 100, 388, 1978; **Haddon, R. C. and Williams, G. R. J.,** *J. Am. Chem. Soc.,* 97, 6582, 1975; **Kollmar, H. and Staemmler, V.,** *J. Am. Chem. Soc.,* 99, 3583, 1977; **Jafri, J. A. and Newton, M. D.,** *J. Am. Chem. Soc.,* 100, 5012, 1978.
35b. **Buenker, R. J. and Peyerimhoff,** *J. Chem. Phys.,* 48, 354, 1968.
36a. **Dewar, M. J. S. and Gleicher, G. J.,** *J. Am. Chem. Soc.,* 87, 3255, 1965; **Dewar, M. J. S. and Kollmar, A.,** *J. Am. Chem. Soc.,* 97, 2933, 1975; **Dewar, M. J. S., and Kormornicki, A.,** *J. Am. Chem. Soc.,* 99, 6174, 1977.
36b. **Allinger, N. L. and Tai, J. C.,** *Theor. Chim. Acta,* 12, 29, 1968.
37. **Chapman, O. L. De La Cruz, D., Roth, R., and Pacansky, J.,** *J. Am. Chem. Soc.,* 95, 1337, 1973.
38. **Carpenter, B. K.,** *J. Am. Chem. Soc.,* 105, 1700, 1983; **Whitman, D. W., Capon, S. W., Grant, E. R., and Carpenter, B. K.,** *J. Am. Chem. Soc.,* in press.
39. **Criegee, R.,** *Angew. Chem.,* 74, 703, 1962; *Bull. Soc. Chim. France,* p. 1, 1963.
40. **Criegee, R. and Moschel, A.,** *Chem. Ber.,* 92, 2181, 1959.
41. **Smirnov-Samkov, I. V.,** *Izv. Akad. Nauk. SSSR. Ser. Khim. Nauk.,* 83, 869, 1952.
42. **Dunitz, J. W.,** *Helv. Chim. Acta.,* 45, 647, 1962.
43. **Longuet-Higgins, H. C. and Orgel, L. E.,** *J. Chem. Soc.,* p. 1969, 1956.
44. **Criegee, R. and Schröder, G.,** *Liebigs Ann. Chem.,* 633, 1, 1959.
45. **Criegee, R., Dekker, J., Engel, W., Ludwig, P., and Noll, K.,** *Chem. Ber.,* 96, 2362, 1963.
46. **Criegee, R. and Louis, G.,** *Chem. Ber.,* 90, 417, 1957; **Criegee, R., Schröder, G., Maier, G., and Fischer, H. G.,** *Chem. Ber.,* 93, 1553, 1960.
47. **Berkoff, C. E., Cookson, R. C., Hudec, J., and Williams, R. O.,** *Proc. Chem. Soc.,* p. 312, 1961.
48. **Maier, G. and Schneider, M.,** *Angew. Chem.,* 83, 885, 1971.
49. **Grubbs, R. H. and Grey, R. A.,** *J. Am. Chem. Soc.,* 95, 5765, 1973.
50. **Kobayashi, Y. et al.,** *Tetrahed. Lett.,* p. 3001, 1975; p. 3819, 1975.
51. **Masamune, S., Machiguchi, T., and Aratani, M.,** *J. Am. Chem. Soc.,* 99, 3524, 1977.
52. **Gerache, M. J., Lemal, D. M., and Ertl, H.,** *J. Am. Chem. Soc.,* 97, 5584, 1975.
53. **Maier, G., Fritschi, G., and Hoppe, B.,** *Tetrahed. Lett.,* p. 1463, 1971.
54. **Gompper, R. and Seybold, G.,** *Angew. Chem. Int. Ed. Engl.,* 7, 824, 1968; **Neuenschwander, M. and Niederhauser, A.,** *Helv. Chim. Acta,* 53, 519, 1970.
55. **Kimling, H. and Krebs, A.,** *Angew. Chem. Int. Ed. Engl.,* 11, 932, 1972.
56. **Masamune, S., Nakamura, N., Suda, M., and Ono, H.,** *J. Am. Chem. Soc.,* 95, 8481, 1973.
57. **Maier, G. and Alzérreca,** *Angew. Chem. Int. Ed.,* 12, 1015, 1973.
58. **Maier, G. and Sauer, W.,** *Angew. Chem. Int. Ed.,* 16, 51, 1977.
59. **Maier, G. et al.,** *Tetrahed. Lett.,* p. 1837, 1978.
60. **Lauer, G., Müller, C., Schulte, K. W., Schweig, A., Maier, G., and Alzérreca,** *Angew. Chem. Int. Ed. Engl.,* 14, 172, 1975.
61. **Delbaere, L. T. J., James, M. N. G., Nakamura, N., and Masamune, S.,** *J. Am. Chem. Soc.,* 97, 1973, 1975.
62. **Maier, G. and Sauer, W.,** *Angew. Chem.,* 89, 49, 1977.
63. **Maier, G., Schäfer, U., Sauer, W., Hartan, H., Matusch, R., and Oth, J. F. M.,** *Tetrahed. Lett.,* p. 1837, 1978.
64. **Maier, G., Pfriem, S., Schäfer, U., and Matusch, R.,** *Angew. Chem.,* 90, 552, 1978.
65. **Irngartinger, H. and Nixdorf, M.,** *Angew. Chem.,* 95, 415, 1983.
66. **Irngartinger, H., Riegler, N., Malsch, K. D., Schneider, K. A., and Maier, G.,** *Angew. Chem.,* 92, 214, 1980.
67. **Heilbronner, E., Jones, T. B., Krebs, A., Maier, G., Malsch, K. D., Pocklington, J., and Schmelzer, A.,** *J. Am. Chem. Soc.,* 102, 564, 1980; **Ermer, O. and Heilbronner, E.,** *Agnew. Chem.,* 95, 414, 1983.
68. **Malatesta, L., Santarella, G., Vallarino, L., and Zingales, F.,** *Angew. Chem.,* 72, 34, 1960.
69. **Blomquist, A. T. and Maitlis, P. M.,** *J. Am. Chem. Soc.,* 84, 2329, 1962.
70. **Maitlis, P. M. and Games, M. L.,** *J. Am. Chem. Soc.,* 85, 1887, 1963.
71. **Dahl, L. F. and Oberhansli,** *Inorg. Chem.,* 4, 629, 1965.
72. **Maitlis, P. M.,** *J. Organometal. Chem.,* 200, 161, 1980.
73. **Maitlis, P. M. and Eberius, K. W.,** *Non-Benzenoid Aromatics,* Vol. 2, Snyder, J. P., Ed., Academic Press, New York, 1971, 376.

74. **Mann, B. E., Bailey, P. M., and Maitlis, P. M.,** *J. Am. Chem. Soc.,* 97, 1275, 1975.
75. **Taylor, S. H. and Maitlis, P. M.,** *J. Am. Chem. Soc.,* 100, 4711, 1978.
76. **Canziani, F., Chini, P., Querta, A., and Martino, A. D.,** *J. Organometal. Chem.,* 26, 285, 1971.
77. **Ranzini, F. and Malatesta, M. C.,** *J. Organometal. Chem.,* 90, 235, 1975.
78. **Fischler, I., Hildenbrand, K. H., and Körner von Hustorf, E.,** *Angew. Chem. Int. Ed.,* 14, 54, 1975.
79. **Sanders, A. and Giering, W. P.,** *J. Am. Chem. Soc.,* 96, 5247, 1974; 97, 919, 1975.
80. **Hübel, W.,** *J. Inorg. Chem.,* 9, 204, 1959.
81. **Schomaker, V.,** *Angew. Chem.,* 72, 755, 1960.
82. **Freedman, H. H.,** *J. Am. Chem. Soc.,* 83, 2194, 1961.
83. **Mingos, D. M. P.,** *J. Chem. Soc. Dalton,* pp. 20, 26, 31, 1976.
84. **Aihara, J.,** *Bull. Chem. Soc. Jpn.,* 51, 28, 1978.
85. **Schmidt, E. K. G.,** *Angew. Chem. Int. Ed.,* 12, 777, 1973; *Chem. Ber.,* 107, 2440, 1974.
86. **Brune, H. A. and Harlbeck, G.,** *Z. Naturforsch.,* 28b, 68, 1973.
87. **Andrews, D. C. and Davidson, G.,** *J. Organometal. Chem.,* 76, 373, 1974.
88. **Lucken, E. A. C., Pozzi, R., and Ramaprasad, K. R.,** *J. Mol. Struct.,* 18, 377, 1973.
89. **Schröder, G.,** *Cyclooctatetraen,* Verlag-Chemie, Weinheim, 1965.
90. **Fray, G. I. and Saxton, R. G.,** *The Chemistry of Cyclooctatetraene and Its Derivatives,* Cambridge University Press, Cambridge, Mass., 1978.
91. **Paquette, L. A.,** *Tetrahedron,* 31, 2855, 1975; *Pure Appl. Chem.,* 54, 987, 1982.
92. **Reppe, W., Schlichting, O., and Meister, H.,** *Liebigs Ann. Chem.,* 560, 93, 1948.
93. **Huisgen, R. and Mietzsch, F.,** *Angew. Chem. Int. Ed.,* 3, 83, 1964; **Huisgen, R., Konz, W. E., and Gream, G. E.,** *J. Am. Chem. Soc.,* 92, 4105, 1970.
94. **Vogel, E., Kiefer, A., and Roth, W. R.,** *Angew. Chem. Int. Ed.,* 3, 442, 1964.
95. **Anet. F. A. L. and Bock, L. A.,** *J. Am. Chem. Soc.,* 90, 7130, 1968; **Anet, F. A. L., Bourn, A. J. R., and Lin, Y. S.,** *J. Am. Chem. Soc.,* 86, 3576, 1964.
96. **Anet, F. A. L.,** *J. Am. Chem. Soc.,* 84, 671, 1962.
97. **Schröder, G. and Oth, J. F. M.,** *Tetrahed. Lett.,* p. 4083, 1966.
98. **Oth, J. F. M. and Gilles, J. M.,** *Tetrahed. Lett.,* p. 6259, 1968.
99. **Oth, J. F. M.,** *Pure Appl. Chem.,* 25, 573, 1971.
100. **Oth, J. F. M., Merenyi, R., Martini, T., and Schröder, G.,** *Tetrahed. Lett.,* p. 3087, 1966.
101. **Schröder, G., Oth, J. F. M., and Merenyi, R.,** *Angew. Chem. Int. Ed. Engl.,* 4, 752, 1965.
102. **Paquette, L. A., Photis, J. M., and Ewing, G. D.,** *J. Am. Chem. Soc.,* 97, 3538, 1975; **Paquette, L. A. and Photis, J. M.,** *J. Am. Chem. Soc.,* 98, 4936, 1976.
103. **Paquette, L. A., Gardlik, J. M., Johnson, L. K., and McCullough, K. J.,** *J. Am. Chem. Soc.,* 102, 5026, 1980.
104. **Gardlick, J. and Paquette, L. A.,** *Tetrahed. Lett.,* p. 3597, 1979; **Paquette, L. A. et al.,** *J. Org. Chem.,* 45, 5105, 1980; *J. Am. Chem. Soc.,* 102, 2131, 1980; 102, 1188, 1980.
105. **Paquette, L. A., Hanzawa, Y., McCullough, K. J., Tagle, B., Swenson, W., and Clardy, J.,** *J. Am. Chem. Soc.,* 103, 2262, 1981.
106. **Warrener, R. N., Anderson, C. M., McCay, I. W., and Paddon-Row, M. N.,** *Aust. J. Chem.,* 30, 1481, 1977; **Warrener, R. N., McCay, I. W., Tan, R. Y. S., and Russell, R. G.,** *Tetrahedron Lett.,* p. 3183, 1979; **Anderson, C. M., McCay, I. W., and Warrener, R. N.,** *Tetrahedron Lett.,* p. 2735, 1970.
107. **Paquette, L. A., Hanzawa, Y., Hefferon, G. J., and Blount, J. F.,** *J. Org. Chem.,* 47, 265, 1982.
108. **Hanzawa, Y. and Paquette, L. A.,** *J. Am. Chem. Soc.,* 103, 2269, 1981.
109. **Bryce-Smith, D., Gilbert, A., and Gazonka, J.,** *Angew. Chem. Int. Ed. Engl.,* 10, 746, 1971.
110. **Paquette, L. A., Meisinger, R. H., and Wingard, R. E., Jr.,** *J. Am. Chem. Soc.,* 94, 9224, 1972.
111. **Paquette, L. A. and Gardlik, J. M.,** *J. Am. Chem. Soc.,* 102, 5016, 1980; **Gardlik, J. M., Paquette, L. A., and Gleiter, R.,** *J. Am. Chem. Soc.,* 101, 1617, 1979.
112. **Ganis, P., Musco, A., and Temussi, P. A.,** *J. Phys. Chem.,* 73, 3201, 1969.
113. **Lyttle, M. H., Streitwieser, A., Jr., and Kluttz, R. Q.,** *J. Am. Chem. Soc.,* 103, 3232, 1981.
114. **Vogel, E., Kiefer, H., and Roth, W. R.,** *Angew. Chem. Int. Ed.,* 3, 442, 1964.
115. **Dewar, M. J. S. and Gleicher, G. J.,** *J. Am. Chem. Soc.,* 87, 685, 1965.
116. **Einstein, F. W. B., Willis, A. C., Cullen, W. R., and Soulen, R. L.,** *Chem. Commun.,* p. 526, 1981.
117. **Streitwieser, A., Jr. and Müller-Westerhoff, U.,** *J. Am. Chem. Soc.,* 90, 7364, 1968; **Hodgson, K. O. et al.,** *Chem. Commun.,* p. 1592, 1971; *J. Organometal. Chem.,* 28, C24, 1971.
118. **Starks, D. F., Parsons, T. C., Streitwieser, A., and Edelstein, N.,** *Inorg. Chem.,* 13, 1302, 1974.
119. **Rausch, G. and Schrauzer, G. N.,** *Chem. Ind.,* p. 957, 1959; **Schrauzer, G. N.,** *J. Am. Chem. Soc.,* 83, 2966, 1961; **Manuel, T. A. and Stone, F. G. A.,** *Proc. Chem. Soc.,* p. 90, 1959; **Nakamura, A. and Hagihara, N.,** *Bull. Chem. Soc. Jpn.,* 32, 880, 1959.
120. **Dickens, B. and Lipscomb, W. N.,** *J. Am. Chem. Soc.,* 83, 4862, 1961.

121. **Ehntholt, D. J. and Kerber, R. C.,** *J. Organometal. Chem.,* 38, 139, 1972.
122. **Paquette, L. A. et al.,** *Tetrahed. Lett.,* p. 2943, 1973; *J. Am. Chem. Soc.,* 97, 4658, 1975.
123. **Brookhart, M. and Davis, E. R.,** *J. Am. Chem. Soc.,* 92, 7622, 1970; **Brookart, M., Davis, E. R., and Harris, D. L.,** *J. Am. Chem. Soc.,* 94, 7853, 1972.
124. **Davison, A., McFarlane, W., Pratt, L., and Wilkinson, G.,** *Chem. Ind.,* p. 553, 1961; *J. Chem. Soc.,* p. 4821, 1962.
125. **Winstein, S., Kreiter, C. G., and Brauman, J. I.,** *J. Am. Chem. Soc.,* 88, 2047, 1966.
126. **Cooke, M. et al.,** *Chem. Commun.,* p. 621, 1971.
127. **Burkoth, T. L. and Van Tamelen, E. E.,** in *Nonbenzenoid Aromatics,* Vol. 1, Snyder, J. P., Ed., Academic Press, New York, 1969, 64; **Van Tamelen, E. E.,** *Acc. Chem. Res.,* 5, 186, 1972.
128. **Banciu, M.,** *Rev. Roum. Chim.,* 18, 657, 1970.
129. **Kemp-Jones, A. V. and Masamune, S.,** in *Topics in Nonbenzenoid Aromatic Chemistry,* Nozoe, T., Breslow, R., Hafner, K., Ito, S., and Murata, I., Eds., Halsted Press, New York, 1973, 121.
130. **Masamune, S. and Darby, N.,** *Acc. Chem. Res.,* 5, 272, 1972.
131. **Prelog, V.,** in *Perspectives in Organic Chemistry,* Todd, A., Ed., Interscience, New York, 1956, 127.
132. **Grob, C. A. and Schiess, P. W.,** *Helv. Chim. Acta,* 47, 558, 1964.
133. **Mulligan, P. J. and Sondheimer, F.,** *J. Am. Chem. Soc.,* 89, 7118, 1967.
134. **Mislow, K.,** *J. Chem. Phys.,* 20, 1489, 1952.
135. **Avram, M., Nenitzescu, C. D., and Marica, E.,** *Chem. Ber.,* 90, 1857, 1957; **Avram, M., Mateescu, G., and Nenitzescu, C. D.,** *Liebigs Ann. Chem.,* 636, 174, 1970.
136. **Cookson, R. C., Hudec, J., and Marsden,** *J. Chem. Ind.,* p. 21, 1961.
137. **Doering, W., von E. and Rosenthal, J. W.,** *J. Am. Chem. Soc.,* 88, 2078, 1966.
138. **van Tamelen, E. E. and Burkoth, T. L.,** *J. Am. Chem. Soc.,* 89, 151, 1967.
139. **Johnson, W. S., Bass, J. D., and Williamson, K. L.,** *Tetrahedron,* 19, 861, 1963.
140. **van Tamelen, E. E. and Greeley, R. H.,** *Chem. Commun.,* p. 601, 1971.
141. **van Tamelen, E. E., Burkoth, T. L., and Greeley, R. H.,** *J. Am. Chem. Soc.,* 93, 6120, 1971.
142. **Masamune, S., Chin, C. G., Hojo, K., and Seidner, R. T.,** *J. Am. Chem. Soc.,* 89, 4804, 1967.
143. **Masamune, S., Hojo, K., Hojo, K., Bigam, G., and Rabenstein, D. L.,** *J. Am. Chem. Soc.,* 93, 4966, 1971.
144. **Farnell, L., Kao, J., Radom, J., and Schaefer, H. F., III,** *J. Am. Chem. Soc.,* 103, 2147, 1981 and further references therein for other theoretical studies, see also **Haddon, H. C., and Raghavachari, K.,** *J. Am. Chem. Soc.,* 104, 3516, 1982.
145. **Sondheimer, F.,** *Pure Appl. Chem.,* 7, 363, 1963; *Proc. R. Soc. A,* 297, 173, 1967; *Acc. Chem. Res.,* 5, 81, 1972; *Chimia,* 28, 163, 1974.
146. **Glaser, C.,** *Ber. Dtsch. Chem. Ges.,* 2, 422, 1869; *Liebigs Ann. Chem.,* 154, 159, 1870.
147. **Eglinton, G. and Galbraith, A. R.,** *J. Chem. Soc.,* p. 889, 1959; *Chem. Ind.,* p. 737, 1956.
148. **Stöckel, K. and Sondheimer, F.,** *Org. Synth.,* 54, 1, 1974.
149. **Schröder, G.,** *Pure Appl. Chem.,* 44, 925, 1974.
150. **Oth, J. F. M., Gilles, J. M., and Schröder, G.,** *Tetrahed. Lett.,* p. 67, 1270; **Oth, J. F. M., Röttele, H., and Schröder, G.,** *Tetrahedron Lett.,* p. 61, 1970.
151. **Schröder, G. and Oth, J. F. M.,** *Tetrahedron Lett.,* p. 4083, 1966; **Schröder, G., Martin, W., and Oth, J. F. M.,** *Angew. Chem. Int. Ed. Engl.,* 6, 4966, 1967.
152. **Jackman, L. M., Sondheimer, F., Amiel, Y., Ben-Efraim, D. A., Gaoni, Y., Wolovsky, R., and Bothner-By, A. A.,** *J. Am. Chem. Soc.,* 84, 4307, 1962.
153. **Sondheimer, F. and Gaoni, Y.,** *J. Am. Chem. Soc.,* 82, 5765, 1960; *Proc. Chem. Soc.,* 299, 1964; **Gaoni, R., Melera, A., Sondheimer, F., and Wolovsky, R.,** *Proc. Chem. Soc.,* p. 397, 1964.
154a. **Bailey, N. A. and Mason, R.,** *Proc. R. Soc. London A,* 290, 94, 1966.
154b. **Chiang, C. C. and Paul, I. C.,** *J. Am. Chem. Soc.,* 94, 4741, 1972.
155. **Gaoni, Y. and Sondheimer, F.,** *Proc. Chem. Soc.,* p. 299, 1964.
156. **Vogel, E., Engels, H. W., Huber, W., Lex, J., and Müllen, K.,** *J. Am. Chem. Soc.,* 104, 3729, 1982.
157. **Loos, D. and Leska, J.,** *Coll. Czech. Chem. Commun.,* 47, 1705, 1982.
158. **Sondheimer, F., Stöckel, K., Clark, T. A., Guss, M., and Mason, R. A.,** *J. Am. Chem. Soc.,* 93, 2571, 1971.
159. **Sondheimer, F. and Gaoni, F.,** *J. Am. Chem. Soc.,* 83, 4863, 1961.
160. **Sondheimer, F. and Wolovsky, R.,** *J. Am. Chem. Soc.,* 84, 260, 1962; **Sondheimer, F., Wolovsky, R., and Amiel, Y.,** *J. Am. Chem. Soc.,* 84, 274, 1962.
161. **Bregman, J., Hirshfeld, F. L., Rabinovich, D., and Schmidt, G. M. J.,** *Acta Cryst.,* 19, 227, 1965; **Hirshfeld, F. L. and Rabinovich, D.,** *Acta Cryst.,* 19, 235, 1965.
162. **Peters, D.,** *J. Chem. Phys.,* 51, 1559, 1566, 1969; **Murrell, J. N. and Hinchliffe, A.,** *Trans. Faraday Soc.,* 62, 2011, 1966.
163. **Gaoni, Y., Melera, A., Sondheimer, F., and Wolovsky, R.,** *Proc. Chem. Soc.,* p. 397, 1964.

164. **Beezer, A. E., Mortimer, C. T., Springall, H. D., Sondheimer, F., and Wolovsky, R.,** *J. Chem. Soc.,* p. 216, 1965.
165. **Oth, J. F. M., Buenzli, J. C., and de Zelicourt, Y. de J.,** *Helv. Chim. Acta,* 57, 2275, 1974.
166. **Stöckel, K., Garratt, P. J., Sondheimer, F., de Zelicourt, Y. de J., and Oth, J. F. M.,** *J. Am. Chem. Soc.,* 94, 8644, 1972.
167. **Gilles, J. M., Oth, J. F. M., Sondheimer, F., and Woo, E. P.,** *J. Chem. Soc. (B),* p. 2177, 1971.
168. **Schröder, G., Neuberg, R., and Oth, J. F. M.,** *Angew. Chem. Int. Ed.,* 11, 51, 1972.
169. **Sondheimer, F. and Ben-Efraim, D. A.,** *J. Am. Chem. Soc.,* 85, 52, 1963.
170. **Calder, I. C., Garratt, P. J., Longuet-Higgins, H. C., Sondheimer, F., and Wolovsky, R.,** *J. Chem. Soc. (C),* p. 1041, 1967.
171. **Woo, E. P. and Sondheimer, F.,** *Tetrahedron,* 26, 3939, 1970.
172. **Wolovsky, R., Woo, E. P., and Sondheimer, F.,** *Tetrahedron,* 26, 2133, 1970.
173. **Du Vernet, R. B., Otsubo, T., Lawson, J. A., and Boekelheide, V.,** *J. Am. Chem. Soc.,* 97, 1629, 1975.
174. **Newkome, G. R and Lee, H. W.,** *J. Am. Chem. Soc.,* 105, 5956, 1983; **Toner, J. L.,** *Tetrahed. Lett.,* 24, 2707, 1983.
175. **Badger, G. M., Elix, J. A., and Lewis, G. E.,** *Aust. J. Chem.,* 19, 1221, 1966.
176. **Badger, G. M., Elix, G. E., and Singh, U. P.,** *Aust. J. Chem.,* 19, 257, 1966.
177. **Ogawa, H., Sadakari, N., Imoto, T., Miyamoto, I., Kato, H., and Taniguchi, Y.,** *Angew. Chem.,* 95, 412, 1983.
178. **Ogawa, A., Fukuda, C., Imoto, T., Miyamoto, J., Taniguchi, Y., Koga, T., and Nogami, Y.,** *Tetrahed. Lett.,* p. 24, 1983.
179. **Sondheimer, F. and Metcalf, B. W.,** *J. Am. Chem. Soc.,* 93, 6675, 1971.
180. **Sondheimer, F., McQuilkin, R. M., and Metcalf, B. W.,** *Chem. Commun.,* p. 338, 1971.
181. **Sondheimer, F. and Wolovsky, R.,** *J. Am. Chem. Soc.,* 81, 4755, 1959.
182. **Sondheimer, F. and Wolovsky, R.,** *J. Am. Chem. Soc.,* 82, 754, 1960.
183. **Hoffmann, R. W.,** *Dehydrobenzene and Cycloalkynes,* Academic Press, New York, 1967.
184. **Levin, R. H.,** in *Reactive Intermediates,* Vol. 1, Jones, M., Jr. and Moss, R. A., Eds., Interscience, New York, 1978, 1.
185. **Fields, E. K.,** in *Organic Reactive Intermediates,* McManus, S. P., Ed., Academic Press, New York, 1973; **Sharp, J. T.,** in *Comprehensive Organic Chemistry,* Vol. 1, Barton, D. and Ollis, W. D., Eds., Pergamon, Oxford, 1979, 455.
186. **Darby, N., Kim, C. U., Salaün, J. A., Shelton, K. W., Takada, S., and Masamune, S.,** *Chem. Commun.,* p. 1516, 1971.
187. **Sondheimer, F., Gaoni, Y., Jackman, L. M., Bailey, N. A., and Mason, R.,** *J. Am. Chem. Soc.,* 84, 4595, 1982.
188. **Raphael, R. and Sondheimer, F.,** *J. Chem. Soc.,* p. 120, 1950.
189. **Nakagawa, M.,** *Pure Appl. Chem.,* 44, 885, 1975.
190. **Nakagawa, M.,** in *The Chemistry of the Carbon-Carbon Triple Bond,* Part 2, Patai, S., Ed., John Wiley & Sons, New York, 1978, 661.
191. **Ojima, J., Katakami, T., Nakaminami, G., and Nakagawa, M.,** *Tetrahed. Lett.,* p. 1115, 1968.
192. **Katakami, T., Tomita, S., Fukui, K., and Nakagawa, M.,** *Chem. Lett.,* p. 225, 1972.
193. **Nomoto, T., Fukui, K., and Nakagawa, M.,** *Tetrahed. Lett.,* p. 3253, 1972; *Bull. Chem. Soc. Jpn.,* 49, 305, 1976.
194. **Fukui, K., Nomoto, T., Nakatsuji, S., and Nakagawa, M.,** *Tetrahed. Lett.,* p. 3157, 1972.
195. **Iyoda, M. and Nakagawa, M.,** *Tetrahed. Lett.,* p. 3161, 1972.
196. **Iyoda, M. and Nakagawa, M.,** *Chem. Commun.,* p. 1003, 1972.
197. **Iyoda, M. and Nakagawa, M.,** *Tetrahed. Lett.,* p. 4253, 1972.
198. **Iyoda, M., Miyazaki, H., and Nakagawa, M.,** *Chem. Commun.,* p. 431, 1972; *Bull. Chem. Soc. Jpn.,* 49, 2306, 1976.
199. **Wolovsky, R. and Sondheimer, F.,** *J. Am. Chem. Soc.,* 87, 5720, 1965.
200. **Untch, K. G. and Wysocki, D. C.,** *J. Am. Chem. Soc.,* 88, 2608, 1966.
201. **Bailey, N. A. and Mason, R.,** *Proc. R. Soc. A,* 290, 94, 1966.
202. **Gaoni, Y. and Sondheimer, F.,** *J. Am. Chem. Soc.,* 86, 521, 1964.
203. **Nomoto, T., Nakatsuji, S., and Nakagawa, M.,** *Chem. Lett.,* p. 839, 1974.
204. **Satake, T., Nakatsuji, S., Iyoda, M., Akiyama, S., and Nakagawa, M.,** *Tetrahed. Lett.,* p. 1881, 1976.
205. **Tomita, S. and Nakagawa, M.,** *Bull. Chem. Soc. Jpn.,* 49, 302, 1976.
206. **Mayer, J. and Sondheimer, F.,** *J. Am. Chem. Soc.,* 88, 602 and 603, 1966.
207. **Nakatsuji, S., Morigaki, M., Akiyama, S., and Nakagawa, M.,** *Tetrahed. Lett.,* p. 1233, 1975.
208. **Sondheimer, F. and Wolovsky, R.,** *J. Am. Chem. Soc.,* 81, 4600, 1771, 1959; **Wolovsky, R.,** *J. Am. Chem. Soc.,* 87, 3638, 1965.

209. **Kabuto, C., Kitahara, Y., Iyoda, M., and Nakagawa, M.,** *Tetrahed. Lett.,* p. 2791, 1976.
210. **Sondheimer, F. and Gaoni, Y.,** *J. Am. Chem. Soc.,* 83, 1259, 1961; 84, 3520, 1962.
211. **Nakatsuji, S. and Nakagawa, M.,** *Tetrahedron. Lett.,* p. 3927, 1975.
212. **McQuilkin, R. M. and Sondheimer, F.,** *J. Am. Chem. Soc.,* 92, 6341, 1970.
213. **Kabuto, C., Kitahara, Y., Iyoda, M., and Nakagawa, M.,** *Tetrahed. Lett.,* p. 2787, 1976.
214. **Akiyama, S., Nomoto, T., Iyoda, M., and Nakagawa, M.,** *Bull. Chem. Soc. Jpn.,* 49, 2579, 1976.
215. **Calder, I. C. and Sondheimer, F.,** *Chem. Commun.,* p. 904, 1966.
216. **Nakatsuji, S., Akiyawa, S., and Nakagawa, M.,** *Tetrahed. Lett.,* p. 3927, 1975.
217. **Loznoff, C. C. and Sondheimer, F.,** *J. Am. Chem. Soc.,* 89, 4247, 1967; **Metcalf, B. W. and Sond-heimer, F.,** *J. Am. Chem. Soc.,* 93, 5271, 1971.
218. **Iyoda, M. and Nakagawa, M.,** *Tetrahed. Lett.,* p. 4745, 1973.
219. **Longuet-Higgins, H. C. and Salem, L.,** *Proc. R. Soc.* A, 251, 172, 1959; A257, 445, 1960; **Coulson, C. A. and Dixon, W. T.,** *Tetrahedron,* 17, 215, 1962; see also Reference 115.
220. **Dewar, M. J. S.,** *Nature (London),* 155, 50, 1945; 155, 141, 1945; 155, 479, 1945.
221. **Doering, W., von E. and Knox, L. H.,** *J. Am. Chem. Soc.,* 79, 352, 1957.
222. **Merling, G.,** *Ber. Dtsch. Chem. Ges.,* 24, 3108, 1891.
223. **Thiele, J.,** *Ber. Dtsch. Chem. Ges.,* 34, 68, 1901.
224. **Stohrer, W. D. and Hoffmann, R.,** *J. Am. Chem. Soc.,* 94, 1661, 1972.
225. **Masamune, S.,** *J. Pure Appl. Chem.,* 44, 861, 1975; **Masamune, S., Sakai, M., Kemp-Jones, A. V., Ona, H., Venot, A., and Nakashima, T.,** *Angew. Chem. Int. Ed. Engl.,* 12, 769, 1973; **Kemp-Jones, A. V., Nakamura, M., and Masamune, S.,** *Chem. Commun.,* p. 109, 1974.
226. **Dewar, M. J. S. and Haddon, R. C.,** *J. Am. Chem. Soc.,* 95, 5836, 1973; **Kollmar, H., Smith, H. O., and Schleyer, P., von R.,** *J. Am. Chem. Soc.,* 95, 5834, 1973; **Hehre, W. J. and Schleyer, P., von R.,** *J. Am. Chem. Soc.,* 95, 5837, 1973.
227. **Balaban, A. T. and Rouvray, D. H.,** *Tetrahedron,* 36, 1851, 1980.
228. **King, R. B. and Rouvray, D. H.,** *J. Am. Chem. Soc.,* 99, 7834, 1977.
229. **Hogeveen, H. and Kwant, P. W.,** *J. Am. Chem. Soc.,* 96, 2208, 1974; *Acc. Chem. Res.,* 8, 413, 1975; *Tetrahed. Lett.,* 1665, 1973; **Hogeveen, H., Kwant, P. W., Postma, J., and Van Duynen, P. Th.,** *Tetrahedron Lett.,* p. 4351, 1974.
230. **Breslow, R.,** *J. Am. Chem. Soc.,* 79, 5318, 1957.
231. **Breslow, R., Groves, J. T., and Ryan, G.,** *J. Am. Chem. Soc.,* 89, 5048, 1967; **Breslow, R. and Groves, J. T.,** *J. Am. Chem. Soc.,* 92, 984, 1970.
232. **Farnum, D. G., Mehta, G., and Silberman, R. G.,** *J. Am. Chem. Soc.,* 89, 5048, 1967.
233. **Krebs, A. W.,** *Angew. Chem.,* 77, 10, 1965; **Closs, G. L.,** *Adv. Alicyclic Chem.,* 1, 1, 1966; **Yoshida, Z.,** *Top. Curr. Chem.,* 40, 47, 1973; **Eicher, T. and Weber, R.,** *Top. Curr. Chem.,* 57, 1, 1975; **West, R.,** *Acc. Chem. Res.,* 3, 130, 1970; **Domnin, I. N. and Lakshin, A. M.,** *Sovremennye Probl. Org. Khim.,* 7, 67, 1982.
234. **Nozoe, T.,** *Prog. Org. Chem.,* 5, 132, 1961.
235. **Dauben, H. J., Gadecki, F. A., Harmon, K. M., and Pearson, D. L.,** *J. Am. Chem. Soc.,* 79, 4557, 1959; **Harmon, K. M. and Harmon, A. B.,** *J. Am. Chem. Soc.,* 81, 865, 1961; **Conrow, K.,** *J. Am. Chem. Soc.,* 81, 5461, 1959.
236. **Volpin, M. E., Kursanov, D. N., Shemyakin, M. M., Maimind, V. I., and Neiman, L. A.,** *Chem. Ind.,* p. 1261, 1958; *Zh. Obshch. Khim.,* 29, 3711, 1959; **Kitaigorodskii, A. I., Struchkov, Y. T., Khotsyanova, T. L., Volpin, M. E., and Kursanov, D. N.,** *Izv. Akad. Nauk SSSR,* p. 39, 1960.
237. **Grimme, W., Hoffmann, H., and Vogel, E.,** *Angew. Chem. Int. Ed. Engl.,* 4, 354, 1965; **Destro, R., Pilati, T., and Simonetta, M.,** *J. Am. Chem. Soc.,* 98, 1999, 1976.
238. **Murata, I., Yamamoto, K., and Kayane, Y.,** *Angew. Chem. Int. Ed. Engl.,* 13, 808, 1974.
239. **Katz, T. J. and Garratt, P. J.,** *J. Am. Chem. Soc.,* 85, 2852, 1963; 86, 5194, 1964; **LaLancette, E. A. and Benson, R. E.,** *J. Am. Chem. Soc.,* 85, 2853, 1963; 87, 1941, 1965; **Boche, G. et al.,** *Angew. Chem. Int. Ed. Engl.,* 8, 594, 1969; 8, 984, 1969.
240. **Schröder, G.,** *Pure Appl. Chem.,* 44, 925, 1975; **Hillenbrand, P., Plinke, G., Oth, J. F. M., and Schröder, G.,** *Chem. Ber.,* 111, 107, 1978.
241. **Doering, W., von E. and De Puy, C. H.,** *J. Am. Chem. Soc.,* 75, 5955, 1953.
242. **Dauben, H. J. and Rifi, M. R.,** *J. Am. Chem. Soc.,* 85, 3041, 1963; **Breslov, R. and Chang, H. W.,** *J. Am. Chem. Soc.,* 87, 2200, 1965; **Staley, S. W. and Orvedal, A. W.,** *J. Am. Chem. Soc.,* 95, 3382, 1973.
242a. **Tidwell, T. T.,** *Angew. Chem. Int. Ed. Engl.,* 23, 20, 1984.
243. **Katz, T. J.,** *J. Am. Chem. Soc.,* 82, 3784, 1960.
244. **Oth, J. F. M., Baumann, H., Gilles, J. M., and Schröder, G.,** *J. Am. Chem. Soc.,* 94, 3498, 1972; **Oth, J. F. M., Woo, E. P., and Sondheimer, F.,** *J. Am. Chem. Soc.,* 95, 7337, 1973.
245. **McQuilkin, R. M., Garrat, P. J., and Sondheimer, F.,** *J. Am. Chem. Soc.,* 92, 6682, 1970.
246. **Garratt, P. J., Rowland, N. E., and Sondheimer, F.,** *Tetrahedron,* 27, 3157, 1971.

247. **Mitchell, R. H., Kloppenstein, C. E., and Boekelheide, V.,** *J. Am. Chem. Soc.,* 91, 4931, 1969.
248. **Paquette, L. A. et al.,** *J. Am. Chem. Soc.,* 93, 168, 1971; 96, 5806, 1974.
249. **Müllen, K., Huber, W., Meul, T., Nakagawa, M., and Iyoda, M.,** *J. Am. Chem. Soc.,* 104, 5403, 1982; **Müllen, K.,** *Chem. Rev.,* 84, 603, 1984; **Vogler, H.,** *Tetrahedron,* 41, 5383, 1985; *Croat. Chim. Acta,* 57, 1177, 1984; *Mol. Struct.,* 51, 289, 1979; **Norinder, U., Wennerström, O., and Wennerström, H.,** *Tetrahedron,* 41, 713, 1985.
250a. **Prakash, G. K. S., Rawdah, T. N., and Olah, G. A.,** *Angew. Chem.,* 95, 356, 1983; (b) **Pagni, R. M.,** *Tetrahedron,* 40, 4161, 1984.
251. **Haddon, R. C.,** *J. Am. Chem. Soc.,* 101, 1722, 1979.
252. **Oth, J. F. M. and Schröder, G.,** *J. Chem. Soc. (B),* p. 904, 1971.
252a. **Schröder, G., Plinke, G., Smith, D. M., and Oth, J. F. M.,** *Angew. Chem. Int. Ed. Engl.,* 12, 325, 1974.
253. **Olah, G. A. and Mateescu, G. D.,** *J. Am. Chem. Soc.,* 92, 1430, 1970; **Olah, G. A. and Staral, J. S. et al.,** *J. Am. Chem. Soc.,* 98, 6290, 1976; 97, 5849, 1975.
254. **Spiesecke, H. and Schneider, W. G.,** *J. Chem. Phys.,* 35, 731, 1961; *Can. J. Chem.,* 41, 966, 1963; *Tetrahed. Lett.,* p. 468, 1961.
255. **Olah, G. A., Staral, J. A., Liang, L. A., Paquette, L. A., Melega, W. P., and Carmody, M. J.,** *J. Am. Chem. Soc.,* 99, 3349, 1977.
256. **Olah, G. A., Liang, G., Paquette, L. A., and Melega, W. P.,** *J. Am. Chem. Soc.,* 98, 4327, 1976.
257. **Oth, J. F. M., Smith, D. M., Prangé, U., and Schröder, G.,** *Angew. Chem. Int. Ed. Engl.,* 12, 327, 1973.
258. **Cresp, T. M. and Sondheimer, F.,** *J. Am. Chem. Soc.,* 97, 4412, 1975; 99, 194, 1977.
259. **Kashitani, T., Akiyama, S., Iyoda, M., and Nakagawa, M.,** *J. Am. Chem. Soc.,* 97, 4424, 1975.
260. **Akiyama, S., Iyoda, M., and Nakagawa, M.,** *J. Am. Chem. Soc.,* 98, 6410, 1976.
261. **Kai, Y., Yasuoka, N., Kasai, N., Akiyama, S., and Nakagawa, M.,** *Tetrahed. Lett.,* p. 1703, 1978.
262. **Nakagawa, M.,** *Angew. Chem. Int. Ed. Engl.,* 18, 202, 1979.
263. **Nakatsuji, S., Akiyawa, S., and Nakagawa, M.,** (a) *Tetrahed. Lett.,* p. 1483, 1978; (b) *Tetrahedron Lett.,* p. 3723, 1977.
264. **Osuka, M., Yoshikawa, Y., Akiyama, S., and Nakagawa, M.,** *Tetrahed. Lett.,* p. 3719, 1977.
265. **Iyoda, M., Akiyama, S., and Nakagawa, M.,** *Tetrahed. Lett.,* p. 4213, 1979.
266. **Yoshikawa, Y., Iyoda, M., and Nakagawa, M.,** (a) *Tetrahed. Lett.,* 22, 1989, 1981; (b) *Tetrahed. Lett.,* 22, 2659, 1981.
267. **Sakano, K., Nakagawa, T., Iyoda, M., and Nakagawa, M.,** *Tetrahed. Lett.,* 22, 2655, 1981.
268. **Iyoda, M., Yoshikawa, Y., and Nakagawa, M.,** *Chem. Lett.,* p. 1501, 1982.
269. **Nakatsuji, S., Akiyama, S., Iyoda, M., and Nakagawa, M.,** *Chem. Lett.,* p. 1777, 1981.
270. **Yoshikawa, Y., Iyoda, M., Nakagawa, M., Nakatsuji, S., and Akiyama, S.,** *Tetrahed. Lett.,* 22, 5209, 1981.
271. **Iyoda, M., Nakagawa, T., Nakagawa, M., and Oda, M.,** *Tetrahed. Lett.,* 23, 5423, 1982.
272. **Hess, B. A., Jr., Schaad, L. J., and Nakagawa, M.,** *J. Org. Chem.,* 42, 1669, 1977; Hess, B. A., Jr., Schaad, L. J., and Agranat, I., *J. Am. Chem. Soc.,* 100, 5268, 1978.
273. **Dewar, M. J. S.,** *Pure Appl. Chem.,* 44, 762, 1975.
274. **Breslow, R., Napierski, J., and Clarke, T. C.,** *J. Am. Chem. Soc.,* 97, 6275, 1975.
275. **Hafner, K. and Kaiser, H.,** *Liebigs Ann. Chem.,* 618, 140, 1958; *Org. Synth. Coll.,* 5, 1088, 1973.
276. **Balaban, A. T., Dinculescu, A., Dorofeenko, G. N., Fischer, G. W., Koblik, A. V., Mezheritskii, V. V. and Schroth, W.,** Pyrylium salts, in *Advances in Heterocyclic Chemistry,* Vol. 2(Suppl.), Academic Press, New York, 1982.
277. **Hafner, K.,** in *The Chemistry of Nonbenzenoid Aromatic Compounds,* Oki, M., Ed., Butterworths, London, 1971.
278. **Hafner, K.,** *Pure Appl. Chem.,* 28, 153, 1971; *Pure Appl. Chem.,* Suppl. 2, 1, 1971; *Angew. Chem. Int. Ed. Engl.,* 3, 165, 1964.
279. **Hafner, K., Donges, R., Gödecke, E., and Kaiser, R.,** *Angew. Chem. Int. Ed. Engl.,* 12, 337, 1973.
280. **Le Goff, E.,** *J. Am. Chem. Soc.,* 84, 1505, 1962; 84, 3975, 1972.
281. **Katz, T. J. and Rosenberger, M.,** *J. Am. Chem. Soc.,* 84, 865, 1962.
282. **Hafner, K., Bangert, K. F., and Orfanos, V.,** *Angew. Chem. Int. Ed. Engl.,* 6, 451, 1967.
283. **Katz, T. J., Rosenberger, M., and O'Hara, R. K.,** *J. Am. Chem. Soc.,* 86, 249, 1964.
284. **Hafner, K.,** *Pure Appl. Chem.,* 54, 939, 1982.
285. **Katz, T. J. and Schulman, J.,** *Am. Chem. Soc.,* 86, 3169, 1964.
286. **Hafner, K. and Krimmer, H. P.,** *Angew. Chem. Int. Ed. Engl.,* 19, 199, 1980.
286a. **Sugihara, Y.,** *J. Am. Chem. Soc.,* 106, 7268, 1984.
287. **Maier, G.,** *Angew. Chem. Int. Ed. Engl.,* 13, 425, 1974.
288. **Hafner, K., Diehl, H., and Süss, H. U.,** *Angew. Chem. Int. Ed. Engl.,* 15, 104, 1976.

289. **Wilner, I. and Rabinowitz, M.,** *J. Am. Chem. Soc.,* 100, 337, 1978; **Wilner, I., Becker, J. Y., and Rabinowitz, M.,** *J. Am. Chem. Soc.,* 101, 395, 1979.
290. **Dauben, H. J., Jr. and Bertelli, D. J.,** *J. Am. Chem. Soc.,* 83, 4659, 1961.
291. **Heilbronner, E., Meier, W., and Meuche, D.,** *Helv. Chim. Acta,* 45, 2628, 1961.
292. **Vogel, E., Königshofen, H., Wassen, J., Müllen, K., and Oth, J. F. M.,** *Angew. Chem. Int. Ed. Engl.,* 13, 732, 1974; **Vogel, E. and Ippen, J.,** *Angew Chem. Int. Ed. Engl.,* 13, 734, 1974; **Vogel, E. and Hogrefe, F.,** *Angew Chem. Int. Ed. Engl.,* 13, 735, 1974; **Paquette, L. A., Browne, A. R., and Chamot, E.,** *Angew Chem. Int. Ed. Engl.,* 18, 546, 1979.
293. **Lammertsma, K., Olah, G. A., Berke, C. M., and Streitwieser, A., Jr.,** *J. Am. Chem. Soc.,* 101, 6658, 1979.
294. **Olah, G. A. and Liang, G.,** *J. Am. Chem. Soc.,* 99, 6045, 1977.
295. **Vogel, E., Runzheimer, H. V., Hogrefe, F., Baasner, B., and Lex, J.,** *Angew. Chem. Int. Ed. Engl.,* 16, 871, 1977.
296. **Oth, J. M. F., Müllen, K., Runzheimer, H. V., Mues, P., and Vogel, E.,** *Angew. Chem. Int. Ed. Engl.,* 16, 872, 1977.
297. **Allinger, N. L. and Gilardeau, C.,** *Tetrahed. Lett.,* p. 1569, 1967.
298. **Dewar, M. J. S. and de Llano, C.,** *J. Am. Chem. Soc.,* 91, 789, 1969.
299. **Müllen, K., Oth, J. M. F., Engels, H. W., and Vogel, E.,** *Angew. Chem. Int. Ed. Engl.,* 18, 229, 1979.
300. **Vogel, E.,** in *Current Trends in Organic Synthesis,* Nozaki, N., Ed., Pergamon Press, Oxford, 1983, 379.
301. **Yoshida, Z. et al.,** *Angew. Chem.,* 96, 75, 1984.
302. **Haddon, R. C., Raghavachari, K., and Whangbo, M. H.,** *J. Am. Chem. Soc.,* 106, 5364, 1984.
303. **Ichikawa, H.,** *J. Am. Chem. Soc.,* 105, 7467, 1983.
304. **Ichikawa, H. and Ebisawa, Y.,** *J. Am. Chem. Soc.,* 107, 1161, 1985.
305. **Haddon, R. C. and Raghavachari, K.,** *J. Am. Chem. Soc.,* 107, 289, 1985.
306. **Herndon, W. C. and Hosoya, H.,** *Tetrahedron,* 20, 3987, 1984.
307. **Glidewell, C. and Lloyd, D.,** *Tetrahedron,* 40, 4455, 1984.
308. **Norrinder, U., Wennerström, O., and Wennerström, H.,** *Tetrahedron,* 41, 713, 1985.
309. **Dewar, M. J. S. and Merz, K. M., Jr.,** *Chem. Commun.,* p. 343, 1985.
310. **Michl, J.,** *Tetrahedron,* 40, 845, 1984.
311. **Stollenwerk, A. H., Kannellakopoulos, B., and Vogler, H.,** *Tetrahedron,* 39, 3127, 1983.
312. **Minsky, A., Meyer, A. Y., and Rabinowitz, M.,** *Tetrahedron,* 41, 785, 1985.
313. **Aihara, J.,** *J. Am. Chem. Soc.,* 107, 298, 1985.
314. **Rösch, N. and Streitwieser, A. Jr.,** *J. Am. Chem. Soc.,* 105, 7237, 1983.
315. **Couitneidge, J. L., Davies, A. G., Lusztyk, E., and Lusztyk, J.,** *J. Chem. Soc. Perkin Trans. II,* p. 155, 1984.
316. **Chan, W. et al.,** *Chem. Commun.,* p. 1541, 1984.
317. **Wilhelm, D., Courtneidge, J. L., Clark, T., and Davies, A. G.,** *Chem. Commun.,* p. 810, 1984.
318. **Stevenson, G. R., Reiter, R. C., and Sedgwick, J. B.,** *J. Am. Chem. Soc.,* 105, 6251, 1983.
319. **Hafner, K. and Thiele, G. F.,** *Tetrahed. Lett.,* 26, 2567, 1985.
320. **Yoshida, Z., Shibata, M., and Sugimoto, T.,** *Tetrahed. Lett.,* 24, 4585, 1983.
321. **Müllen, K. et al.,** *Tetrahedron,* 39, 1575, 1983.
322. **Mortensen, J. and Heinze, J.,** *Tetrahed. Lett.,* 26, 415, 1985.
323. **Sugihara, Y., Wakabayashi, S., and Murata, I.,** *J. Am. Chem. Soc.,* 105, 6718, 1983.
324. **Regitz, M. and Eisenbarth, P.,** *Chem. Ber.,* 117, 1991, 1984; **Fink, J. and Regitz, M.,** *Tetrahed. Lett.,* 25, 1711, 1984; *Chem. Ber.,* 118, 2255, 1985; **Michels, G., Fink, J., Maas, G. and Regitz, M.,** *Tetrahed. Lett.,* 26, 3315, 1985; **Vogelbacher, U. J., Eisenbarth, P. and Regitz, M.,** *Angew. Chem. Int. Ed. Engl.,* 23, 708, 1984.
325. **Grassi, M., Mann, B. E. and Spencer, C. M.,** *Chem. Commun.,* p. 1169, 1985.
326. **Minkin, V. I. and Minyaev, R. M.,** *Nonclassical Structures of Organic Compounds* (in Russian), Izd. Rostovskogo Univ., Rostovon-Don, 1985; **Minkin, V. I. et al.,** *J. Mol. Struct. (Theochem.),* 110, 241, 1984; *Usp. Khim.,* 54, 86, 1985.
327. **Stezowski, J. J., Hoier, H., Wilhelm, D., Clark, T. and Schleyer, P. von R.,** *Chem. Commun.,* p. 1263, 1985.
328. **Bernhard, W., Brügger, P., Schönholzer, P., Weber, R. H. and Hansen, H. J.,** *Helv. Chim. Acta,* 68, 429, 1985.

Chapter 5

BRIDGED ANNULENES

I. INTRODUCTION

In Chapter 4 the brilliant work of Sondheimer concerning [4n]- and [4n + 2] annulenes was presented in detail. Until 1964 from the list of known annulenes (or derivatives) a notable example was missing, namely the [10]annulene. The spectacular results obtained between 1950 and 1964 in confirmation of Hückel's rule led to the idea that [10]annulene should also be an aromatic compound if the geometry allows the planarity of the molecule. Inspection of stereomodels indicates two possible configurations for [10]annulene, namely, the *all-cis* isomer (**1**) and the di-*trans*-isomer (**2**).

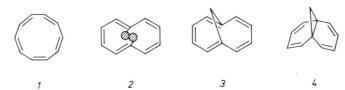

| 1 | 2 | 3 | 4 |

The *all-cis* isomer (**1**) is strongly strained (C–C–C angles of 144°) and an interesting question arose, namely, if the delocalization energy (DE) of the 10π system could overcome the strain energy. It was shown later (Masamune 1969, 1971) that **1**, which exists at low temperatures, behaves as a polyolefin (see Chapter 4).

The di-*trans* isomer **2** suffers from strong interactions between internal hydrogen atoms (at C-1 and C-6).

It was the brilliant idea of Professor Emanuel Vogel (Cologne) to substitute these two hydrogens by a methylene bridge in order to observe if the intriguing molecule thus resulted, 1,6-methano[10]annulene (**3**) gains planarity and aromatic character.

This was the beginning of a large and fascinating chapter on aromatic bridged annulenes, developed step by step with a remarkable intuition and perseverence, especially by Vogel and co-workers ("Cologne annulenes" or "Vogel's annulenes") and Boekelheide and co-workers ("Boekelheide's annulenes").

II. CARBON BRIDGED [10]ANNULENES

A. Syntheses
1. 1,6-Methano[10]Annulene

The first synthesis of **3** started from the idea that the strained propellane **4** which contains two norcaradiene units could be unstable and could easily be converted into its valence isomer, the methano-bridged[10]annulene **3**. An earlier attempt to synthesize **4** through direct carbene addition to naphthalene had failed.[1] The successful route to **3**, followed by Vogel in 1964,[2] involved the addition of dichlorocarbene to the tetrasubstituted central double bond of tetrahydronaphthalene **5**; after reductive substitution of the chlorine atoms, an extra unsaturation was introduced by bromine addition followed by treatment with KOH in ethanol. The product was directly 1,6-methano[10]annulene (**3**), m.p. 28 to 29°C.

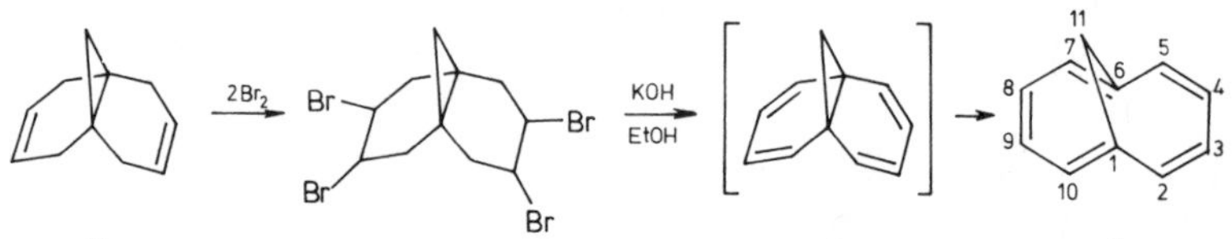

Later, **5** was converted directly into **7** (in modest yields) by the Simmons-Smith reaction[3] and **7** was dehydrogenated to **3** with 2,3-dichloro-5,6-dicyanobenzoquinone, (DDQ) in quantitative yield,[4a] or with MnO$_2$.[4b]

Recently a novel and simple synthesis of **3** was announced by Banwell.[5] Treatment of **6** with 5-*M* equivalents of potassium *t*-butoxide in dimethylsulfoxide (DMSO) affords **3** (13% yield) along with 2-methylazulene (10%).

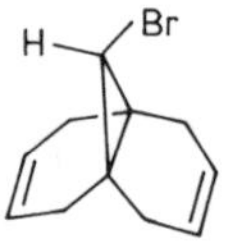

The treatment of **8** with silver nitrate in methanol also affords **3** in small yields.[6] Very recently, an entirely different approach to the synthesis of **3** and of higher bridged [4*n* + 2]annulenes, called the "building block approach" ("Baukastenprinzip") was developed by Vogel et al. using as key intermediate, cycloheptatriene-1,6-dicarbaldehyde (**10**). The initial tedious[7] synthesis of **10** was replaced by a rational five-step synthesis starting with the industrially available cycloheptatriene (**9**).[8]

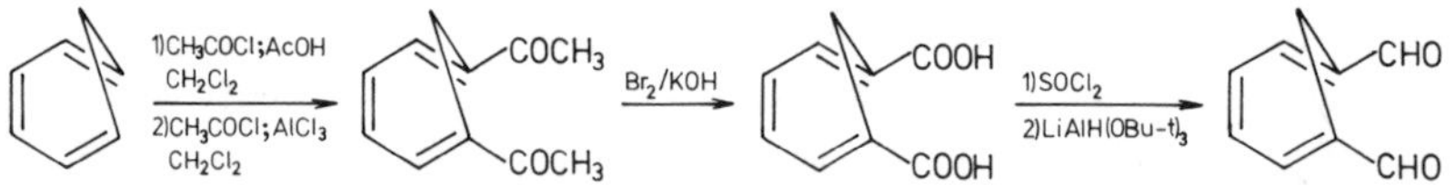

The synthetic usefulness for bridged annulenes of the "homophthalaldehyde" **10** is reminiscent of the well-known annelation ability of phthalic anhydride.

Conversion of **10** into **3** includes: (1) two successive olefinations with Wittig reagents methylene-triphenylphosphorane and chloromethylene-triphenylphosphorane leading to **11** and (2) "cyclodehydrohalogenation" of **11** (a 10π-electrocyclic process combined with HCl elimination by heating in dimethylformamide.[9-11]

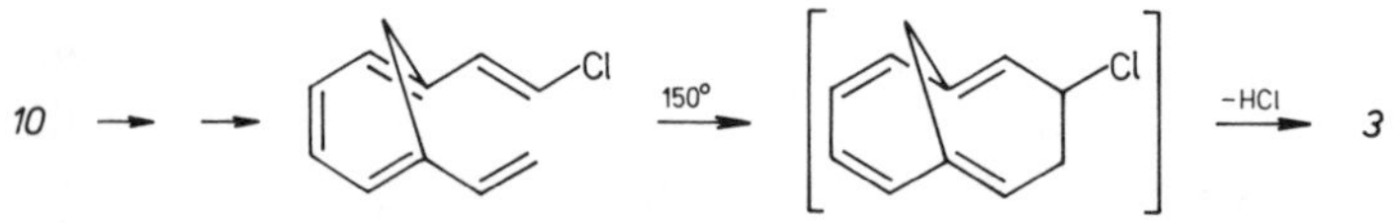

2. 1,5-Methano[10]Annulene (Homoazulene)

Considerable efforts directed towards the synthesis of 1,5-methano[10]annulene (**12**), isomeric with **3**, initially failed;[12-14] the preparation of **12** was accomplished by Masamune in 1976 to 1977[15,16] on the basis of the remarkable finding that the skeleton of 1,5-methano[10]annulene could be constructed in a single-step synthesis (in spite of the enormous amount of torsion). Condensation of cycloocta-2,4,6-trienone with the anion of methyl 4(dimethylphosphinyl)-2-butenoate affords **13** which underwent a rapid base-catalyzed isomerization to **14**.[15]

3

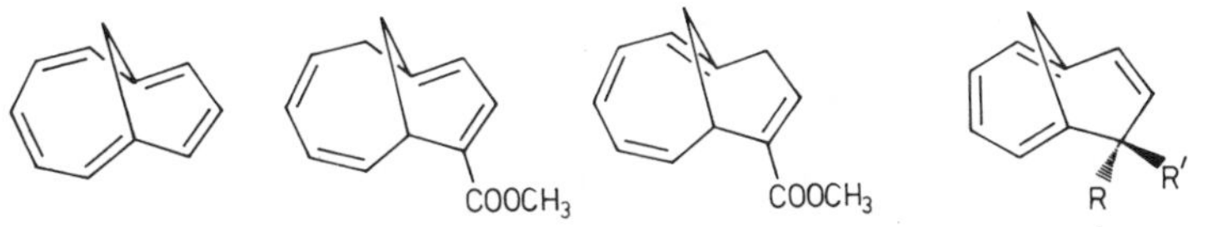

12   13   14   15

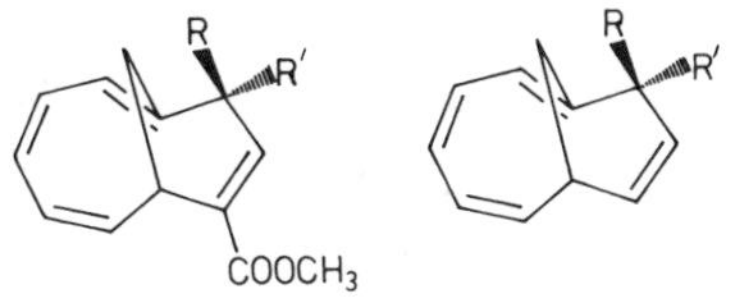

16 R=OH; R'=H   17

a , R=COOCH$_3$; R'=OH
b , R=CH$_2$OH ; R'=OH
c , R; R'=O
d , R=OH; R'=H
e , R=H ; R'=OBz
f , R=H ; R'=OH
g , R=H ; R'=O$_2$CNHC$_6$H$_4$NO$_2$

a , R=OBz ; R'=H
b , R=OH ; R'=H

Oxygenation with ^{3}O$_2$ of the enolate anion resulted from **13** or **14** followed by reduction with triethylphosphite affords a 1:1 mixture of **15a** and **16**. The successive conversion of **15a** into **15b** and **15c** was initially followed by treatment with PCl$_5$, then by reduction with lithium diisopropylamide. The monochloro-derivative of **12** obtained in the last-mentioned step was accompanied by small amounts of **12** (which could be identified by v.p.c.-mass spectrometry and ^{1}H-NMR in a reaction mixture[15]). An improved synthesis (and the full characterization) of **12** followed in the next year, 1977, using the following steps: stereospecific reduction of **15c** to **15d** (endo), conversion of **15d** into the isomeric mixture of benzoates **15e** and **17a**, hydrolysis to **15f** and **17b**. Reaction of the alcohol **15f** with *p*-nitrophenylisocyanate yielded the carbamate **15g** which on pyrolysis (300°C) afforded the crystalline **12**.[16] Another synthesis of **12** due to Scott in 1981[17a] includes as essential steps: (1) intramolecular cyclization with ring expansion of diazomethylketone **18** obtained from dehydrocinamic acid, (2) regioselective cyclopropanation of **19**, (3) introduction of a new double bond via Bamford-Stevens reaction of the tosylhydrazone of **20**, (4) bond breaking of the cyclopropanic ring in the propellane **21** with lead tetraacetate, and (5) two successive eliminations affording **12**:

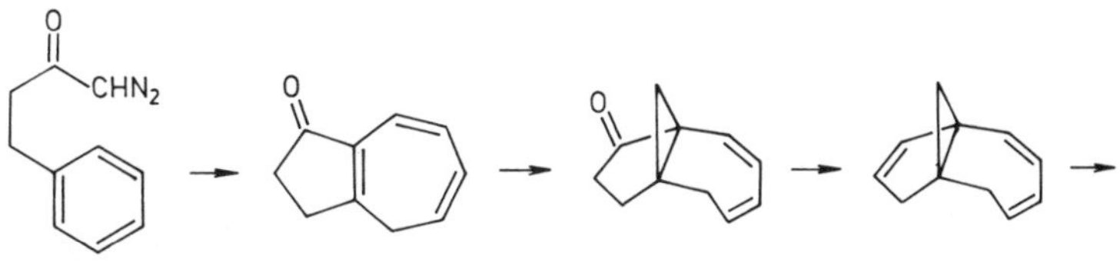

18   19   20   21

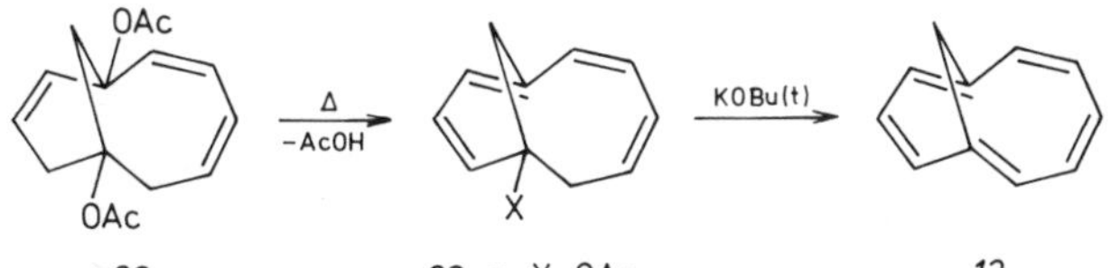

22   23 a , X=OAc   12
                   b , X=OH
                   c , X=OMs

It should be added that starting from the ketone **20** a methoxy-1,5-methano[10]annulene was synthesized in 1978 also by Scott.[17b]

3. Derivatives of 7bH-Cyclopent[c,d]Indene

Tricyclic carbon bridged [10]annulenes, derivatives of 7b*H*-cyclopent[c,d]indene (e.g., the methyl derivative·**24**, R = CH$_3$) were recently synthesized by Rees and co-workers (*vide infra*); in these compounds the skeleton of [10]annulene is preserved and the three inner hydrogens are replaced by a one-carbon tridentate bridge.

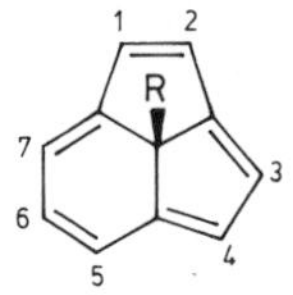

    24 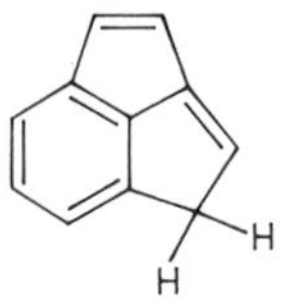   **25** 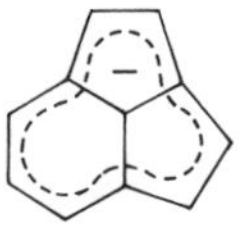   **26** 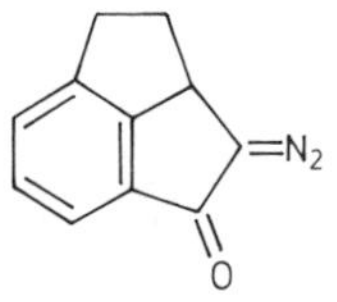   **27**

Some hydrogenated derivatives of 1*H*-cyclopent[*c,d*]indene were previously known[18-20] but dehydrogenation attempts failed. Hafner and Eilbracht[21,22] succeeded in the synthesis of 1*H*-cyclopent[*c,d*]indene (**25**) and of its 12π-anion **26** from the diazoketone **27** via the ring-contracted acid **28**. No trace of tautomers **24**, R = H or **29** was observed.

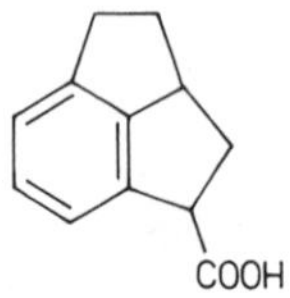

    28 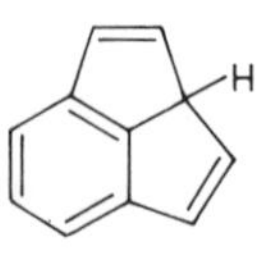   **29** 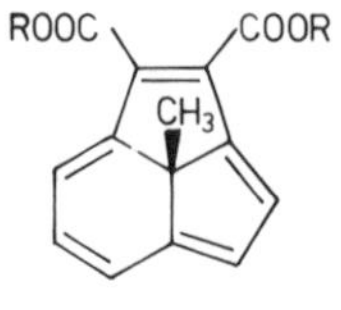   **30** *a*,R=CH$_3$ **31**
                                                    b,R=H

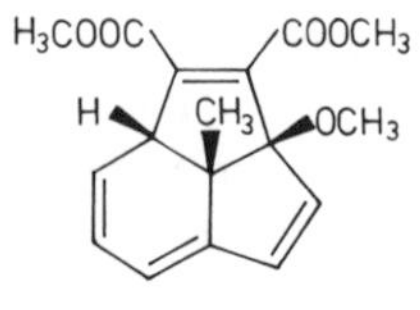

The first derivative of **24** was the diacid **30b** obtained by Rees[23] through methanol elimination from the cycloadduct **31** of 3-methoxy-3a-methyl-3a*H*-indene (**32**)[24] with dimethylacetylene dicarboxylate. 7-Methyl-7b*H*-cyclopent[*c,d*]indene (**24**; R = CH$_3$) was obtained[25] by bisdecarbonylation (in the presence of rhodium(I) complexes) of the dialdehyde corresponding to **30**. An improved synthesis of **24**, R = CH$_3$, suitable for large-scale preparation includes:[26] (1) cycloaddition of **32** with 2-chloroacryloyl chloride affording **33**, (2) conversion of **33** into ketone **34**, (3) introduction of a new double bond through Bamford-Stevens reaction, and finally, (4) methanol elimination from **35**.

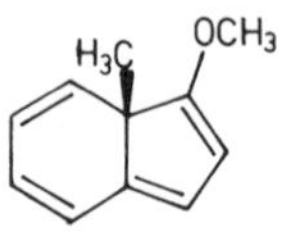

    32 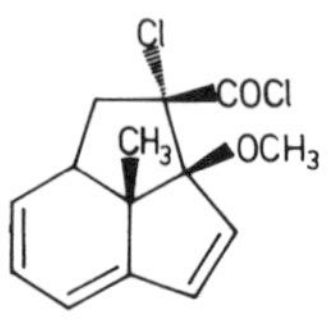   **33** 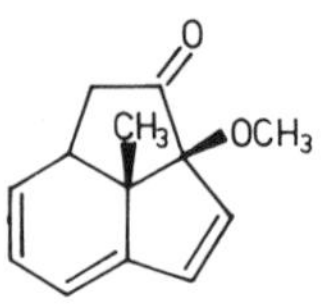   **34** 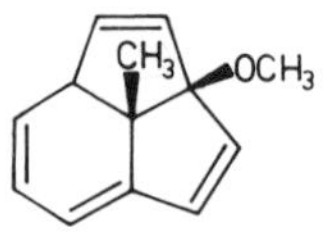   **35**

Another versatile synthesis of **24**, R = CH$_3$, which avoids the cycloadditions of labile **32** was announced by the same group.[27] The dienone **36** (obtained from 7-methoxyindan-1-one) is phenylselenated, then oxidized, and the formed trienone converted to **37** with lithium methyl vinyl ether; methylation of **37** is followed by hydrolysis and cyclization to **38** (intramolecular aldol condensation); reduction of **38** to the epimeric mixture of alcohols and dehydration of this mixture affords **39**; the final elimination of methanol from **39** yielding **24**, R = CH$_3$ is catalyzed by *p*-toluenesulfonic acid.

38 39

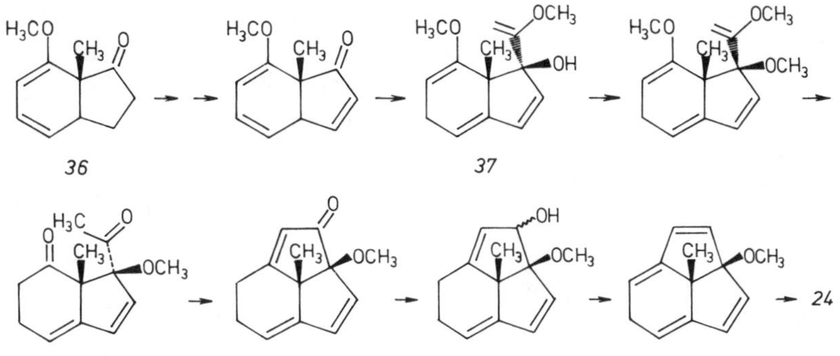

Another route to **24**, $R=CH_3$, involving the single-pot conversion of ketone **34**[26] includes treatment with trimethylchlorosilane and sodium iodide; subsequent reduction of the formed **40a** to **40b** and dehydration afforded **24**, $R=CH_3$.[28]

40 *41* *42* *43*

a, R,R′=O
b, R=H; R′=OH

a, R=COOMe; R′=Me
b, R=CHO ; R′=Me
c, R=CHO; R′=H
d, R=H ; R′=Me

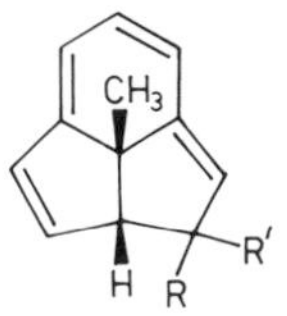

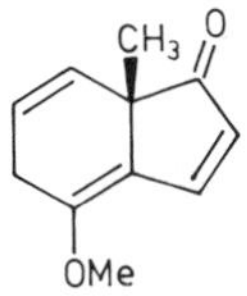

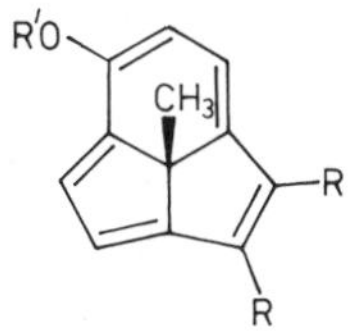

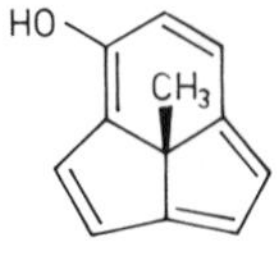

Recently, a 5-hydroxyderivative of **24**, $R=CH_3$ was synthesized, also by Rees and co-workers,[29] in a multistep reaction sequence including as essential stages: (1) conversion of 4-methoxyindanone to the trienone **41**,[24,27] (five steps); (2) formation of the tricyclic skeleton of **42a** (three steps) by means of dimethyl acetylenedicarboxylate; (3) conversion into the dialdehyde **42b** then demethylation to **42c**; (4) decarbonylation (Rh-(I) complexes) to **42d**; (5) formation of **43** via a bromoderivative on treatment with BBr_3 then diazabicycloundecene.

From the cycloadduct **44**[24,27] (obtained from 6-methoxyindanone) the substituted hydroxy-derivative **45** was obtained[29] (along with two other products) on H_2SO_4 treatment.

44 *45* *46*

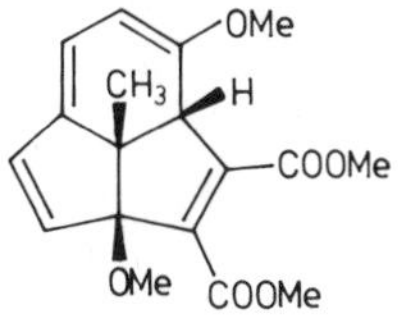

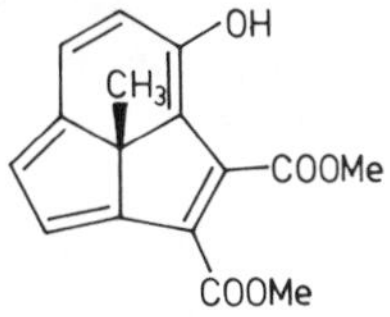

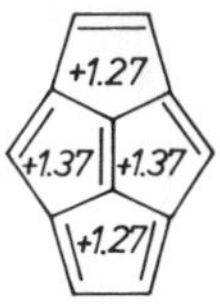

4. The Pyranene Problem

The intriguing pyranene molecule **46**, which possesses a periphery of [10]annulene and contains only five-membered rings, is of interest to theoretical and synthetic organic chemists. First calculations (going back to 1954[30] and 1960[31]) suggested a high degree of thermodynamic stability (RE of 4.19β) as well as the inclusion of the central bond in resonance stabilization. Calculation of ring currents[32] (Pople method) suggested diamagnetism and aromaticity (the values of ring currents in comparison with benzene are noted in the formula). However, different calculations[33] showed that pyranene cannot be aromatic due to: (1) steric strain, (2) high values of free valences in all positions, and (3) the presence of an unoccupied bonding orbital. Thus, the P index of **46** is zero,[33] the calculated resonance energy per π-electron (REPE) is -0.036β,[34] the topological resonance energy (TRE-PE) is -0.045β,[35] and the resonance energy (from theory of conjugated circuits) is -1.075 eV.[36]

On the basis of the above data, it does not appear strange that, in spite of optimistic suggestions for the synthesis of **46**,[37] all attempts to obtain pyranene **46**, e.g., from bicyclo[3.3.0]octane derivatives,[31] have failed.

B. Structure and Physical Characterization

1. 1,6-Methano[10]Annulene

Concerning the structure of 1,6-methano[10]annulene, three hypotheses should be taken into consideration: (1) an annulenic structure with five fixed double bonds, (2) a system with fluctuating double bonds **3** ⇄ **47** (directly or through **4**), and (3) a delocalized 10 π-system with **3** and **47** as mesomeric structures of methano[10]annulene **48**.

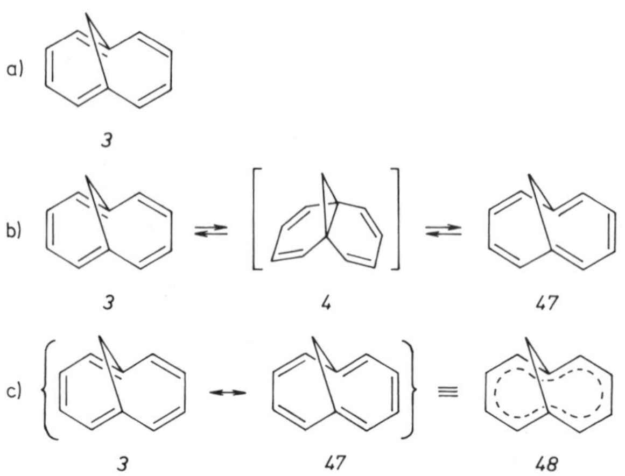

This "fascinating and very subtle structural problem"[38] was elucidated using the whole range of physical methods (see also Table 1).

a. Results of Calculations

The first calculations[39] indicated a resonance energy of 0.869 eV (PPP) and 0.830 eV (SPO); these values are close to that of cyclodecapentaene but much smaller than for naphthalene. Recent determinations[39b] of resonance energies for **3** (17.2 kcal · mol^{-1}) and **12** (6.5 kcal · mol^{-1}) present evidence for the differences of planarity as well as for favorable C-1—C-6 transanular interactions (homoconjugation) in **3**. The aromaticity of **3** results also from a topological approach.[39c]

The geometry of **3** was calculated by π-SCF-MO and force field methods.[40-43] *Ab initio* and semiempirical MO calculations of charge distribution and bond orders[44] are in agreement with observed chemical reactivity. After calculations using the geometry of the crystallographic data, **3** possesses a low or medium aromaticity[44] in opposition to ring current results.

The stabilities of the annulenic and norcaradienic forms in some 11-substituted derivatives of **3** were calculated by the EHMO method[45] and were found to be in agreement with experiments in all cases except the parent hydrocarbon. Recent empirical force field calculations[43] confirmed the annulenic structure with C_{2v} symmetry. The bond lengths indicate the aromatic character of the perimeter.[43] The calculated strain of **3** is 30.72 kcal/mol^{-1}.[43]

Table 1
PHYSICAL CHARACTERIZATION OF SOME CARBON-BRIDGED [10]ANNULENES

Compound	Mp or bp (°C)/pressure (Torr)(Color)	¹H-NMR and/or ¹³C-NMR spectra (δ = ppm, J Hz)	Electronic spectra λ_{max} (nm)	Ref.
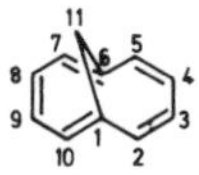 **3**	28—29 (Colorless)	¹H-NMR: -0.5(s; CH$_2$); 6.8—7.5(m; 8H; A$_4$B$_4$); ¹³C-NMR: 38.4(C-11); 126.1(C-3,4,8,9); 128.7(C-2,5,7,10); 114.6(C-1,6); J_{13C-H} = 142 (CH$_2$) (in CS$_2$/d$_6$-acetone)	256; 259; 298; 350—400	2 46 47 52
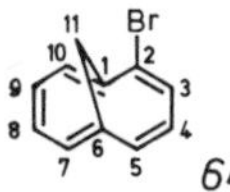 **64**	87—88/0.05 (Yellow)	-0.5(CH$_2$; AB); 6.6—7.9(m; 7H aryl)	244; 266; 313; 395; 405; 415	46 73
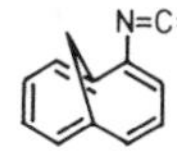 **86c**	69—72/0.01 (Yellow)	-0.6(CH$_2$; AB); 6.5—7.8(m; 7H aryl) (in CCl$_4$)		94
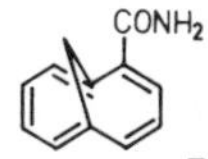 **73**	108—109	-0.41(s; CH$_2$); 6.38(s; NH$_2$); 6.9—7.7(m; 6H aryl); 7.85(m; 1H-α to carbonyl) (in CDCl$_3$)		82
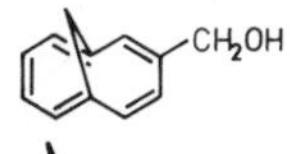	125—127/0.01	-0.42(s; bridge CH$_2$); 2.78(s; OH); 4.67(s; CH$_2$O); 6.9—7.5(m; 7H aryl (in CDCl$_3$)	259; 298; 363; 371; 380; 402; (in dioxane)	108
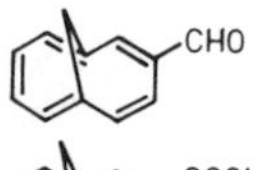	117—119/0.05 (Yellow)	-0.25(s; CH$_2$); 6.95—7.85(m; 6H aryl); 7.98(s; H-2); 9.97(s; CHO) (in CDCl$_3$)	272; 321; 387; 398; 407 (in dioxane)	108
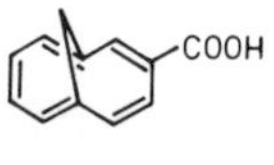	148—149 (Yellow)	-0.22(s; CH$_2$); 6.88—7.63(m; 5H aryl); 7.99(d; H-4); 8.37(s; H-2); 12.6(s; COOH) (in CDCl$_3$)	266; 313; 380; 391; 401; 410; (in dioxane)	108
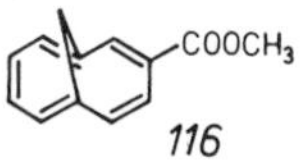 **116**	36—37 (Yellow)	-0.35(s; CH$_2$); 3.84(s; CH$_3$); 6.73—7.50(m; 5H aryl); 7.82(d; H-4); 8.17(s; H-2) (in CCl$_4$)	265; 312; 370; 380; 391; 401; 411 (in cyclohexane)	108
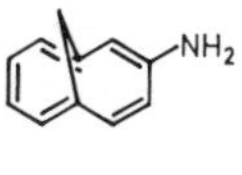	114—115/0.03 (Yellow)	-0.45(d; J = 9) and -0.27(d; J = 9) −CH$_2$; 3.50(s; NH$_2$); 6.16(d; H-4); 6.29(s; H-2); 6.55—7.50 (m; 5H aryl) (in CCl$_4$)	272; 326; 398; 403; 414; 426; (in cyclohexane)	108
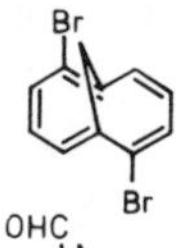 **60**	87 (Yellow)	-0.3(s; CH$_2$); 6.8—7.9 (6H-ABC)	234; 277; 327; 419	46 73
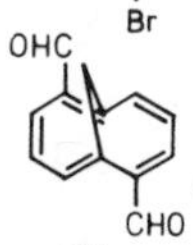 **82**	107 (Yellow)	-0.13(s; CH$_2$); 7.30—8.60(m; H-3-5; 8-10; AB); 10.15(s; CHO) (in CDCl$_3$)	272; 356; 404; (in CH$_2$Cl$_2$)	86
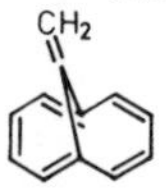	85—86 (Pale yellow)	3.2(s; CH$_2$); 7.2(8H aryl; two identical A$_2$B$_2$)	258; 300; 360-410 (in cyclohexane)	102
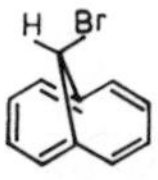 **97**	122—123 (Yellow)	1.9(s; bridge H); 6.7—7.4(m; 8H aryl)	258; 295; 386; 395; 405 (in cyclohexane)	101
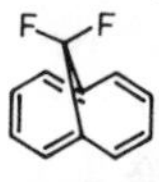 **105** X=F	122—123 (Yellow)	¹H-NMR: 6.98(H-3,4,8,9) and 7.17(H-2,5,7,10)-AA'BB' (in CCl$_4$); ¹⁹F-NMR: 126.2(CCl$_3$F-external standard)	253; 293; 380; 389; 398; 409; (in cyclohexane)	104

Table 1 (continued)

PHYSICAL CHARACTERIZATION OF SOME CARBON-BRIDGED [10]ANNULENES

Compound	Mp or bp (°C)/pressure (Torr)(Color)	¹H-NMR and/or ¹³C-NMR spectra (δ = ppm, JHz)	Electronic spectra λ_{max} (nm)	Ref.
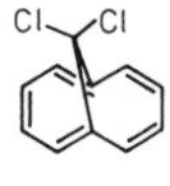 **105** X=Cl	104—107 (Yellow)	6.94(8H)-AA′BB′ (in CDCl₃)	256; 280; 388 (in cyclohexane)	104
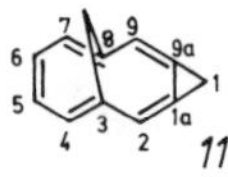 **112**	65—67/0.1 (Yellow)	−0.82 and 0.02 (bridge CH₂; AB); 2.40 and 3.54(cyclopropanic H; AB); 7.1(H-5,6) and 7.3(H-4,7)-AA′BB′ 7.4(s; H-2,9) (in CCl₄)	268; 302; 377; 387; 396; 406; (in dioxane)	108
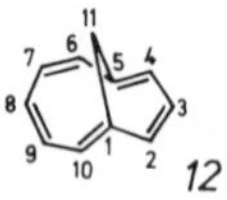 **12**	Thermally unstable (orange)	¹H-NMR: −1.23(dt; H-11); −0.74(d; H-11); 6.85(t; 1H); 7.23—7.4(m; 3H); 7.53(d; 2H); 8.12(d; 2H); ¹³C-NMR; 34.5; 124.7; 128.1; 129.5; 132.8; 143.3; 160.4 (in d₆-acetone)	279; 482 (in hexane)	16 17
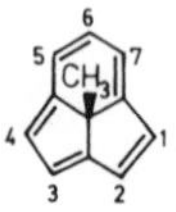 **24** (R=CH₃)	8.5—11 (Yellow)	−1.67(CH₃); 7.53—7.83(H-5,6,7; AB₂); 7.89—7.92(H-1-4; AB) (in CDCl₃)	249; 282; 335; 398; 439; 450; (in ethanol)	25
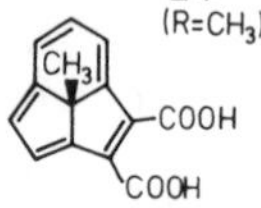 **30b**	195—197 (Orange)	−1.25(CH₃); 7.7—8.5(5H aryl) (in CDCl₃)	480	23
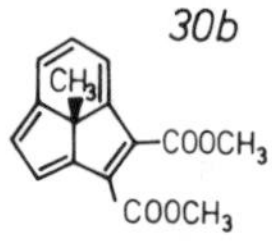 **30a**	Oil (Yellow)	−1.34(CH₃); 3.99 and 4.04 (s; ester CH₃); 7.68(H-5,6); 8.06 (H-7); 8.08(H-4); 8.22(H-3) (in CDCl₃)	217; 262; 305; 336; 471 (in ethanol)	23

b. UV Spectrum

The UV spectrum of **3**[2,46,47] (Table 1) is characteristic for a conjugated aromatic system and excludes the structure **4**; the spectrum could be considered as a bathochromically shifted spectrum of naphthalene. A more recent analysis of electronic spectra of **3**[48] and of its higher 14π and 18π homologues has shown important transanular interactions between bridgehead carbon atoms.

c. Photoelectron Spectrum

The photoelectron (PE) spectrum of **3** resembles that of naphthalene.[49a] The utility of PE spectroscopy in structure determination of bridged annulenes was reviewed.[49b]

d. MCD Studies

Though the absolute polarizations cannot be derived from a simple qualitative analysis, nevertheless, Magnetic Circular Dichroism (MCD) proved to be a suitable tool for differentiating various 1,6-methano[10]annulenes.[50] In a recent detailed MCD study[51] of 2- and 3-substituted 1,6-methano[10]annulenes it was shown that the B terms of the L_b transitions are strongly dependent on the electronic effect of substituents. The MCD spectroscopy was used on this occasion for the first time to establish the MO sequence in an aromatic perimeter; from this orbital ordering, a strong C-1 to C-6 transanular interaction in **3** was assessed.

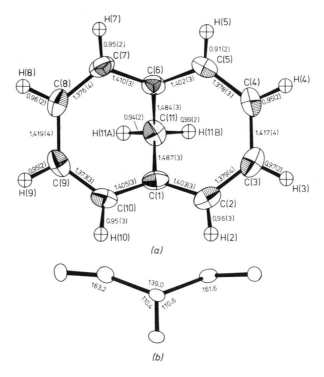

FIGURE 1.   X-ray data for 1,6-methano[10]annulene (**3**): (a) numbering scheme and bond lengths (Å); (b) angles between least-squares planes. (From Bianchi, R., Pilati, T., and Simonetta, M., *Acta Cryst.*, B36, 3146, 1980. With permission.)

e. NMR Spectra

The ^{1}H-NMR spectrum of **3** consists of an A_4B_4 system (8H) in the range $\delta = 6.8$ to 7.5 ppm and a sharp singlet at $\delta = -0.5$ ppm (2H of CH$_2$). The downfield position of the olefinic protons and the considerable shielding of the methylenic protons are clear proofs of diamagnetic ring current and implicitly of the structure with delocalized π-electrons. In agreement with such a structure, the ^{1}H-NMR[46] and ^{13}C-NMR spectra[52a] are temperature independent. It could also be added that the $J_{{}^{13}C-H}$ coupling constant of 142 Hz for the CH$_2$ protons rules out the norcaradienic structure **4** (for which a coupling constant of about 160 Hz was expected).[46] The utility of ^{13}C-NMR in assigning either a propellane structure or the corresponding bridged annulenic one was reviewed.[52b] Recently quantitative correlations of ^{13}C-NMR shifts with π-electron densities were reported.[52c]

f. X-Ray Analysis

The first X-ray analysis, performed to 1,6-methano[10]annulene-2-carboxylic acid[53] indicated a bridge angle of 99.6° and a C_{10} perimeter which is almost planar. The bond lengths (1.38 to 1.42 Å) are typically aromatic and lie in a more narrow interval than for naphthalene (1.36 to 1.42 Å). More recently, the X-ray analysis of unsubstituted **3** has become available.[54] From these data (see Figure 1) the aromatic structure of **3** is undoubtedly proved. From some crystal structure data[55] of derivatives with skeleton of **3** the existence of C-1 to C-6 interactions was considered.

g. Electron Diffraction

From an electron diffraction study[56] the C-1 to C-6 distance in **3** was found to be 2.257 Å, suggesting the absence of significant direct electronic interactions between these carbon atoms.

h. ESR Spectra

The ESR spectra of radical anions of **3** and of deuterated derivatives of **3** indicated[57] spin populations in agreement with predictions made on the basis of the HMO model, including the electron-repelling effect of the bridge. ESR spectra of radical anions of **3** have a marked temperature dependence[58] which was attributed to geometry changes.

i. Dipole Moments

The dipole moment of **3** is 0.39 D[59] and is oriented from the ring towards the methylenic group. Contributions of about 0.8 D of the aromatic ring and of 0.4 D of the CH_2 group to the total dipole moment was estimated.

The "enthalpy of formation", determined by combustion calorimetry, is $\Delta H_f^{298} = 75.2 \pm 0.6$ kcal $\cdot$ mol^{-1}.[60]

Some "mass spectral" characteristics of **3** were also reported.[61]

j. Valence Tautomerism

The results of *ab initio* calculations have shown that for 1,6-methano[10]annulene the annulenic form **3** is clearly preferred over the norcaradienic (propellanic) one **4**.

3, R=R'=H ; 49, R=R'=CH₃   4, R=R'=H ; 50, R=R'=CH₃
51, R=CH₃ ; R'=CN   52, R=CH₃; R'=CN ; 53, R=R'=CN

The energy difference between **3** and **4** was calculated to be between 6.24 and 15.6 kcal $\cdot$ mol^{-1} [62] or 4.5 kcal $\cdot$ mol^{-1},[63] thus similar to that for cycloheptatriene and norcaradiene. From these data a concentration of about 0.05% **4** results at room temperature, the valence tautomerism of **3** being characterized by an asymmetric double-well potential.[63] A mapping of the reaction path for conversion of 1,6-methano[10]annulenes into their cyclized propellanic forms using X-ray data was performed by Dunitz et al.[55] underlining the essential role of the distances between bridgehead atoms 1 and 6.

According to theoretical and experimental investigations, bridge substituents attached to C-11 can shift the valence tautomeric equilibrium to the side of the norcaradienic isomer. Thus, calculations suggested that in the case of dimethyl compounds the norcaradienic form **50** might be more stable than **49**.[43,45,62] However, ^{13}C-NMR data show that **49** is slightly more stable than **50** ($\Delta H^\circ = 0.2$ kcal $\cdot$ mol^{-1} in favor of **49**) but the bis-norcaradiene **52** is clearly more stable than its valence tautomer **51**.[52] X-ray analyses confirmed the propellanic forms (in the crystal) for dimethyl-**50**[64] and for methyl-cyano-substituted derivatives-**52**.[65,66] Recently[67] the 11,11-dicyanoderivative **53** was obtained from 11-cyano-1,6-methano[10]annulene; ^{13}C-NMR, ^{1}H-NMR, UV-spectra, as well as X-ray analysis showed that **53** possesses (in solution and in solid state) exclusively the propellanic structure[67] in agreement with results of calculations.[62a]

The above-described equilibria, probably the first valence tautomerism between an aromatic compound and an olefin, are determined by the shorter 1,6-distance in **3** and by the interaction of Walsh orbitals of the cyclopropanic ring with the π-orbitals of butadiene moieties in **4**; the bridge substituents influence considerably these factors.

2. 1,5-Methano[10]Annulene (Homoazulene)

The electronic spectrum of **12** ($\lambda_{max} = 482$ nm) is very similar to that of azulene.[16] The bathochromic shift relative to 1,6-methano[10]annulene finds an analogy in the comparison of the spectra of azulene and naphthalene.[15] The ^{1}H-NMR spectrum (CDCl$_3$, δ ppm): 6.8 to 8.1 (8H arom) and −0.7 and −1.2(CH$_2$)[16] indicates a diamagnetic ring current similar to 1,6-methano[10]annulene. These data contradict the earlier prediction[40] that **12** will exhibit bond alternation (due to the important deviation from coplanarity) and lack of aromaticity.

In order to emphasize the similarity between **12** and azulene the name ''homoazulene'' was proposed by Scott[16] for **12**.

3. Derivatives of 7bH-cyclopent[c,d]Indene

The ^{1}H-NMR spectrum of 7b-methyl-7b*H*-cyclopent[*c,d*] indene (**24**,R = CH$_3$)(δ ppm): −1.67(CH$_3$); 7.53 to 7.83 (AB$_2$ System, H-5,6,7); and 7.89 to 7.92 (AB system, H-1-4) indicates the existence of a diamagnetic ring current.[24] In a MNDO-SCF-MO study[68] the aromatic structure of **24** (R = CH$_3$; R′ = H) with C$_s$ symmetry was found to be slightly lower in energy than the bond alternating form. The aromaticity of the diacid **30b** was also proved by its ^{1}H-NMR spectrum[22] as well as by X-ray analysis (perimeter bonds of 1.35 to 1.44 Å).[22]

24 , R=CH$_3$ (R′=H)
30b, R=CH$_3$ (R′=COOH)

C. Chemical Properties
1. 1,6-Methano[10]Annulene — Reactions and Derivatives
a. Thermal Behavior

1,6-Methano[10]annulene (**3**) is an unusually stable hydrocarbon showing no tendency to polymerization and remaining unchanged on prolonged heating to 150 to 200°C.[69] On flash pyrolysis at 500°C, **3** undergoes a complete isomerization to 7*H*-benzocycloheptene (**54**), proceeding probably through a Berson-Willcott rearrangement.[38,69]

3   4   54

b. Diene Syntheses

The Diels-Alder reaction of **3** with maleic anhydride proceeds with difficulty, affording at 150°C a 1:1 adduct **55** of the norcaradienic form;[69] structure **55** was proved by X-ray analysis.[70]

55   56, R=CH$_3$; C$_6$H$_5$ 57, R=CH$_3$; C$_6$H$_5$

Diene reaction of **3** with 4-substituted-1,2,4-triazoline-3,5-diones affords mono- (**56**) and bis-adducts (**57**)[71a] through *anti*-attack of the dienophile. Similar adducts were obtained from 11-substituted-1,6-methano[10]annulenes.[71b] As Ginsburg noted:[72] "it is more usual to obtain a *bis*-adduct than a mono-adduct for the wise molecules of Köln are aware of their aromaticity and do not want to lose it."

c. Aromatic Substitutions

From the numerous characteristic aromatic substitutions of **3** the bromination was thoroughly investigated. Bromination with 1 mol of bromine at 0°C or with *N*-bromosuccinimide in CH_2Cl_2 yielded regioselectively a monobromo-derivative **64** via the 1,4-addition product **58**.[46,73] At $-75°C$ with 2 mol of bromine the tetrabromo-adduct **59** with cyclopropanic ring is obtained;[73,74] this gives rise to dibromo-derivatives **60** and **61** on treatment with potassium *t*-butoxide.[73]

Scheme 1

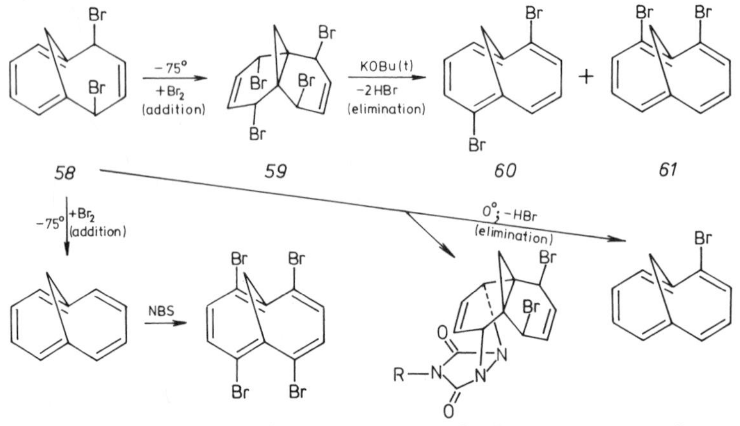

The dibromo-derivative **60** could also be obtained from **3** with two equivalents of *N*-bromosuccinimide whereas on treatment with 5.5 equivalents of *N*-bromosuccinimide the tetrabromo compound **62** was obtained.[75]

The structures of bromination products were proved spectrally[46,73] through X-ray analysis (for **60**),[46] as well as chemically, e.g., conversion of **61** into the corresponding di-acid, see below[73] or trapping of **58** as adduct **63**; X-ray analysis indicated[9,74] that the attack of bromine on **3** occurs from the more hindered *syn*-position (in contrast to additions of dienophiles). Formation of **58** from **3** was recently discussed in terms of MO theory.[76]

A mixture of three tetrahalo-adducts of the type **59** was obtained on bromination of **3** with bromine in the presence of triethylbenzylammonium chloride,[77] suggesting that **58** and **59** are formed through intermediate σ-complexes and not by synchronous addition of bromine.[74]

Other substitution reactions of **3** are (1) acetylation with acetic anhydride and $SnCl_4$ affording **65**;[46] (2) nitration with cupric nitrate in acetic anhydride affording a mixture of 2-nitroderivative (**66**), 3-nitroderivative, and 2,7-dinitroderivative;[46,77] (3) nitration with collidine/nitronium trifluoromethanesulfonate affording exclusively **66**;[77] (4) sulfonation with 0.9 equivalents of SO_3 affording exclusively **67** or with four equivalents of SO_3 affording only **68**;[78] and (5) protiodetritiation which was studied kinetically.[79]

2-Substituted 1,6-methano[10]annulenes can be further substituted, e.g., by bromination with *N*-bromosuccinimide.[80] Similarly, sulfonation of the 2-methyl-derivative of **3** affords **69** whereas that of **70** affords, along with the 5-perisubstituted compound, about 35% **71** formed through *ipso*-attack.[78]

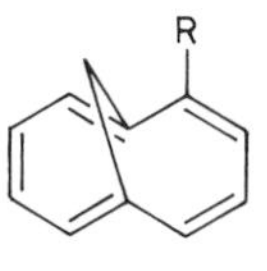

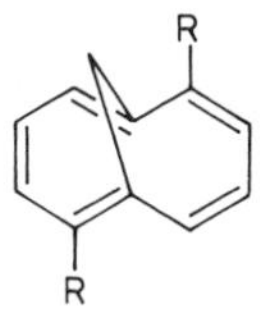

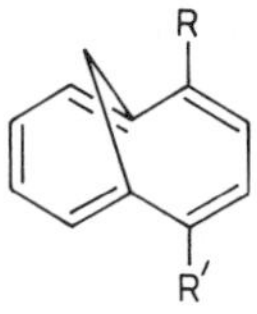

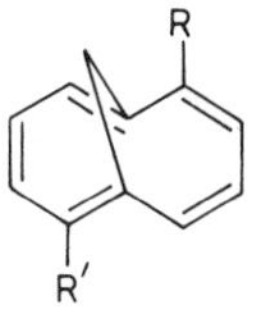

65, R=$COCH_3$; *72*, R=COOH
66, R=NO_2 ; *73*, R=$CONH_2$
67, R=SO_3H ; *74*, R=CN

68, R=SO_3H
70, R=CH_3

69, R=CH_3
R′=SO_3H

71, R=CH_3
R′=SO_3H

d. Derivatives

In Table 1 some physical characteristics of bridged [10]annulenes are collected. The above-mentioned bromo-derivatives are important key intermediates for the syntheses of other substituted 1,6-methano[10]annulenes. Thus, the bromo-derivative **64** could be easily converted, via the Grignard reagent, into the carboxylic acid **72**.[46] The same acid results on oxidation of **65**.[81] From this acid (as well as from **3** with chlorosulfonylisocyanate followed by hydrolysis) the amide **73** then the nitrile **74** could be obtained.[82] Starting from the same acid **72** the homologous acid **75** could be obtained through usual reactions[83,84] (Table 1).

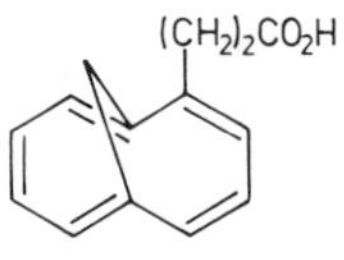

75

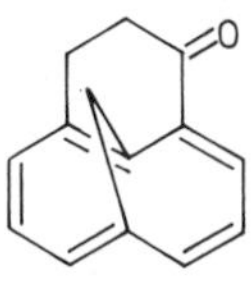

76

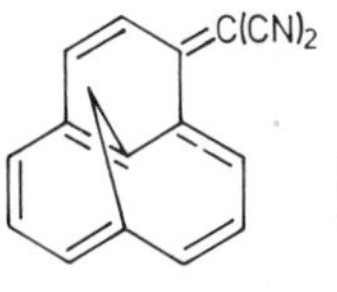

77

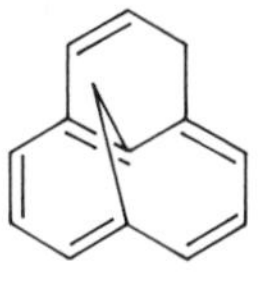

78

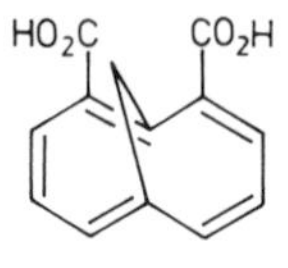

79

Interesting tricyclic derivatives, e.g., **77** (a dicyano-fulvene with a homophenalene skeleton) and **78** (homophenalene) could be obtained from **75** via the ketone **76**.[83,84] On the other hand, the dibromides **60** and **61** were converted into the corresponding dicarboxylic acids; from the acid **79** an intramolecular anhydride could be obtained.[69,81] From dibromide **60** annulenophanes of the type **80** were obtained,[85] whereas from dibromide **61** the disulfide **81a** and diselenide **81b** were synthesized.[9] Charge transfer complexes of **81** with 7,7,8,8-tetracyanoquinodimethane have interesting electrical properties.[9]

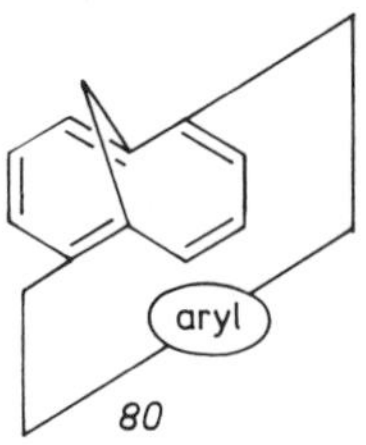

80

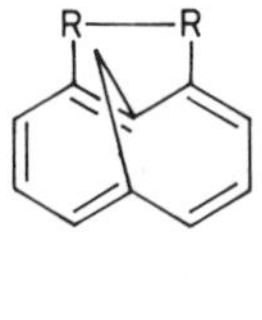

81, *a*, R=S
b, R=Se

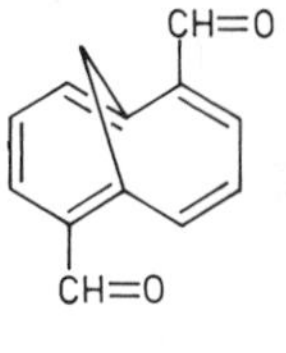

82

Another key intermediate in the syntheses of various 1,6-methano[10]annulene derivatives is the 2,7-diformyl derivative **82** obtained also from dibromide **60**.[86]

Numerous derivatives were obtained by Neidlein and co-workers starting from **82**, e.g., **83** (tetrathiofulvalene homologues),[86-88] **84**,[83] and **85**.[89-92] Some of these compounds, e.g., **83b**, **85a**, and **85c** are push-pull (or vinylogous push-pull) stabilized 1,6-methano[10]annulenes, presenting interesting electronic spectra.

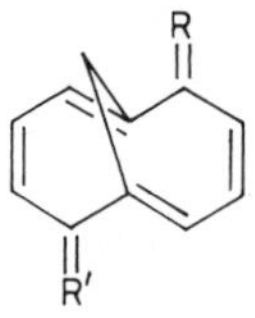

83

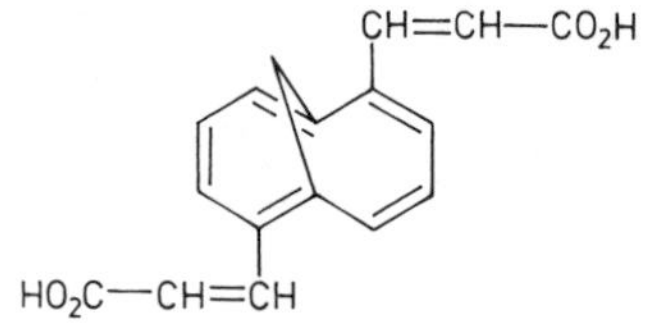

84

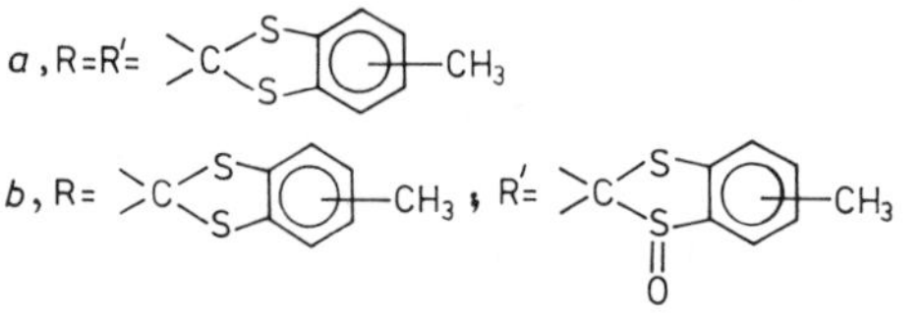

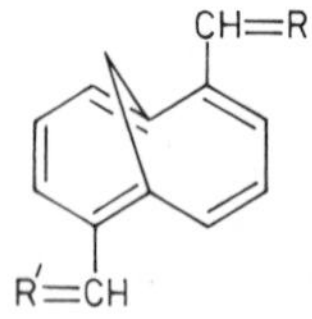

a, R=R'= (structure) —CH₃

b, R= (structure) —CH₃ , R'= (structure) —CH₃

a, R=R'= (structure) —CH₃ or (structure)

b, R=R'= (structure)

c, R= (structure) ; R'= (structure) —CH₃

85

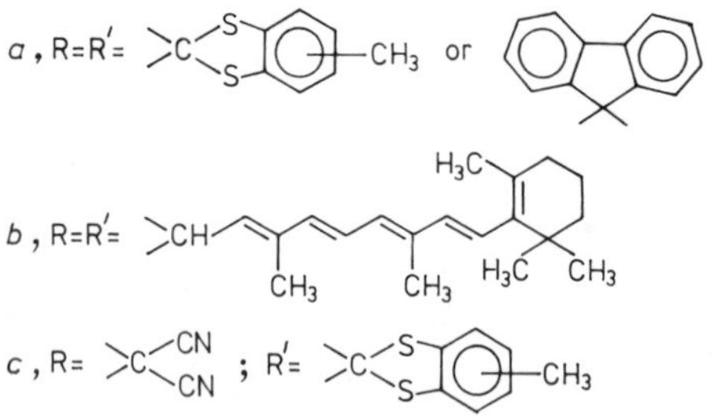

1,6-Methano[10]annulene (**3**) is easily deuterated to the 2,5,7,10-tetradeuterioderivative.[69] Together with other reactions described above, this isotopic exchange indicates the preferential substitution of **3** in the 2-position (corresponding to the α-position of the naphthalene moiety), in agreement with theoretical calculations.[39] ^{13}C-NMR data indicate a remarkable similarity with naphthalene for blends of inductive and resonance effects.[93]

An earlier synthesis of 3-substituted-1,6-methano[10]annulene involved the reaction of **64** with lithium piperidide[94] when a 1:1 mixture of 2- and 3-piperidinoderivatives was obtained, presumably through the intermediate aryne (for other methods of generation of 3-substituted derivatives, see later).

Among the derivatives of **3** which could not be obtained directly through electrophilic substitution the 2-amino- and 2-hydroxy-derivatives (α-naphthylamine and α-naphthol analogs) deserve special mention. The air-sensitive amine **86a**, the first amino-derivative of an annulene, with no trace of its imino tautomer, was obtained from a classical sequence through **86b**, **86c** including Curtius degradation.[94]

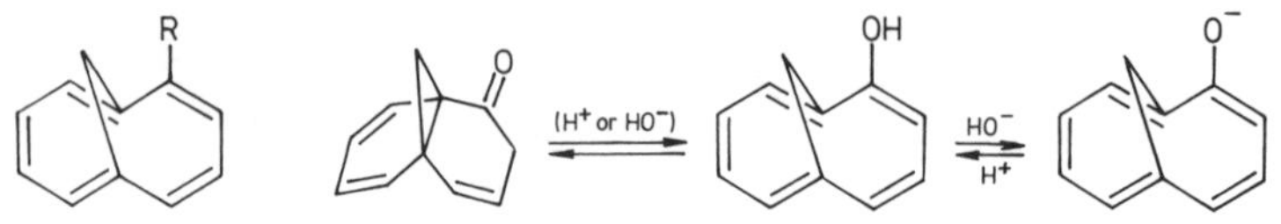

86   87   88   89

a, R = NH₂
b, R = COCl
c, R = N=C=O
d, R = OBu(t)

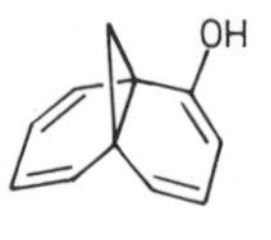

87A 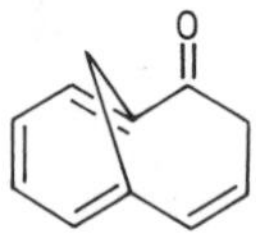   88A

Hydrolysis of the amine affords a mixture of ketone **87** and enol **88** from which the ketone could be crystalized.[94] The ^{1}H-NMR spectra indicated the norcaradienic (propellanic) structure for **87** and the annulenic one for **88**.[94] In principle, four tautomers and/or valence isomers are possible, but the propellanic enol **87A**, unlike **88**, does not benefit from any aromatic delocalization; hence is less stable than its propellanic ketonic tautomer **87**. On the other hand, because of this delocalization, the annulenic enol **88** is more stable than its annulenic ketonic tautomer **88A**. Both **87** and **88** are converted by alkali into the yellow enolate anion **89**.[94] Another route to the **87** ⇌ **88** mixture includes the conversion of **64** into **86d** via the corresponding Grignard reagent followed by acidolysis of the butyl ether.[95]

Starting from this keto-enol tautomeric mixture different quinomethides, e.g., **90**[96] (donor-acceptor stabilized derivatives of 1,6-methano[10]annulene) as well as push-pull stabilized quinodimethanes, e.g., **91**[96] were obtained.

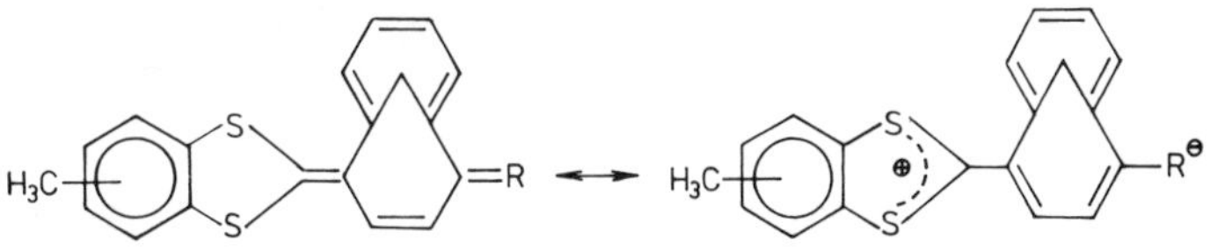

R = O   90   R⁻ = O⁻

R = C(CN)₂   91   R⁻ = C(CN)₂⁻

The ¹H-NMR spectra of these derivatives are interesting since the bridge CH₂ protons are a probe for the detection of the ring current while the CH₃ protons are a probe for the push-pull effect.

Interesting dimers, e.g., **94** were obtained (probably through carbene **93**) on pyrolysis of the sodium salt of tosylhydrazone **92**.[97,98] The aldehyde corresponding to **92** was obtained starting from the acid **72** (*vide supra*).

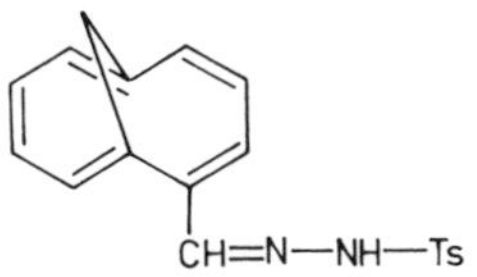

92 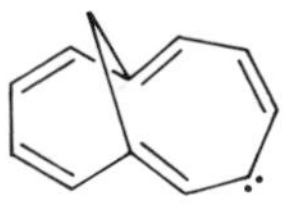   93 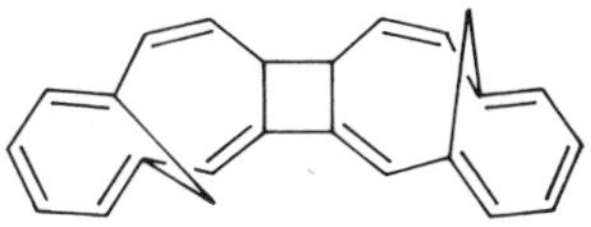   94

e. Metal Complexes

UV irradiation of a solution of **3** and hexacarbonylchromium in hexane yielded complex **95** characterized by ¹H-NMR spectrum.[99]

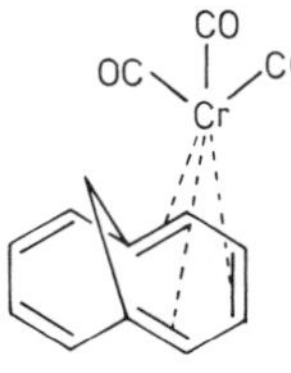

95 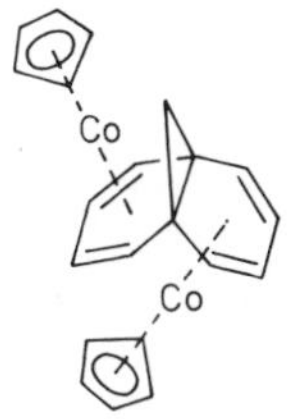   96a 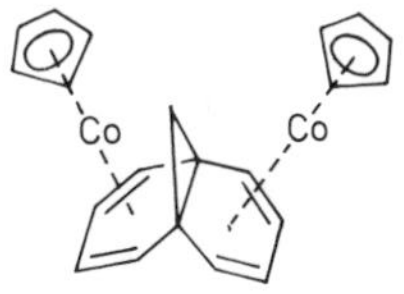   96b

More recently complexes **96a,b** were obtained on treatment of **3** with CpCo(C₂H₄)₂.[100]

f. Derivatives with Functionalized Bridge

The chemistry of the formally allylic CH₂ bridge group in **3** was thoroughly investigated. Because direct bridge substitution in **3** is not possible, the synthesis of such derivatives starts with substituted precursors. Thus, the 11-bromo-derivative **97** was obtained using the following sequence in which the aromatization is the last step.[101]

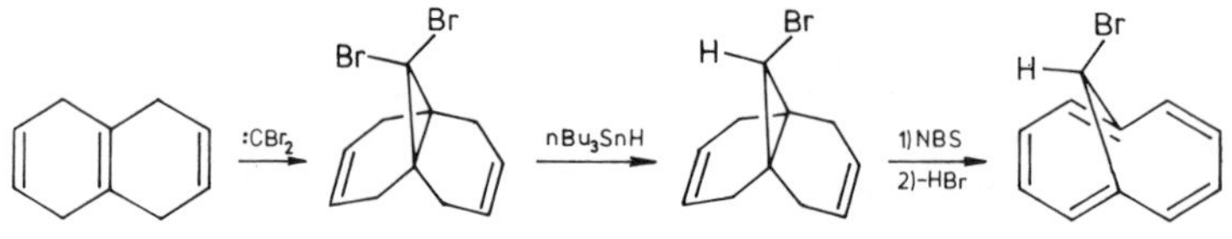

The versatility of **97** results from its different conversions.[38,69,102]

Scheme 2

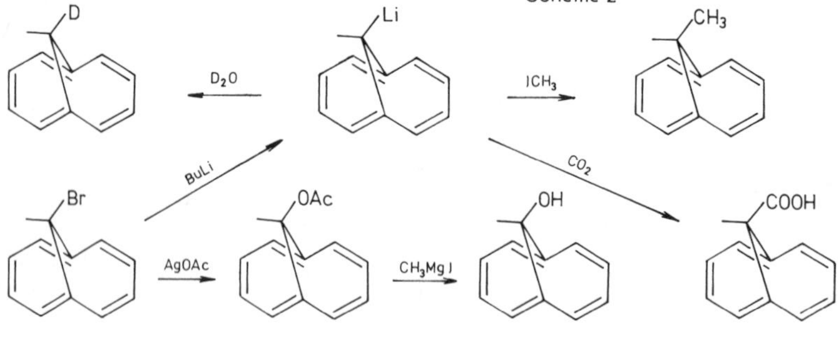

The study of bridge-substituted derivatives of **3** led to the idea of replacement of the CH_2 group by a $C=O$ or $C=CH_2$ group in order to produce an additional flattening of the annulenic perimeter and thus an increase of aromaticity. Initial classical attempts to oxidize alcohol **98** to the ketone **99** failed.[69,102] Ito's elegant synthesis of **99** through [6 + 4] cycloaddition of butadiene and tropone[103] led to the reinvestigation of this oxidation. It was found that oxidation of **98** with oxidizers free from metal ions (Pfitzner-Moffat reagent or Corey's DMSO complex) affords **99**.

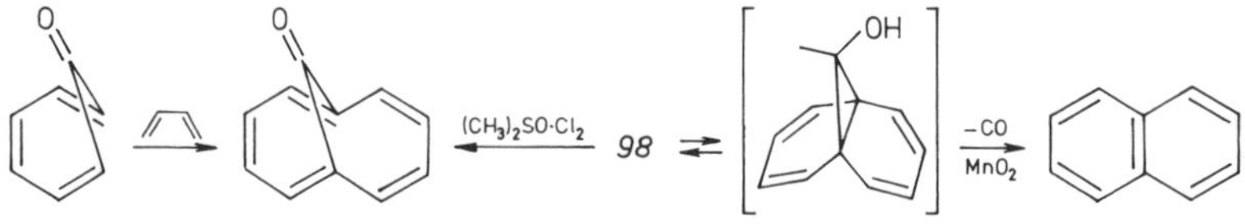

In the presence of metal ions the propellanic cyclopropanol valence tautomer of **98** undergoes cyclopropanic bond breaking followed by bridge carbon loss affording naphthalene in a reaction reminiscent of other cyclopropanols.

Contrary to expectations, the carbonyl group of **99** failed to react with nucleophiles.[9]

The dicyanoderivative **100** was obtained through olefination of the tetraenic ketone **101** followed by aromatization.[9] The 11-methylene derivative **102** was synthesized[102] by Hofmann degradation of the quaternary base **103**, obtained in turn from nitrile **104** after aromatization and modification of the cyano group.

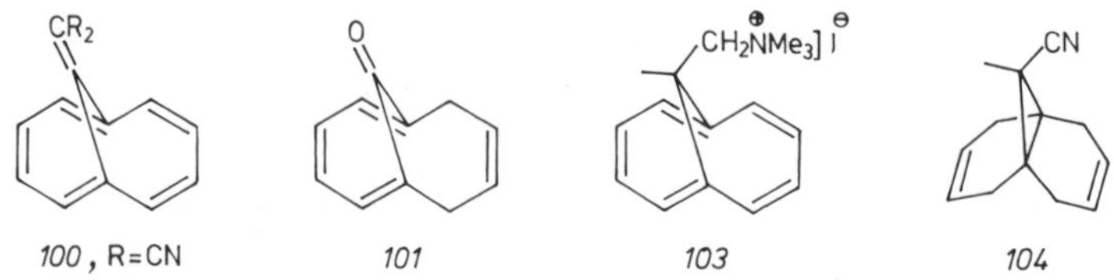

100 , R=CN
102 , R=H   101   103   104

The study of derivatives **99, 100** and **102** indicated that their geometry does not appreciably differ from that of **3**, and that the bridge and perimeter π-systems do not interact, most probably owing to geometric factors.[102]

Another interesting group of bridge-substituted derivatives is that of 11,11-dihalocompounds. The compounds **105** were obtained using the classical synthesis of 1,6-methano[10]annulene applied to dihalo-adduct of 1,4,5,8-tetrahydronaphthalene (**5**).[104]

5 105 X=F,Cl,Br 106

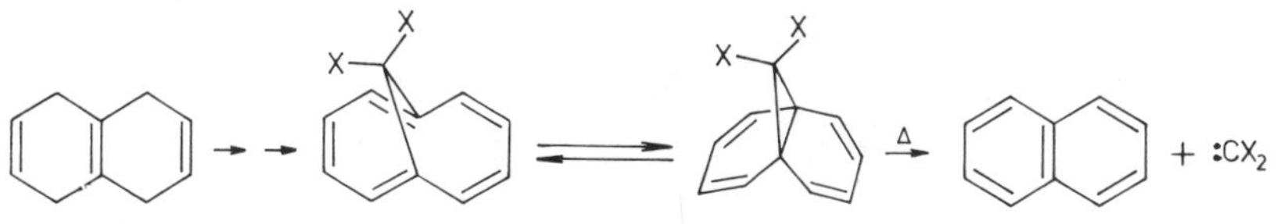

The yields in which the derivatives **105** were obtained are 90% for X = F, 5% for X = Cl, and 2% for X = Br[38,104] (for comparison the yield was 70% for the parent hydrocarbon **3**). Thus, it seems that the *gem*-difluorine atoms accentuate the aromaticity of **3**, compound **105**, X = F being "the most benzene-like of all the bridged [10]annulenes yet prepared".[38]

In striking contrast to the parent hydrocarbon (which affords thermally 7*H*-benzocycloheptene) the compounds **105** undergo thermal fragmentation, through the corresponding norcaradienic valence tautomers **106**, into naphthalene and dihalocarbenes (which can be trapped with olefins,[104] or detected by means of mass spectrometry.[61,105] The above fragmentations resulted in a practical method for structure determination of substituted 1,6-methano[10]annulenes. Thus, the structures of 2- or 3-nitroderivatives of **105**, X = F were proved by heating at 250°C and by identifying α- or β-nitronaphthalene, respectively.[38]

g. Formation of Benzocyclopropene and Related Compounds from 1,6-Methano[10]Annulene

Benzocyclopropene (**107**) was obtained in 1965[101] in the pyrolysis of Diels-Alder adduct **108** of **3** with methyl acetylenedicarboxylate. Interestingly, the aromatic bond common to both rings is a good C=C dienophile affording Diels-Alder adducts, e.g., **109** (with α-pyrone) which eliminate CO_2 regenerating 1,6-methano[10]annulene.[38]

107 108 109 110 111

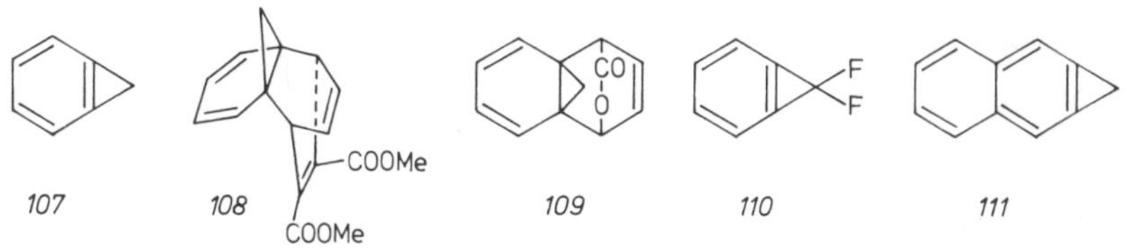

1,1-Difluorobenzocyclopropene (**110**), which was obtained analogously to **107**,[106] proved to be one of the most stable benzocyclopropenes yet synthesized. Another compound similarly synthesized is 1*H*-cyclopropa[*a*]naphthalene (**111**).[107]

112 113 114 115, R=CH_2OH
116, R=$COOCH_3$

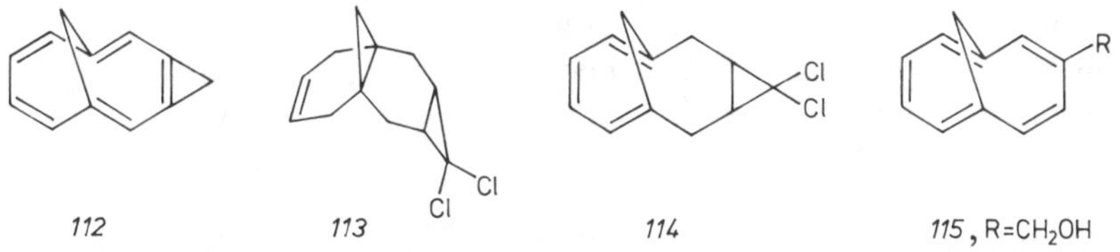

A related compound, 1*H*-3,8-methanocyclopropa[10]annulene (**112**) was obtained from the dichlorocarbene adduct **113** through bromine addition/HBr elimination conducting to **114** followed by treatment with six equivalents of potassium *t*-butoxide.[108] ^{1}H-NMR data show that the 10 π-electron system of **112** is not markedly different from that of the parent compound **3**.[108] Very interestingly, starting with **112** many 3-substituted-1,6-methano[10]annulenes became available[108] because, on hydrolysis with Ag[+] ions, **112** affords the alcohol **115** from which other compounds, e.g., **116** or CHO-, COOH-, NH₂-substituted derivatives could be obtained,[108] (Table 1).

In this connection it must be added here that numerous 3- and 3,4-substituted-1,6-methano[10]annulenes **117** to **119** could be recently synthesized[9,109] via the "cycloheptatriene-1,6-dicarbaldehyde route" (see the syntheses of **3**).

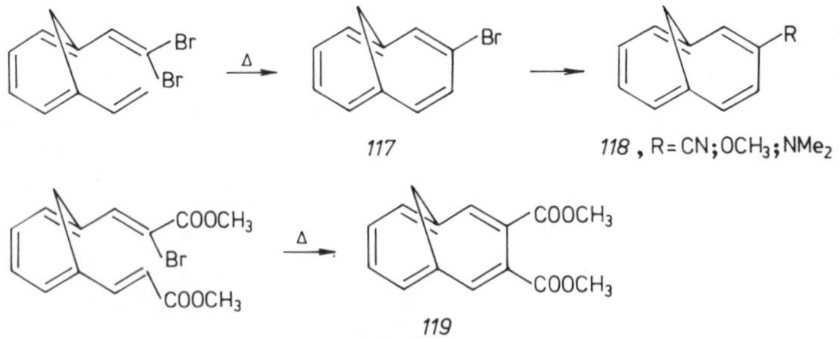

117   *118* , R= CN; OCH₃; NMe₂

119

The ester **116** as well as its 2-substituted isomer give mono- and bis-Diels-Alder reactions stereospecifically (*anti*-attack of the dienophile) but devoid of regioselectivity for the substituted or unsubstituted diene moieties.[110]

2. 1,5-Methano[10]Annulene (Homoazulene)

Homoazulene (**12**) presents a remarkable thermal stability since it remains unchanged on flash pyrolysis up to 300°C.[111] The pyrolysis at 435°C gives rise to five isomeric products, **120** to **124**.

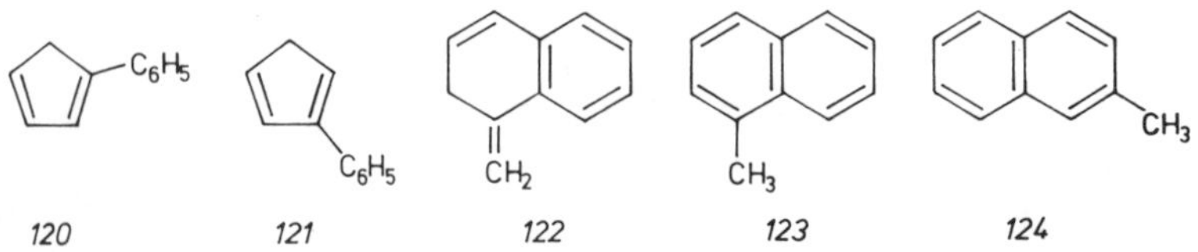

120   *121*   *122*   *123*   *124*

The suggested mechanism comprises:[111] (1) an unusual ground-state di-π-methane rearrangement to **125** (source of **120** and **121**) and (2) a parallel allowed electrocyclic process affording **126** then **127** (sources of **122** to **124**).

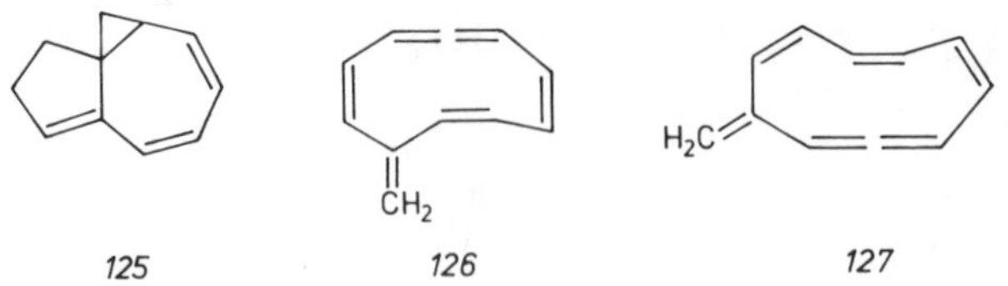

125   *126*   *127*

A chemical analogy of **12** with azulene was found in [2 + 2]cycloaddition reactions with tetracyanoethylene leading to the 1:1 adduct **128** and with dimethylacetylene dicarboxylate leading to the homoheptalene ([4*n*]annulene) derivative **130** formed probably via **129**.[112]

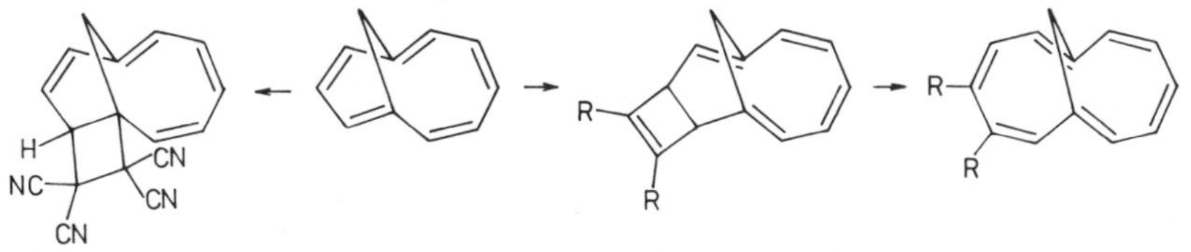

128	*12*	*129* R=COOCH₃	*130*

3. Derivatives of 7bH-Cyclopent[c,d]Indene

7b-Methyl-7bH-cyclopent[c,d]indene (**24**, R = CH₃) is an aromatic compound which presents characteristic aromatic substitutions. Whereas nitration (copper II nitrate/acetic anhydride) affords a mixture of all possible mononitroderivatives, acetylation and formylation (with dichloromethyl *n*-butyl ether and tin-IV chloride) are more selective[26] (75 and 93% 5-substituted products and mono-2-substituted products, respectively).

The hydroxy-derivative **131** of **24**, R = CH₃ is unstable: the tautomeric equilibrium is completely shifted towards the keto form **40a**[28] (the synthesis of ketone **40a** from **34** was mentioned above[26]). A subsequent MNDO-SCF-MO study[68] indicated that the ketone **40a** is calculated to be more stable than annulenol **131** by 12.4 kcal · mol⁻¹.

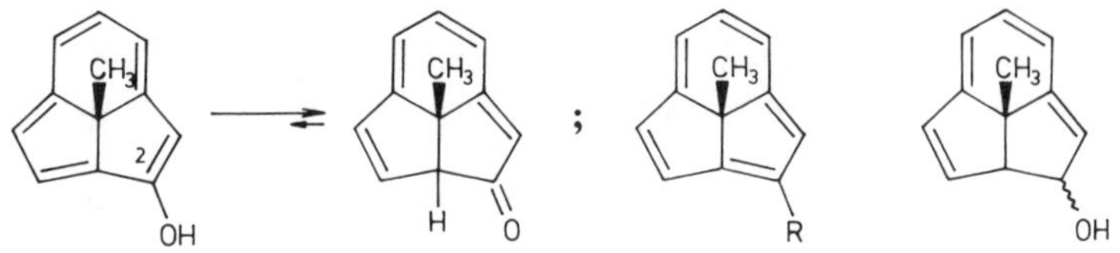

131	*40a*	*132 a* R=O⁻ M⁺	*133*
		b R=OSiMe₃	
		c R=OMe	

Starting from the ketone **40a** different oxygenated derivatives of **24**, R = CH₃, e.g., **132** and **133** were obtained.[28]

Contrary to 2-hydroxy-derivative **131**, the 5-hydroxy-derivatives **43**, **42c**, and **45** exist only in the enolic form[29] (after spectral data and lack of ketonic reactivity), and seem to be the first isolable higher annulenols.

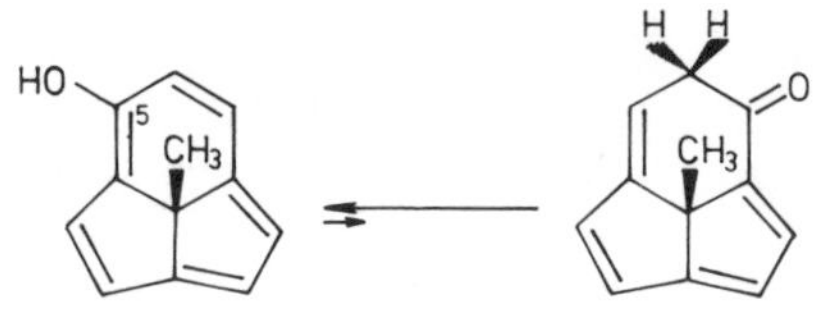

43	*134*

The above results are in agreement with theoretical predictions[68,113] that the 5-hydroxy-derivative of **24**, R = CH₃ could be more stable than its ketonic tautomer **134**.

Other reactions of **24**, R = CH₃ are the hydrogenation to **135**, the thermal rearrangement to the 2aH-isomer **136** (400°C, flash pyrolysis) or to the more stable 1H-isomers **137** and **138** (600°C, flash pyrolysis).[25]

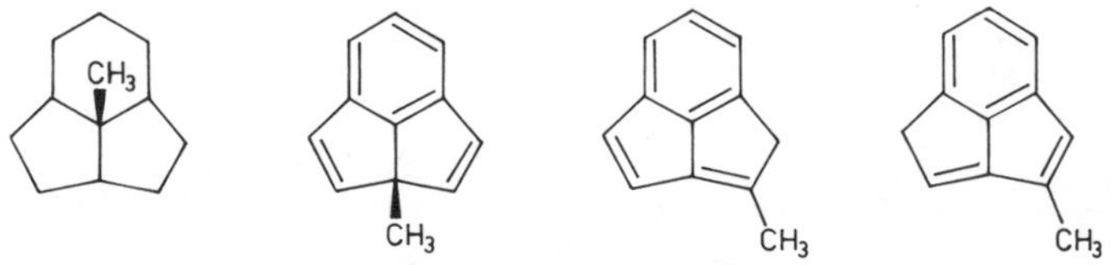

135	*136*	*137*	*138*

D. Conclusions

The synthesis of carbon-bridged [10]annulenes has enriched our knowledge of aromaticity and has confirmed Hückel's rule for [10]annulene.

The standard representative, 1,6-methano[10]annulene (**3**) has typical aromatic behavior which parallels that of naphthalene. As noted by Vogel:[114] "while the aliphatic bridge moiety may be looked upon as a *Schönheitsfehler*, it has the benefit of adding to the chemical versatility of the hydrocarbon."

Interestingly, some furanosesquiterpenes, e.g., *Spiniferin* (**139**) isolated recently by Cimino from a Mediterranean sponge (*Pleraphysilla spinifera*) include a 1,6-methano[10]annulene moiety,[91] and the 1,6-methano[10]annulene derivatives **140** possess a β-adrenergic blocking activity.[115]

139

140 , R=H;Me;i-Pr
R=Pr;t-Bu;cyclohexyl

The more recent derivatives of 7b*H*-cyclopent[*c,d*]indene are also interesting aromatic compounds resembling Boekelheide's tetracyclic bridged annulenes (see later) whereas homoazulene resembles azulene.

The derivatives of 7b*H*-cyclopent[*c,d*]indene, **24** with R = Et and R = *i*−Pr were synthesized (analogously to **24**, R = Me, see Section II.A.3) by Gibbard et al.[397] The same authors described the preparation and reactions (especially aromatic substitutions) of **24**, R = CH$_2$C$_6$H$_5$.[398]

The PE spectrum of **24**, R=Me, was published by Bischof et al.[399] The full paper of preliminary notes[28,29] concerning phenol-keto tautomerism in hydroxy-derivatives of **24** has been published by Lidert et al.[400]

A charge density topological approach to the equilibrium dinorcaradiene ⇄ [10]annulene in various 11,11-disubstituted 1,6-methano[10]annulene was presented by Gatti et al.[401]

Andréa et al.[402] discussed helium-photoelectron spectra as well as CNDO/S and MNDO-MO calculations of 1,6-methano[10]annulene and substituted derivatives.

The total synthesis of (±) *Spiniferin* (**139**), a naturally occurring 1,6-methano[10]annulene, was described by Marshall and Conrow.[403]

III. CARBON-BRIDGED [12]ANNULENES

A. Syntheses
1. Bicyclic-Bridged [12]Annulenes
a. 1,7-Methano[12]Annulene (146)

The 12 π-electron analog of 1,6-methano[10]annulene, namely 1,7-methano[12]annulene (**146**), was synthesized in 1974 by Vogel et al.[116] in a reaction sequence comprising: (1) a double dibromocarbene addition to **7**, (2) the acetolysis of **141** accompanied by double ring enlargement to **142**, (3) conversion of **142** in two steps into the diol **144**, (4) conversion of **144** into the olefin **145**, and (5) the final dehydrogenation to **146** by means of DDQ.

141   142   143

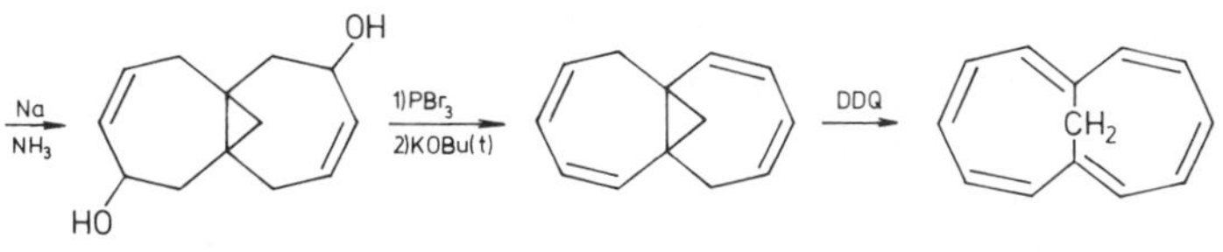

144   145   146

The synthesis of substituted **146** derivatives such as **130** through cycloadditions to homo-azulene was mentioned before (see Section II.C.2).

b. 1,6-Methano[12]Annulene (153)

1,6-Methano[12]annulene (**153**) was obtained by Vogel and co-workers[117] again starting from **7**; after cyclopropanation with ethyl diazoacetate followed by hydrolysis, the acid **147** was converted into the aldehyde **148**; the Bamford-Stevens reaction of the tosylhydrazone of **148** affords the strange cage compound **151** (the diene adduct of the initial cyclobutenic hydrocarbon reacting in the norcaradienic form (**150**). Pyrolysis of **151** affords **152** (retro-diene reaction accompanied by cyclobutene → butadiene isomerization) which on dehydrogenation gives rise to **153**.

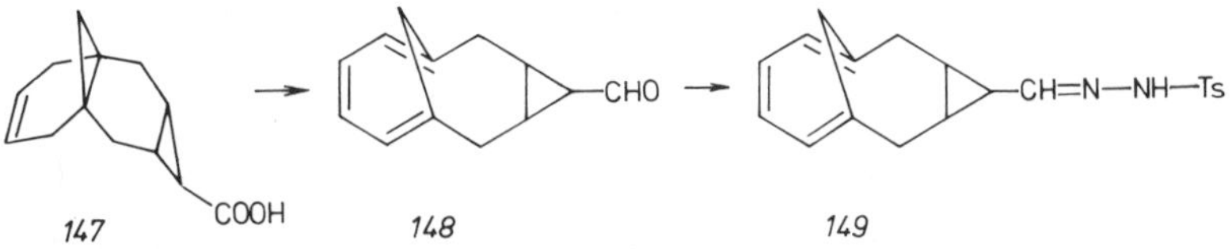

147   148   149

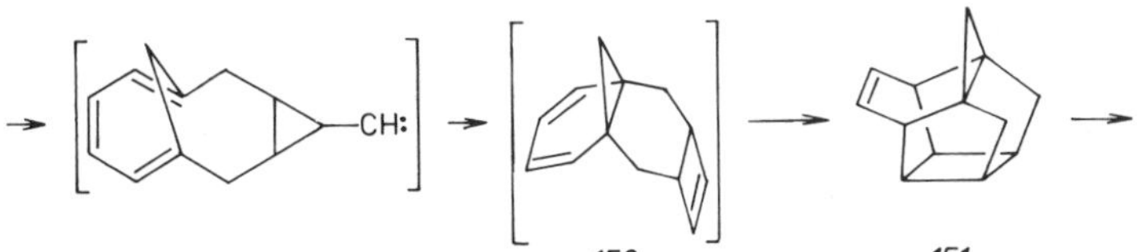

150   151

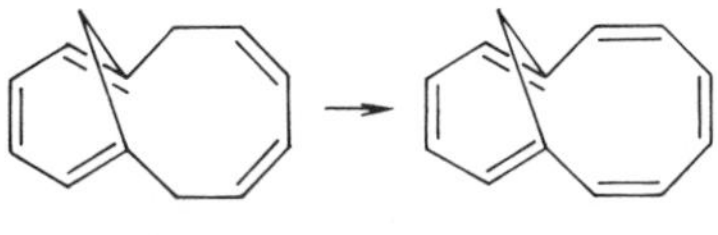

152   153

2. Tricyclic-Bridged [12]Annulenes

There are three possible tricyclic-bridged [12]annulenes with five- to eight-membered rings, namely **154**, **155**, and **156**.

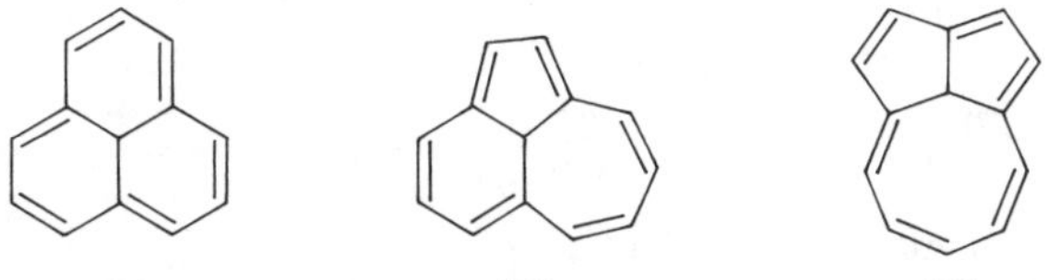

154   155   156

Though none of these hydrocarbons were yet obtained the isomeric compounds **157**[118-120] and **158**[121-124] are known as well as the ions of **154** and **155** (see Section VIII). MO calculations[123] indicated that annelated derivatives of **155**, e.g., **159** and **160** should be sufficiently stable to be isolated.

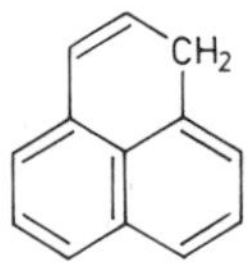

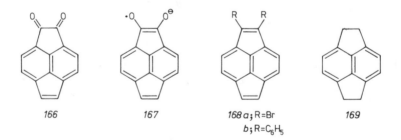

157 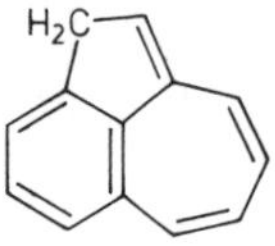   158 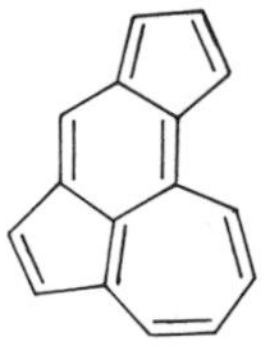   159 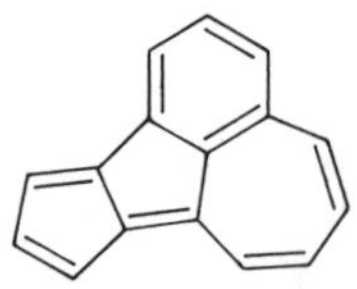   160

3. Tetracyclic-Bridged [12]Annulenes

Five isomeric etheno-bridged tetracyclic systems with [12]annulene perimeter containing five- to seven-membered rings (**161** to **165**) are possible.

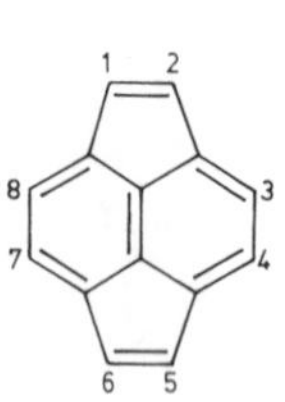

161 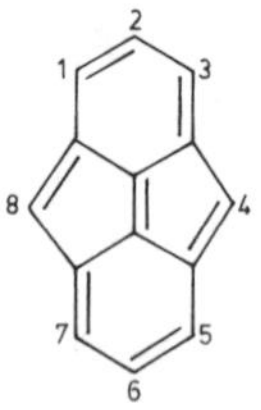   162 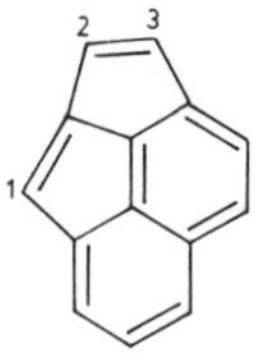   163 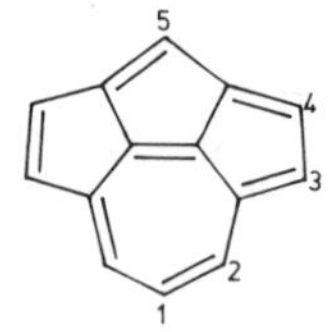   164 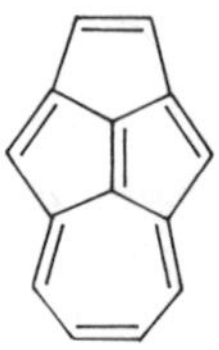   165

From these, only pyracylene (**161**), some derivatives with skeletons **162**, **163**, and **164**, as well as a dibenzoderivative of **164** were synthesized.

a. Pyracylene* (161) and Derivatives

Some benzo-derivatives of **161**[125-128] as well as a dihydrobenzopyracylene[129] were known before the parent compound.

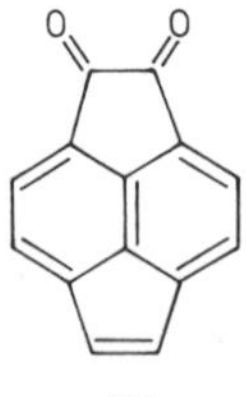

166   167   168 a; R=Br   169
                          b; R=C₆H₅

168 a; R=Br
b; R=C_6H_5

Attempts to synthesize pyracylene itself go back to 1958 when Anderson[130] obtained only dihydropyracylene on oxidation of pyracene with chloranil, and only polymers on pyrolysis of pyracene diacetate. A 1,2-diphenyl-5,6-dihydropyracylene was obtained in 1960.[131] In 1966 Trost[132] synthesized the quinone **166** and from it the anion-radicals **167** with pyracylene skeleton[133] which were investigated through ESR.[134] The first neutral derivative of **161**, namely the 1,2-dibromopyracylene (**168a**) was obtained[135] by bromination of pyracene (**169**) to the tetrabromo-derivative **170** followed by debromination with potassium iodide.

---

*   This name was suggested by Brown[404] by analogy with acenaphthylene and acenaphthene. Some authors name the same compound "pyracyclene."[416]

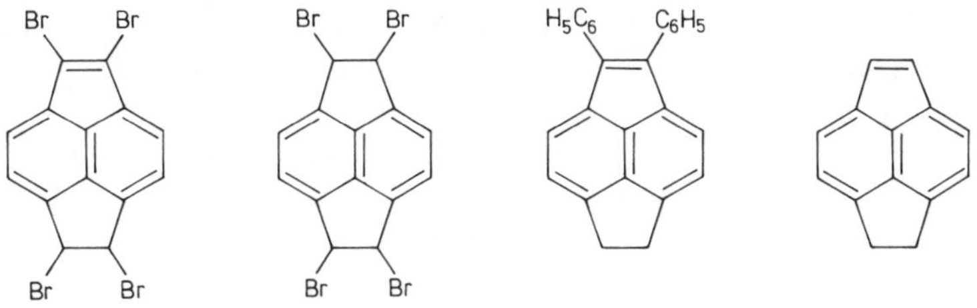

170　　　　　171　　　　　172　　　　　173

Pyracylene (**161**) was synthesized by Trost and Bright in 1967[136] through iodide ion stepwise debromination of the tetrabromo-derivative **171** (obtained on bromination of pyracene **169**). Later this synthesis was optimized and extended to the preparation of 1,2-diphenylpyracylene (**168b**) via the dihydro-derivative **172**.[137]

An attempt to synthesize pyracylene through dehydrogenation of **169** with chloranil or with DDQ failed in the second step, suggesting that **161** is not appreciably stabilized relative to its isolated dihydroderivative **173**.[137]

b. Derivatives of Dibenzo[cd, gh]Pentalene

4,8-Dihydrodibenzo[*cd, gh*]pentalene (**176**) was obtained[138] in a reaction sequence where the essential step is the photochemical ring contraction (Wolff rearrangement) of the diazoketone obtained form the quinone **174** (yield 2%).

Later,[139] the pentalenoquinone **177**, the first neutral conjugated derivative of dibenzo[*cd, gh*]pentalene, was obtained starting from **176**.

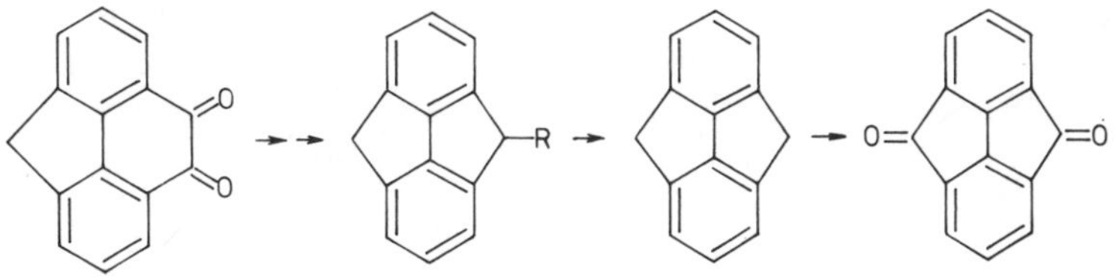

174　　　　175, a, R=CONHBu-t　　　176　　　　177
　　　　　　　　b, R=CO₂H

c. Derivatives of 1H-Cyclopent[b,c]Acenaphthene

Some hydrogenated derivatives with the skeleton of **163** have been known for a long time, e.g., the ketone **179** obtained on cyclization of the acid **178**.[19]

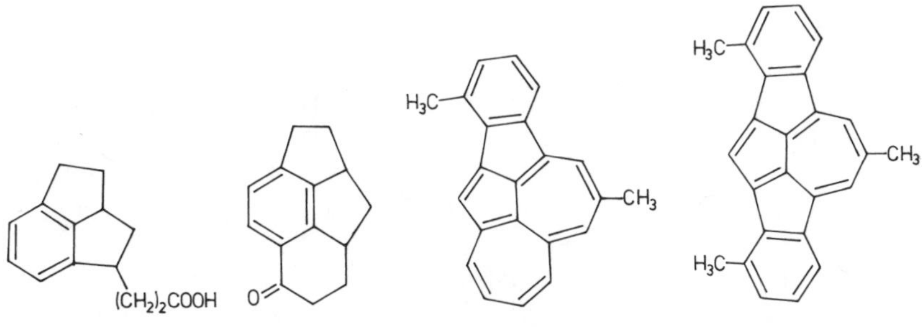

178　　　　　179　　　　　180　　　　　181

d. Derivatives of Dicyclopent[cd,ij]azulene

The dibenzoderivative **181**, the first conjugated thermally stable derivative of the yet unknown dicyclopent[*cd,ij*]azulene (**164**), was obtained by Hafner[140] from methyl-substituted benzo-pentalenoheptalene **180** and *N,N*-diethyl-1-propynylamine. Two consecutive valence isomerizations and a subsequent elimination of diethylamine were proposed in order to explain the reaction course.[140]

1,2-Dihydrodicyclopent[*cd,ij*]azulene (**185**) was obtained also by Hafner,[141] as a thermally unstable compound starting from 5-methyl-1,2-dihydrocyclopent[*c,d*]azulene (**183a**) via **183b, 184a,b**. Interestingly, the synthesis of **183a** started from 4,8-dimethylazulene (**182**) following a similar route as the conversion **183a → 185**.[141] The dehydrogenation of **185** has not yet been reported.

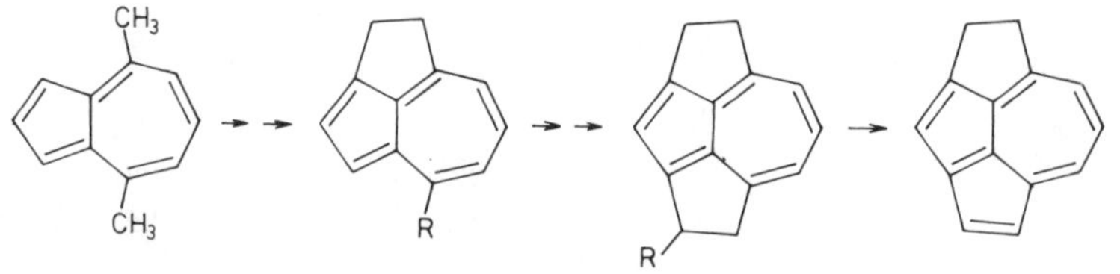

182	*183 a*, R=CH₃ *b*, R=CH₂CONMe₂	*184 a*, R= =NMe₂ *b*, R=-NMe₃	*185*

B. Structure and Physical Characterization
1. Bicyclic-Bridged [12]Annulenes
a. 1,7-Methano[12]Annulene (146)

1,7-Methano[12]annulene (**146**) is a red-brown compound unstable in air, with m.p. 4 to 6°C.[116] The UV spectrum (cyclohexane), λ_{max}: 254 (ϵ 54000); 263(51000; 300(1000sh); 425 nm(240) is characteristic for a polyene and resembles the UV spectrum of dehydro[12]annulene.[116] The ¹H-NMR spectrum of **146** (CCl₄; δ ppm): 5.2 (m; H-3,4,5,9,10,11); 5.5 (m; H-2,6,8,12); 6.0 (s, CH₂) clearly indicates paratropicity in striking opposition to 1,6-methano[10]annulene (see Section II.B.1) The temperature dependence of the ¹H-NMR spectrum provides evidence for the valence tautomerism **146a ⇌ 146b** which leads above −60°C to a C₂ᵥ symmetry.[116]

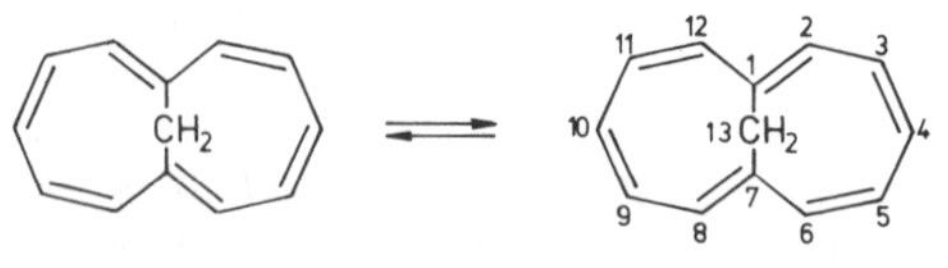

146a	*146b*

The same conclusion could be also drawn from the ¹³C-NMR spectrum (d₈-THF; δ ppm): 34.1(C-13); 129.7(C-4); 131.6(C-2,6); 134.1(C-3,5), and 146.2(C-1).[116] Whereas at −118°C the molecule has a "frozen" structure, at room temperature the double bond fluctuation is fast; the activation energy is 5 kcal · mol⁻¹ for the process **146a ⇌ 146b**.[116] Force field calculations confirmed the twisted conformation with C₂ᵥ symmetry and bond alternation; the total steric energy is 30.1 kcal · mol⁻¹.[43] An X-ray analysis of the 4,10-dibromo-derivative of **146**[142] as well as a study of crystal and molecular structure of 13,13-difluoro-1,7-methano[12]annulene[143] were reported.

In conclusion, the 1,7-methano[12]annulene indicates, in striking contrast to the non-bridged [12]annulene, all the usual characteristics of a $4n\pi$ system.

b. 1,6-Methano[12]Annulene (153)

1,6-Methano[12]annulene is a relatively stable orange-red solid with m.p. 30 to 31°C.[117] The ¹H-NMR spectrum (CCl₄; δ ppm): AA′BB′-system at 6.17 (H-3,4) and 5.73 (H-2,5);

Table 2
CALCULATED DATA FOR TETRACYCLIC BRIDGED [12]ANNULENES

Compound	DE/m[a]	P[b]	RE[c]	REPE[d]	TRE[e]	RE[f]
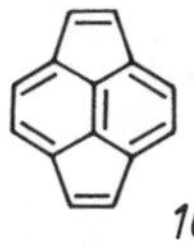 **161**	0.3186	0.000	0.254	0.018	0.009	0.767
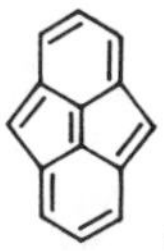 **162**	0.3182	0.000	0.246	0.018	—	—
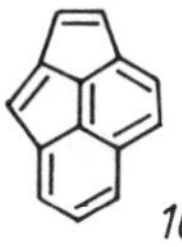 **163**	0.3232	0.042	0.332	0.024	—	0.640
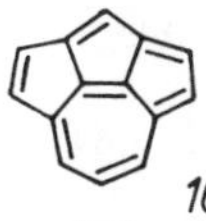 **164**	0.3106	0.014	0.118	0.008	—	−0.155
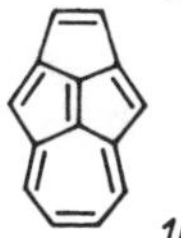 **165**	0.3016	0.000	−0.034	−0.002	−0.011	−0.505

[a] Delocalization energy per bond (m = 17).[33]
[b] Stability index.[33]
[c] Resonance energy (β-values).[34]
[d] Resonance energy per π-electron (β-values).[34]
[e] Topological resonance energy (β-values).[35]
[f] Resonance energy (eV).[36]

5.5 (m; H-7 to 12); AB-system at 2.29 (H-13a) and 7.00 (H-13b) indicates no significant temperature dependence.[117] Comparison of these data with those of a suitable model, **186**, permitted the assessment of paratropicity in **153**.[117] From the analysis of NMR coupling constants, the existence of a cycloheptatrienic moiety was proved, hence the exclusiveness of the valence tautomer **153**. On the other hand, from the large chemical shift difference between H-13a and H-13b the preference of the conformation **153a** was proved.[117]

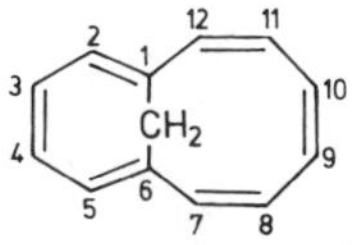

153 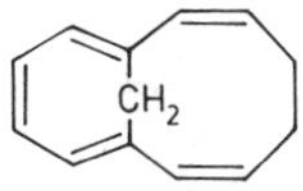   *186* 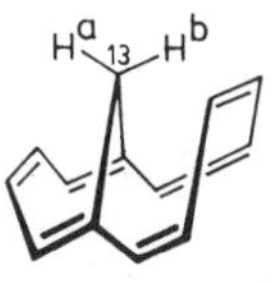   *153a*

2. Tetracyclic-Bridged [12]Annulenes
a. Results of Calculations

Though the first MO calculations of pyracylene due to Brown in 1951[144] suggested a high reactivity of C-1 to C-2 bond, little was done until 1965 to 1966 when important contributions of Zahradnik and co-workers appeared[33,145] (Table 2). With the exception of hydrocarbon **163** (stability index P = 0.042), which appears to possess some stability but might polymerize easily, the remaining hydrocarbons from Table 2 are predicted to be very unstable.

Table 3

PHYSICAL DATA OF PYRACYLENE AND OF SOME DERIVATIVES [136,137]

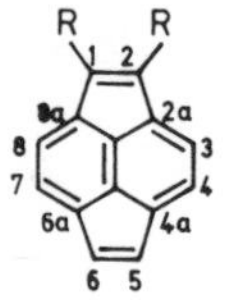

R=	¹H-NMR (δ ppm)				UV-VIS λ_{max} (nm)
	H-1,2	H-3	H-4	H-5,6	
H (in CCl₄)	6.01	6.52	6.52	6.01	218; 326; 332; 341; 385; 405; 408; 427
Br (in CDCl₃)	—	6.60	6.57	6.11	222; 261; 303; 317; 331; 339; 346; 355; 363; 391
C₆H₅ (in CCl₄)	—	6.61	6.61	6.06	217; 242; 291; 338; 353; 371; 393; 418; 548
5,6-Dihydro pyracylene (in CCl₄)	7.04	7.65	7.32	3.49	—

Small values for RE and REPE were found for **161** to **165** by Hess and Schaad[34] (Table 2). More recent data of Randić[36] (Table 2) using the "conjugated circuits" concept suggested for **161** and **163** prevailing aromatic character, for **164** and **165** dominating antiaromatic character, whereas for **162** exclusive antiaromatic character.

Concerning the pyracylene molecule **161**, conflicting theoretical results were reported. Thus, Lo and Whitehead[146] calculated a stabilization energy per C–C bond of 0.3103 eV in comparison to 0.3355 eV for naphthalene; Coulson and Mallion[147] calculated a paratropism for all rings in **161** or at least for the five-membered rings of **161** whereas more recent results of calculations, performed by six different methods, span the range from a strong paratropism to a marginal diatropism.[148]

SCF-MO calculations for molecule **162**[149] suggested a triplet ground state and the lack of appreciable bond fixation in the peripheral skeleton.

Calculations of ring currents for **164** and **165** indicated paramagnetic (minus sign) and diamagnetic (plus sign) ring currents as indicated in formulas (in comparison with benzene).[32]

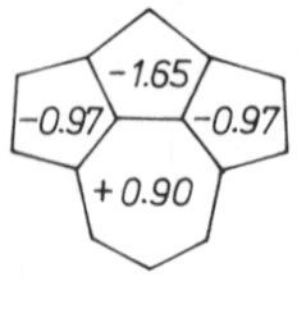

164 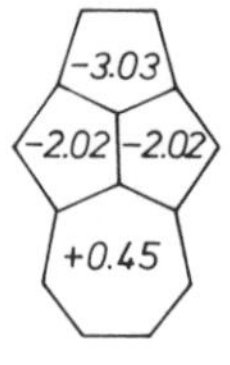   *165* 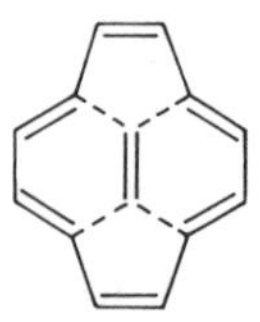   *187*

Other theoretical papers concerning hydrocarbons **161** to **165** were published.[150-154]

b. Experimental Results

Pyracylene (**161**) was obtained only in solution; attempts to isolate the pure compound led to polymerization. Somewhat more stable than **161** are the 1,2-dibromo- (**168a**) and 1,2-diphenyl-(**168b**) derivatives. Some physical data are collected in Table 3. The ¹³C-NMR spectrum of **161**, in excellent agreement with the predicted one, indicates the following signals (δ ppm): 124.8 (C-3,4,7,8); 131.5 (C-8b,8c); 132.4 (C-1,2,5,6); 142.0 (C-2a,4a,6a,8a). From the NMR data (and by comparison with 5,6-dihydropyracylene, Table 3) a paramagnetic ring current is undoubtedly proved. Thus, pyracylene is an antiaromatic compound as suggested by Craig's rule. Though containing 14π electrons, pyracylene can be described as a perturbed planar [12]annulene having an isolated central double bond as indicated by formula **187**.

The 4,8-dihydrodibenzo[*cd,gh*]pentalene (**176**) is a solid with m.p. 137.5 to 138.3°C, whose X-ray analysis demonstrates its planarity.[139] Interestingly, in the mass spectrum of **176** the important part (64%) of the total ion current is given by the ions with m/e 178, 176, 89, 88, corresponding to the molecular ion of **176** and of dibenzo[*cd,gh*]pentalene (probably appearing under electron impact) and, respectively, to the doubly charged molecular ions of the above hydrocarbons.[139]

C. Conclusions

The bicyclic-bridged[12]annulenes **146** and **153** are paratropic systems with $4n\pi$-electrons. Whereas the symmetrical **146** presents a valence tautomerization through double bond fluctuation, the isomeric compound **153** exists in a single form.

No tricyclic-bridged [12]annulene with five- to eight-membered ring was yet synthesized.

From tetracyclic-bridged [12]annulenes **161** to **165** only pyracylene **161** could be obtained in solution as an unstable paratropic species. Although possessing 14π-electrons, pyracylene behaves as a [12]annulene perturbed by a central double bond.

IV. CARBON-BRIDGED [14]ANNULENES

A. Introduction

Different types of carbon-bridged [14]annulenes are known at present. They will be presented after the following classification:

1. Tricyclic-bridged [14]annulenes and derivatives (Vogel's type)

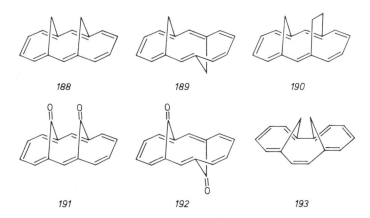

188   189   190

191   192   193

2. Tetracyclic-bridged [14]annulenes and derivatives

    a. Vogel's type

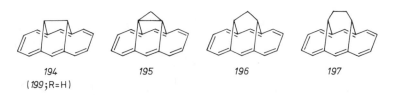

194   195   196   197
(*199*;R=H)

    b. 15,16-Dialkyl-15,16-dihydropyrenes (Boekelheide's type) and isomers

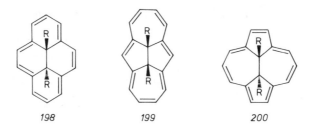

198   199   200

c.   Pyrene and isomers (ethene-bridged [14]annulenes)

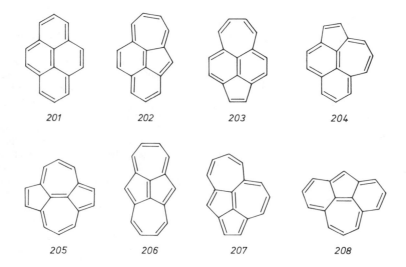

201   202   203   204

205   206   207   208

So far no related system is known such as **209** with smaller rings than five-membered ones.

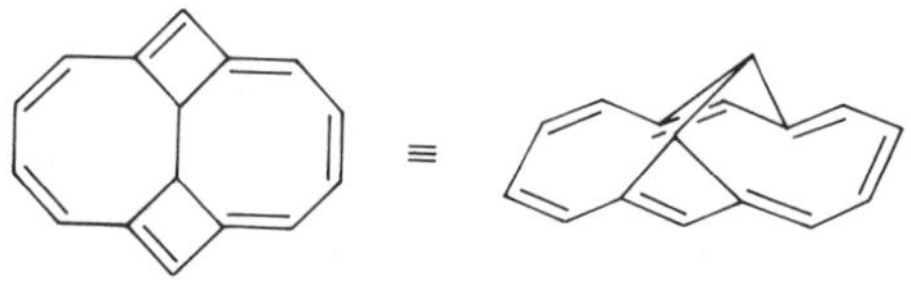

≡

209

B. Syntheses

The synthesis and the aromatic character which were proved for 1,6-methano[10]annulene raised the question if a generalized "homologous" series of bridged [4n + 2]annulenes formally derived from the acene series can be explored. The solution to this problem was provided by Vogel.

The existence of more than one bridge in such a molecule permits the existence of *syn*- and *anti*-isomers such as **188** (*syn*) and **189** (*anti*).

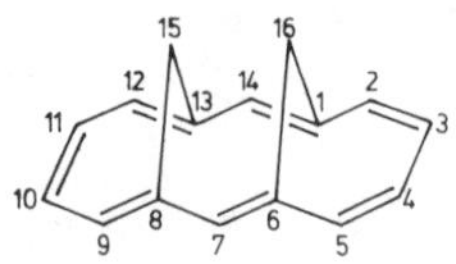

188 (syn) 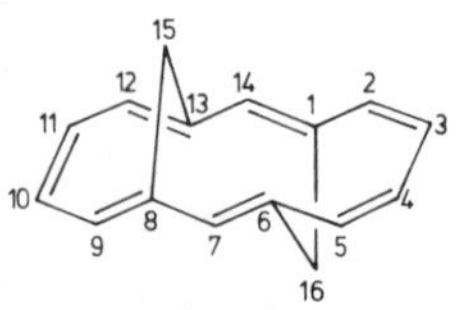   189 (anti)

Inspection of stereomodels reveals that the *syn*-isomers (e.g., **188**) could possess a nearly planar carbon perimeter with a parallel orientation of the $2p_z$ orbitals belonging to carbon atoms 6, 7, 8 and 1, 14, 13. On the contrary, the *anti*-isomers belonging to carbons 6, 7, 8 and 1, 14, 13 which are noticeably skewed. One can thus predict that only the *syn*-bridged annulenes (e.g., **188**) could present aromatic character.

1. Tricyclic-Bridged [14]Annulenes and Derivatives (Vogel's Type)

a. Syn-1,6:8,13-Bismethano[14]Annulene (188)

After unsuccessful attempts to synthesize **188**,[38,155] this compound was obtained in 1975 following a new synthetic concept.[156] Reaction of 1*H*-3,8-methanocyclopropa[10]annulene (**112**) with bromine provided the tetrabromoderivative **210** (stereochemistry proved by X-ray analysis[11]). Reaction of **210** with potassium iodide followed by heating (producing an allowed cyclopropane ring opening in the intermediate dibromoderivative) afforded **211a** with *syn*-configuration of the two CH_2 groups. Successive conversions of **211a** into **211b** and **211c** followed by a Wittig reaction with thiodimethylenebis(triphenylphosphonium bromide) yielded **212** (with traces of **188**). Final heating of **212** with triphenylphosphane in benzene produced **188** via the arene episulfide **213**.[156]

112 210 211,a,R=Br b,R=CN c,R=CHO

212 213 188

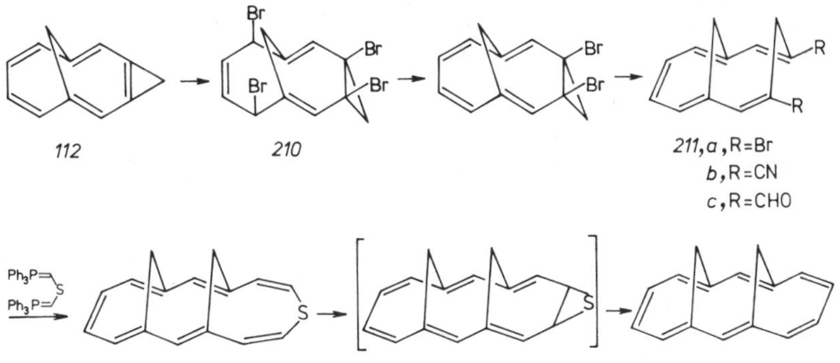

Later it was shown[9] that the conversion **211c** → **188** could be conveniently performed in an analogous fashion to the conversion of cycloheptatriene-1,6-dialdehyde (**10**) into 1,6-methano[10]annulene (**3**) (see Section II.A.1). Similarly, the acetylene **214** was thermally converted into **188** probably via the nonisolable cyclic cumulene **215**[109]

214 215 188

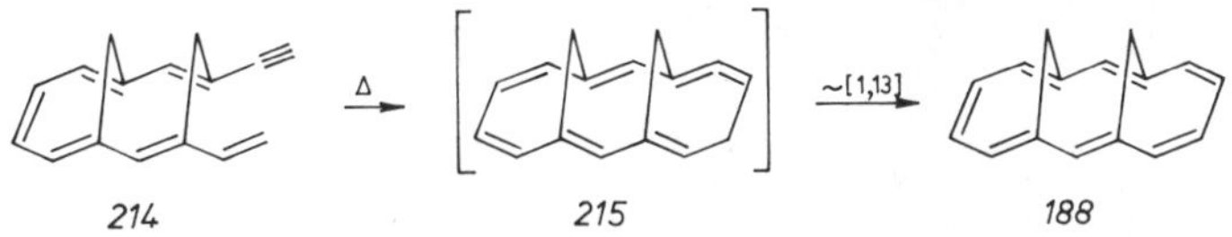

The above dialdehyde **211c** was recently obtained by a "homologization sequence" in Vogel's "building block approach" synthesis.[8] Thus, **10** is olefinated with the bifunctional Wittig-Horner reagent **216**, the diester **217** is reduced with diisobutylaluminium hydride, and the intermediate diol is oxidized with DDQ.[8]

10 216 217

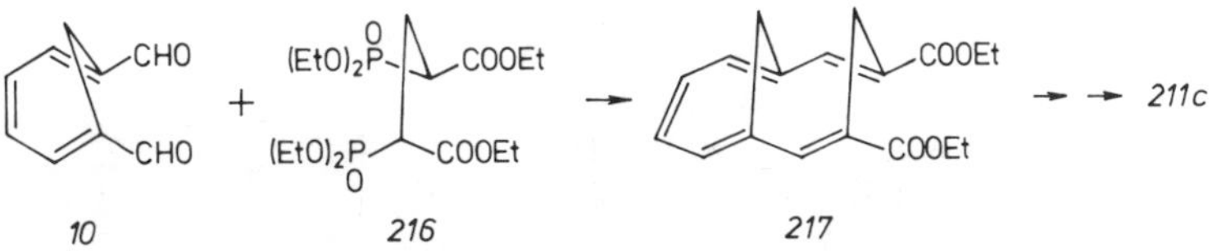

b. Anti-1,6:8,13-Bismethano[14]Annulene (189)

This compound was obtained starting from the hexahydroanthracene (**218**) through intermediates **219** to **222**. Though dehydrogenation of **222** failed,[157] bromination followed by treatment with sodium iodide in acetone afforded **189**.[155]

218 219 220 221

222 223 189

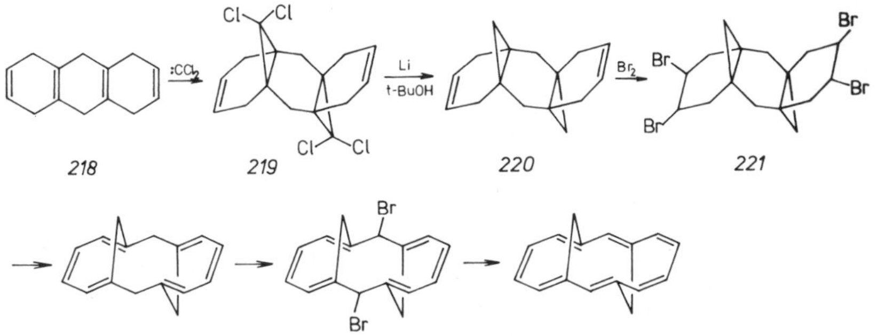

c. Syn-1,6-Ethano-8,13-Methano[14]Annulene (190)

This compound was synthesized in 1980 by Vogel[158] using the "building block approach". The dialdehyde **224** was obtained from **10** analogously to **211c** (see above). Olefination of **224** followed by "cyclodehydrochlorination" of the resulted **225** afforded **190**.

224 225 190

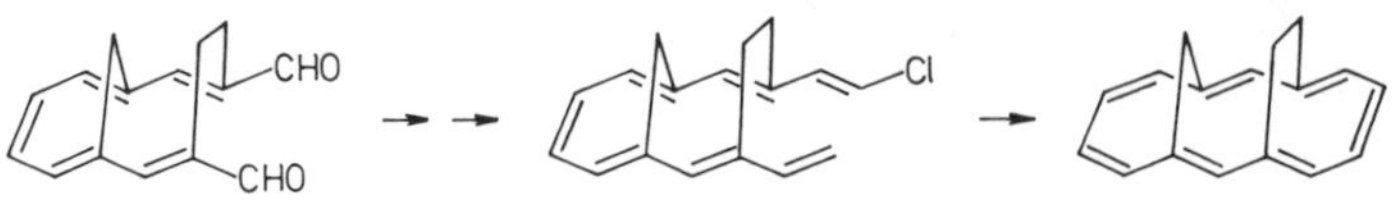

d. 15,16-Dioxo-Syn-1,6:8,13-Bismethano[14]Annulene (191)

This compound was obtained in 1981[159] using a multistep reaction sequence including: (1) the conversion of the diazoketone **226** (obtained from **218**) into **227**, (2) bromination of **227** to **228**, (3) thermolysis of the crude **228** to the symmetrical α,α′-dibromoketone **229**, (4) dehydrogenation to **230**, (5) conversion to the trihydroxyderivative **231** through reduction of **230** followed by solvolysis of the bromines, (6) cleavage of **231** with periodic acid in acetone/water to the desired **191**.

226 227 228

229 230 231 191

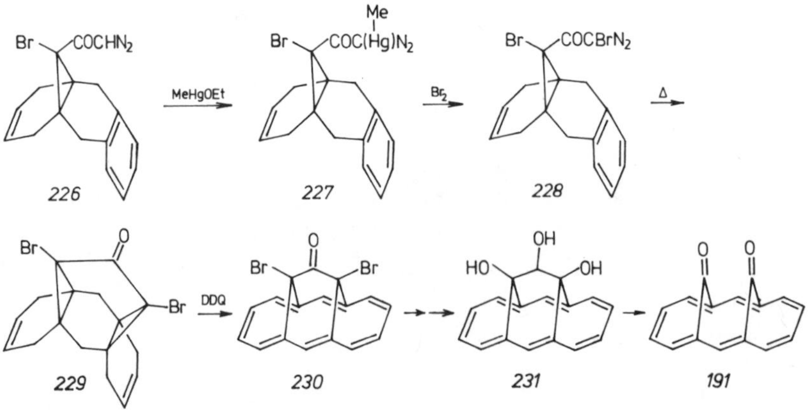

e. 15,16-Dioxo-Anti-1,6:8,13-Bismethano[14]Annulene (192)

This compound was synthesized starting also from **218**.[160] Addition of sodium chlorodifluoroacetate (difluorocarbene-latent carbonyl group) yielded a mixture of mono- and bis-adducts. The bis adduct **232** was converted to **233** (bromine addition/hydrogen bromide elimination) which was hydrolyzed further with sulfuric acid to diketone **234a**. A new bromination followed by dehalogenation of **234b** with sodium iodide in acetone afforded **192**.

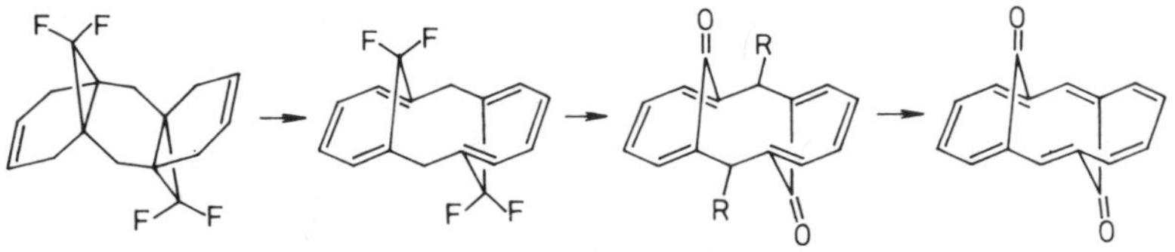

| 232 | 233 | 234 a, R = H | 192 |
| | | b, R = Br | |

f. Syn-1,6:7,12-Bismethano[14]Annulene (193)

This compound was recently synthesized by Vogel[109] and is the first representative of a bridged [14]annulene formally related to phenanthrene (instead of anthracene). The key intermediate, **235a**, formerly obtained by a tedious synthesis,[101] becomes available in large amounts after Okazaki's method,[161] including a photochemical reaction of benzocyclopropene. Conversion of **235a** into the mononitrile **235b** was followed by the treatment with tetrakis(triphenylphosphine)nickel(O) affording the dimeric product **236a**. The corresponding dialdehyde, **236b**, was converted to **193** on treatment with WCl$_6$/n-butyllithium (Sharples reagent).

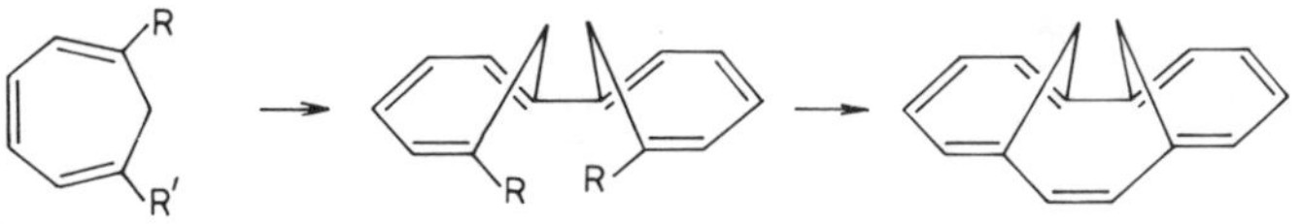

| 235 a, R = R' = I | 236 a, R = CN | 193 |
| b, R = I; R' = CN | b, R = CHO | |

2. Tetracyclic-Bridged [14]Annulenes and Derivatives (Vogel's Type)
a. 1,6:8,13-Ethanediylidene[14]Annulene (194)

This compound was obtained in 1972[162] after a multistep reaction sequence including: (1) the conversion of dibromocarbene adduct **273a** into the corresponding acid **237b** then into the diazoketone **237c**; (2) intramolecular cyclization of **237c** to **238**; (3) aromatization on treatment with DDQ; (4) Favorskii rearrangement of **239** to **240a**; (5) conversion into the S-acylxanthate **240b**, and (6) irradiation of **240b** in the presence of tri-n-butyltin hydride affording **194**.

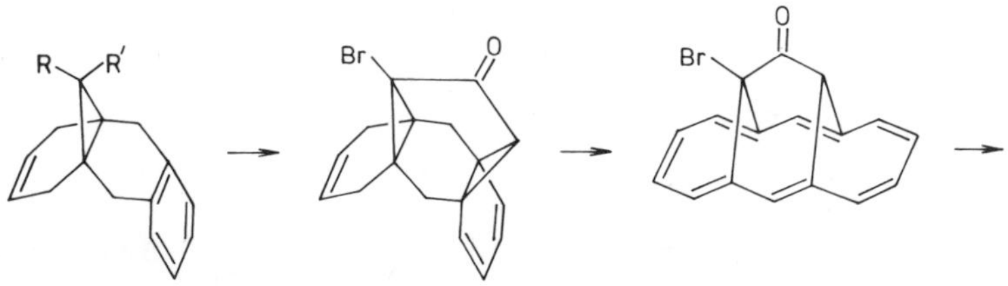

237, a, R = R' = Br	238	239
b, R = Br; R' = COOH		
c, R = Br; R' = COCHN$_2$		

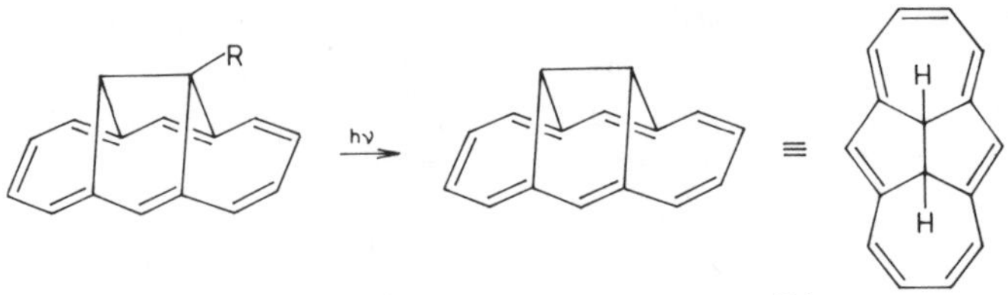

240, a, R=COOH
b, R=CO-S-CS-OEt

194

b. 15,16-Methylene-1,6:8,13-Ethanediylidene[14]Annulene (195)

This compound was obtained serendipitously instead of an aldehyde when the tosylate **241** was heated with sodium carbonate in DMSO.[11]

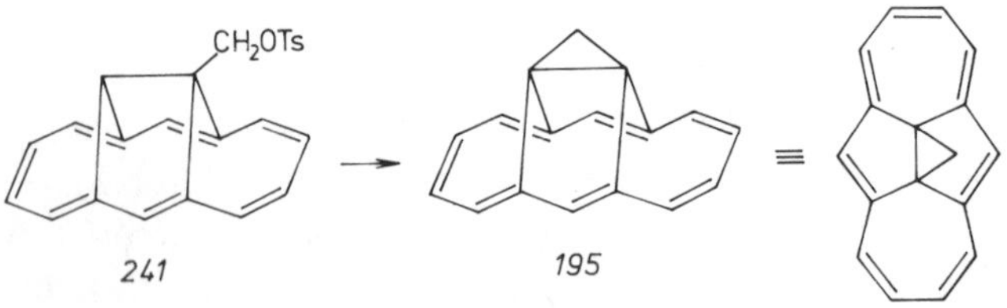

241   *195*

c. 1,6:8,13-Propanediylidene[14]Annulene (196)

This compound was obtained in 1970[163] after a reaction sequence including: (1) the conversion of the dibromocarbene adduct **242a** to **242c** via **242b**; (2) stepwise dehydrogenation of **242c** to **243** then to **244**; (3) intramolecular cyclization of diazoketone **245** obtained from *endo*-acid **244** to the polycyclic ketone **247**; (4) aromatization of the carbon perimeter; (5) a final Huang-Minlon reduction of **248** to **196**.

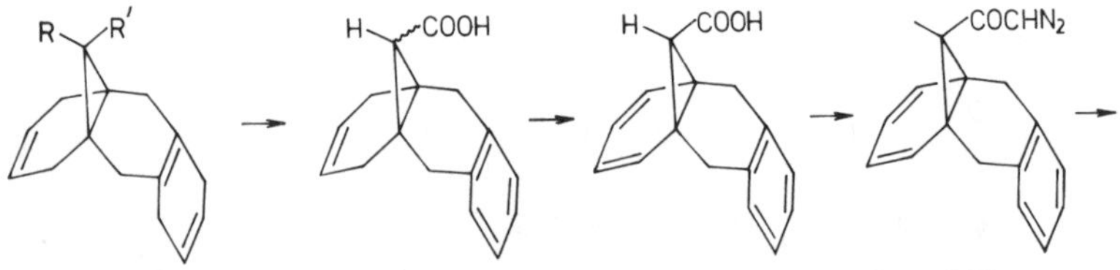

242, a, R=R'=Br
b, R=H; R'=Br
c, R=H; R'=COOH

243   *244*   *245*

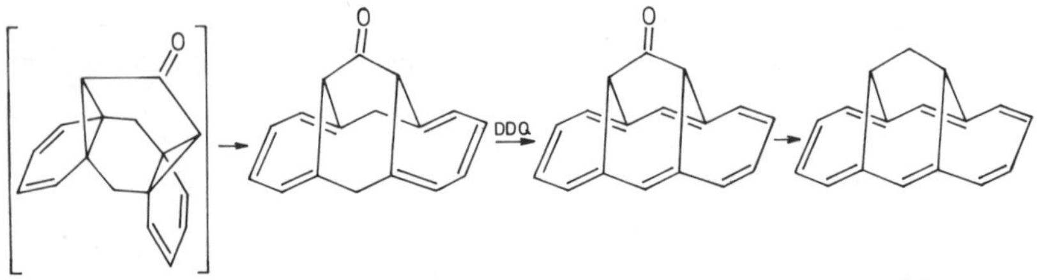

246   *247*   *248*   *196*

d. 1,6:8,13-Butanediylidene[14]Annulene (197)

This compound was synthesized[164] from the above ketone **247**. Ring expansion with diazomethane afforded **249** which was dehydrogenated by DDQ to **250**. The final Huang-Minlon reduction afforded **197**.

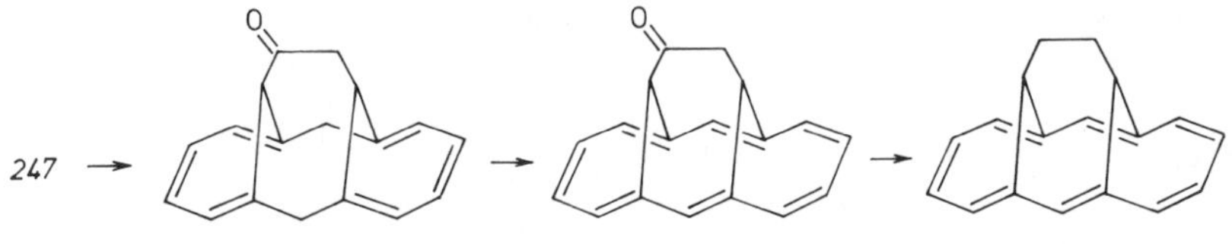

247 → 249   250   197

e. 15,16-Dialkyl-15,16-Dihydropyrenes* (Boekelheide's Annulenes) and Isomers

The first attempts to synthesize the *trans*-15,16-dimethyldihydropyrene (**254**), an aromatic molecule bearing substituents within the cavity of the 14-π-electron cloud, go back to 1961:[165] the metacyclophane **252**, obtained by Wurtz reaction from dibromide **251** or from 2,6-bis(bromomethyl)toluene could not be dehydrogenated to cyclophanediene **253** (the valence tautomer of **254**). Attempts to methylate the dianion of pyrene[165] also failed.

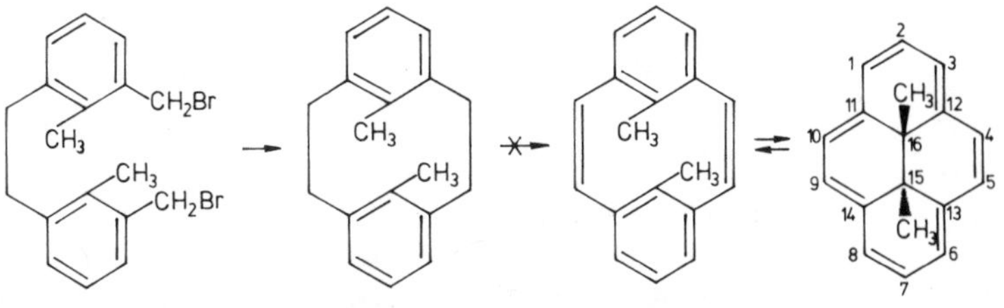

251   252   253   254

In 1963 Boekelheide[166] succeeded in synthesizing 2,7-diacetoxy-*trans*-15,16-dimethyl-15,16-dihydropyrene (**260**), the first derivative with the skeleton of **254**. The 17-step synthesis starting with p-cresol is summarized in the scheme below.

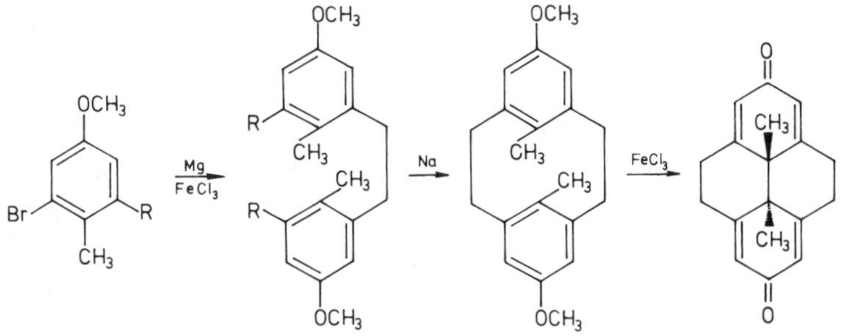

255, R=Br;CN;   256, R=Br;CN;   257   258
COOH;COOMe;   CH₂NH₂;CH₂OH;
CH₂OH;CH₂Br   CH₂Br;CH₂I

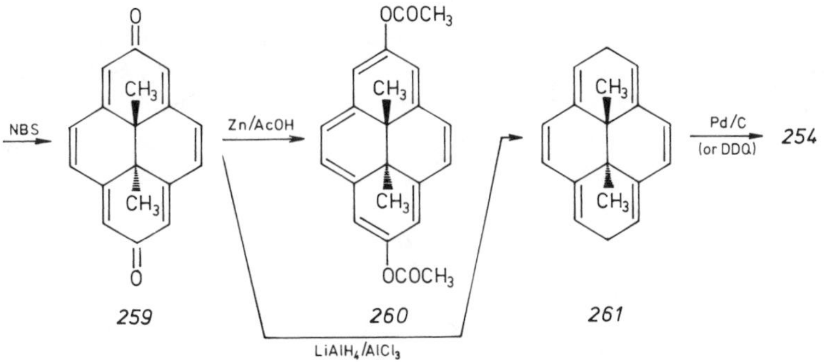

259   260   261   254

LiAlH₄/AlCl₃

---

* Though now *Chemical Abstracts* calls these compounds *trans*-10b,10c-dihydropyrenes, we preserve the original nomenclature in order to avoid confusions.

In the next year, 1964, Boekelheide announced[167] the synthesis of **254** from the above quinone **259** in two steps (see Scheme 3). Later, the above syntheses were optimized.[168]

f. 1,3,6,8-Trans-15,16-Hexamethyldihydropyrene (265)

This compound was obtained in seven steps starting with 2,4,6-trimethylanisol (**262**) and following the same final steps as in the synthesis of **254**.[169,170]

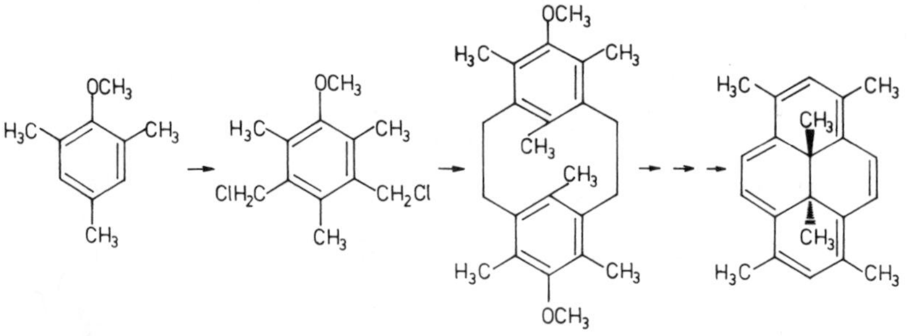

262   263   264   265

g. Trans-15,16-Diethyldihydropyrene (268)

This compound could not be obtained using the same procedure as for **254**.[171] Consequently, in the intermediate **255**, R = COOMe the C–CH_3 group was replaced by an ethyl group by sequential conversions into –CH_2CN, –CH_2–CH_2–$N(CH_3)_2$, –CH_2–CH_2–$\overset{+}{N}(CH_3)_3$, –CH=CH_2, and –CH_2–CH_3, obtaining **266** then the desired cyclophane **267** from which **268** as well as its diacetoxyderivative **269** were obtained in a manner similar to the methyl series.[171]

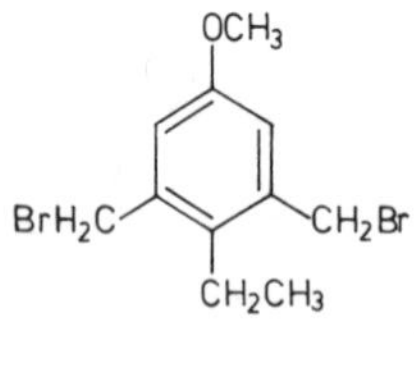

266 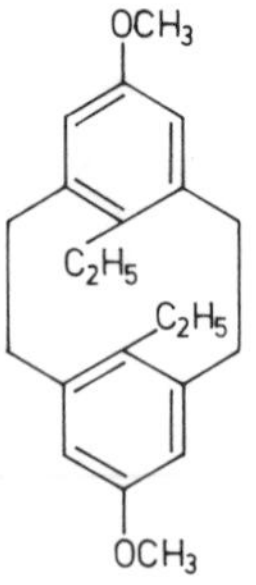   267 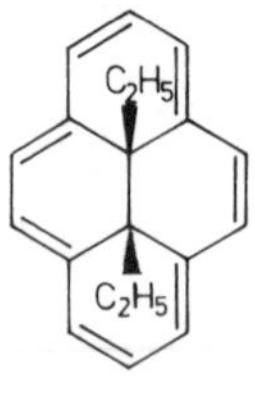   268 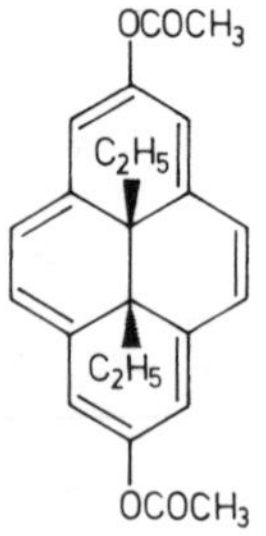   269

h. Trans-15,16-Di-n-Propyldihydropyrene (271)

In 1970 Boekelheide and Hylton[172] devised a general method for synthesizing *trans*-15,16-dialkyldihydropyrenes including a novel optimized four-step conversion of 4-alkylanisole into 3,5-dibromo-4-alkylanisole (**270**) (nitration, reduction to NH_2, bromination, and elimination of the 2-NH_2 group). Interestingly, the yields of the Wurtz dimerization reaction do not decrease with the increasing size of internal substituents. Through this method the new di-*n*-propyl derivative **271** was obtained.[172]

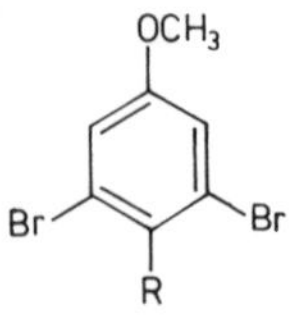

270 , R=C_2H_5 ; C_3H_7 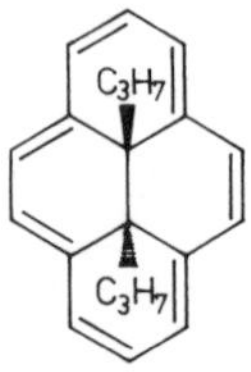   271

An entirely different route to *trans*-15,16-dialkyl dihydropyrenes was developed in 1970 by Boekelheide.[173] The essential steps of the reaction sequence are (1) Wittig synthesis of the substituted *trans*-stilbene **272**, (2) photochemical isomerization into the *cis*-stilbene **273a**, (3) cyclization of **273b** to [2.2]metacyclophan-1-ene **274**, and (4) subsequent dehydrogenation with DDQ.

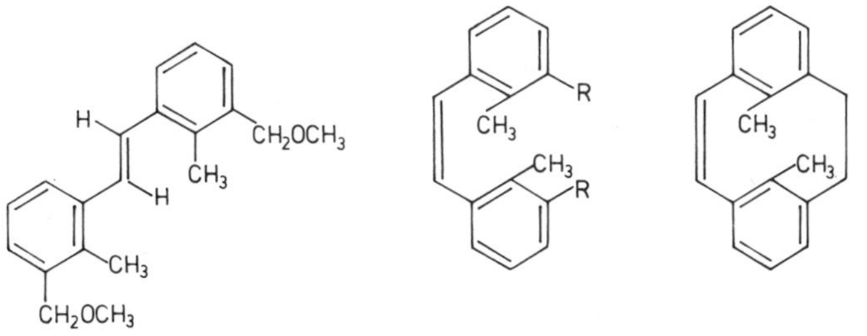

272

273,*a*,R=CH$_2$OCH$_3$
  b,R=CH$_2$Br

274

A new, more efficient synthesis of *trans*-15,16-dimethyldihydropyrene was described in 1971 by Boekelheide.[174,175] This synthetic approach is based on two key steps, namely the Stevens rearrangement of bis-sulfonium salts (e.g., **277**) and the Hofmann elimination from the bis-sulfonium salts (e.g., **279**) (see the following scheme). The reaction mixture in the condensation of **275a** with **275b** contains seven parts *anti*-sulfide **276a** and one part *syn*-sulfide **276b**; the *anti*-isomer is converted to **277** by means of $(CH_3O)_{\frac{1}{2}}CH(BF_4)^-$; the reaction mixture from the Stevens rearrangement need not be separated into isomers but could be further converted directly to **279** then to **280**. The product of Hofmann elimination is the *anti*-methacyclophanediene **280** which on heating or irradiation[176] undergoes the valence tautomerization to **254**.

Scheme 4

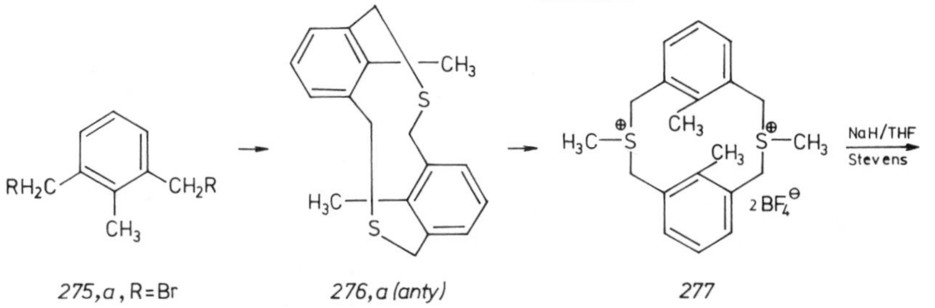

275,*a*,R=Br
   b,R=SH

276,*a (anty)*

277

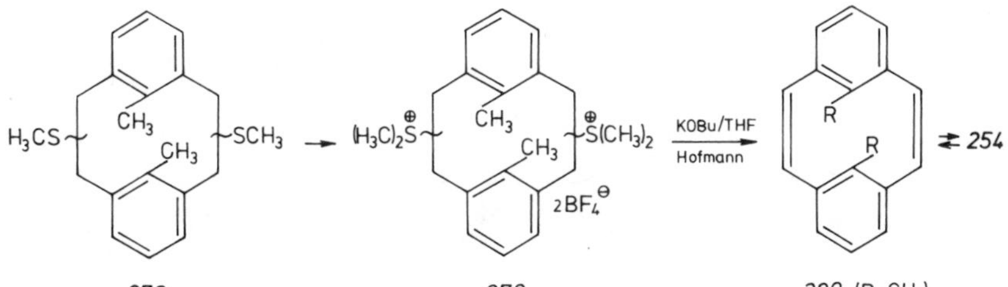

278

279

280,(R=CH$_3$)

The generality of the above sequence in replacing a sulfide linkage by a carbon-carbon double bond was proved by the synthesis of (1) the parent compound, *trans*-15,16-dihydropyrene (**282**) starting from the unsubstituted disulfide **281**[174,175] and (2) of *cis*-15,16-dimethyldihydropyrene (**283**) from the *syn*-disulfide **276b** (*vide supra*).[175]

281   282   276b(syn)   283

Similarly, dihydropyrenes with functionality in the cavity of the π-electron cloud, e.g., **284**,[177] as well as the *trans*-15n-butyl,16-methyl-dihydropyrene (**285**)[178] were obtained starting with suitably 1,2,3-substituted benzenes.[177,178] However, the analogous attempted synthesis of 15-methyl-15,16-dihydropyrene failed, the reaction products being pyrene and methane.[179]

284,*a*,R=-CH$_2$OCH$_3$ 285   286
b,R=-CH$_2$-CH=CH$_2$

Recently *trans*-15-phenyl-16-methyldihydropyrene (**286**), an unusual dihydropyrene containing an aromatic π-cloud within and perpendicular to a second one was synthesized after the same reaction scheme.[180] Some syntheses of 2,7-di-*t*-butyl derivatives of 15,16-dialkyl-dihydropyrenes were described by Tashiro and Yamato.[181,182]

i. Isomers of 15,16-Dimethyl-15,16-Dihydropyrene

The synthesis of **199** was announced as a preliminary communication[11] without experimental details.

199 (R=CH$_3$)   200 (R=CH$_3$)   287

The isomeric hydrocarbon **200** resulted on treatment of azupyrene (**205**) with lithium in THF follwed by alkylation of the resulting dianion with dimethylsulfate.[183] Alkylation of the same anion with 1,3-dichloropropane afforded the interesting derivative **287**.[184]

j. Pyrene and Isomers

From the pyrene isomers recently enumerated[185] only eight contain a central tetrasubstituted ethene bridge and include five- to seven-membered rings, namely **201** to **208** (see Section IV.A). Analogous compounds, including four- or eight-membered rings, seem to be yet unknown.

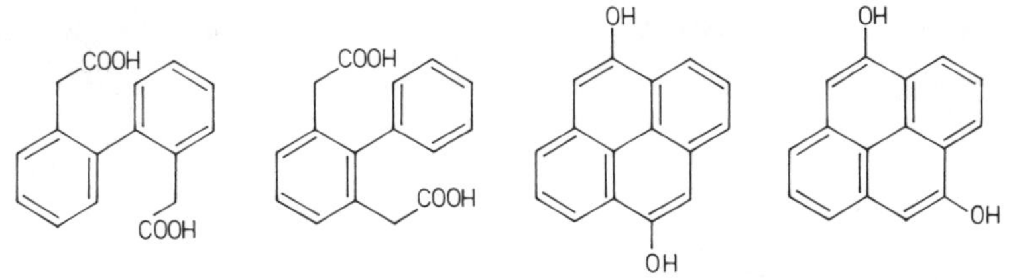

288   289   290   291

Pyrene (**201**)[186] is found in coal tar[187] and results in cracking processes of the petroleum industry. An early synthesis[188] converts diphenyldiacetic acids **288** and **289** into dihydroxypyrenes **290** and, respectively, **291** by cyclization, then converts these compounds into **201** by zinc distillation. Similarly, the 1,5-naphthalene-dipropionic acid was converted to **201**.[189] Other syntheses of pyrene involve tetracyclic ketones **292** and **293** obtained from tricyclic ketones **294**[190] and **295**,[191] respectively. Zinc reduction of tetracyclic diketone **296** obtained from perinaphthane also affords pyrene.[192]

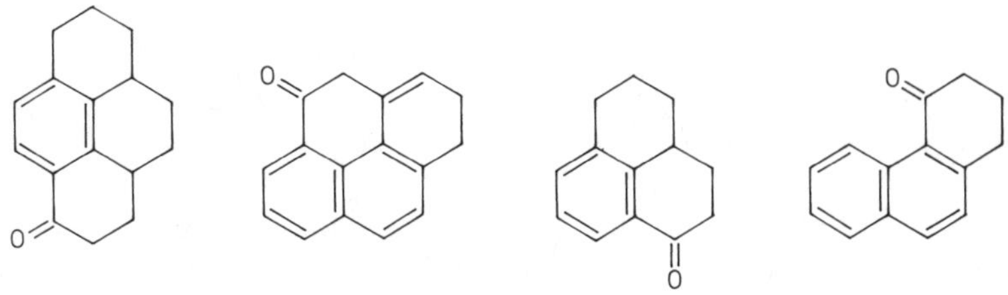

292   293   294   295

Wurtz reaction of *m*-xylylene dibromide affords the *m*-cyclophane **297**, which is converted into pyrene on dehydrogenation catalyzed by Pd/C.[193]

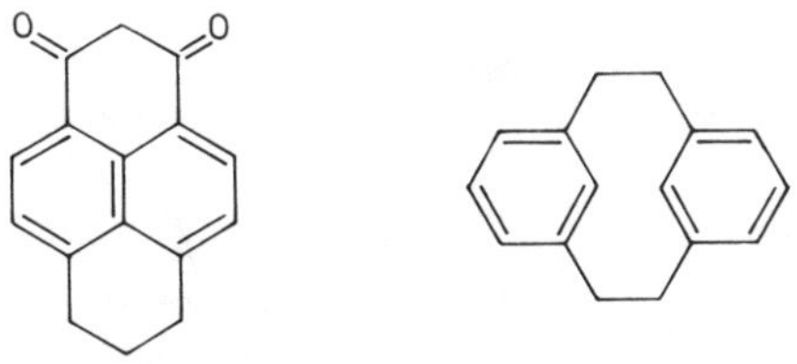

296   297

Cyclohepta[*bc*]acenaphthylene (**202**) was obtained in 1955[194] from acenaphthene-butyric acid (**298**) in four usual steps.

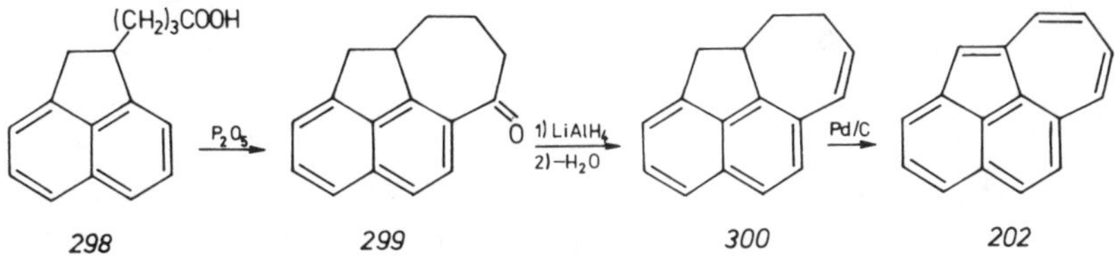

298   299   300   202

Cyclohepta[*fg*]acenaphthylene (Acepleiadylene) (**203**) was first obtained by Boekelheide and Vick[195] in the Pd/C-catalyzed disproportionation of acepleiadiene (**302**),* a compound

---

* The utilization of the stem "pleiad" to indicate the presence of the seven-membered ring joined to the naphthalene nucleus was suggested by Fieser.[405] The name "acepleiadiene" for **302** as well as that of "pleiadiene" for the compound without the five-membered ring were proposed by Boekelheide.[196]

obtained previously from 1,4-acepleiadanedione **301**.[196] Later, **203** was also obtained on treatment of acepleiadane **304** with chloranil.[197]

301 302 203 303 304

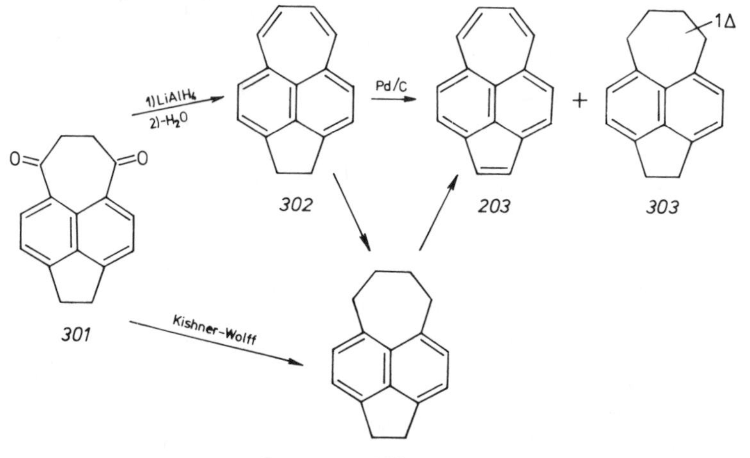

Napht[*cde*]azulene (Cyclohepta[*klm*]benz[*e*]indene (**204**) was prepared in 1958[197] starting from the hydroxyester **305** which could be obtained through a Reformatsky reaction. From both unsaturated acids obtained on dehydration of **305** the tetracyclic ketone **306** was prepared through usual steps. The alkene **307** obtained from **306** was converted to **204** by means of chloranil.

305 306 307 204

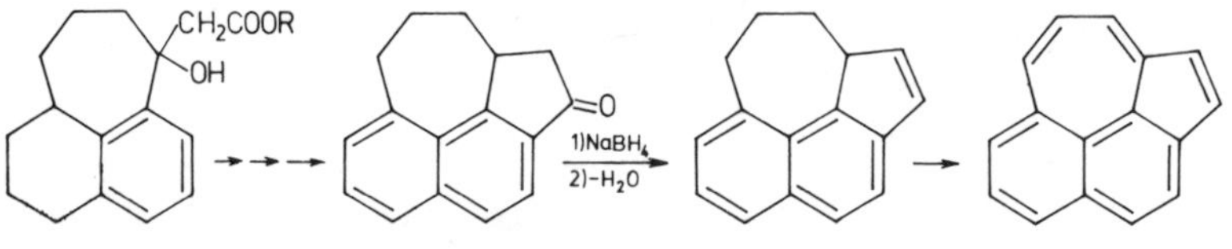

Dicyclopenta[*ef,kl*]heptalene (azupyrene, pyraceheptylene) (**205**) — The first synthesis of **205** (a nonalternant isomer of pyrene), announced by Anderson in 1968[198] and presented in detail some years later,[199,200] used as key intermediate the ketone **308** obtained from the cyclization of 4(1-indanyl)-butanoic acid. A Reformatsky reaction followed by polyphosphoric acid treatment yielded the tetracyclic ketone **309**. Treatment of the hydrocarbon **310** (obtained on reduction of ketone **309**) with ethyl diazoacetate afforded a ring-enlarged ester **311**. Saponification of **311** followed by treatment with Pd/C at 350°C (decarboxylation and dehydrogenation) yielded **205**.

308 309 310 311

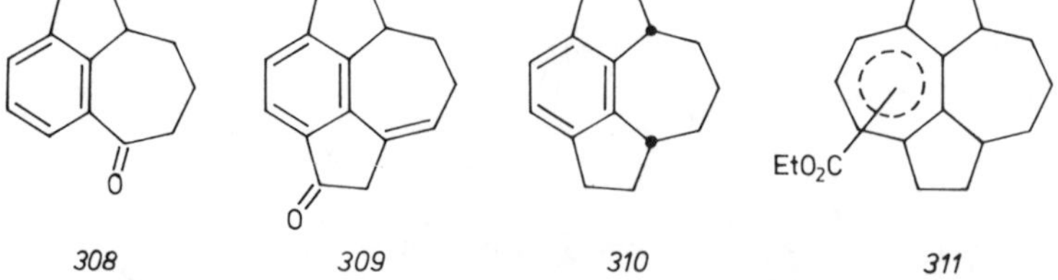

Other syntheses of **205** were devised by Jutz and Schweiger. In one of these the Ziegler-Hafner azulene synthesis was utilized starting with enamines **312** and **313**.[201]

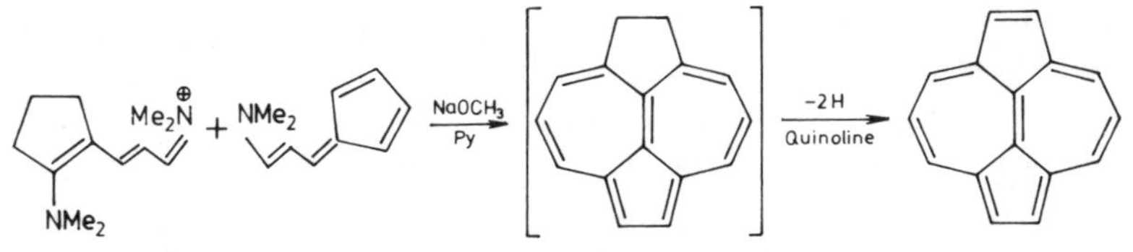

312   313   205

In the second improved one, the cyclopentenoazulene **315** was obtained by a Hafner synthesis from **312** via **314**; the cyclization of **316** to the tetracyclic skeleton, a pericyclic 10π reaction, was performed thermally and the final dehydrogenation of **317** was achieved by means of chloranil.[202]

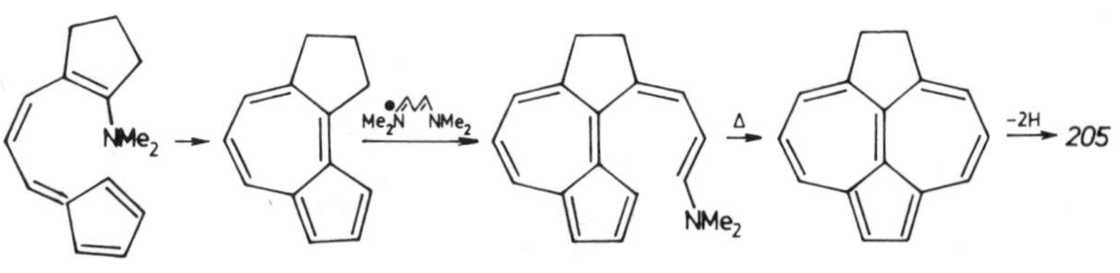

314   315   316   317

A facile synthesis of disubstituted azupyrenes (e.g., **319**) was described in 1976 by Hafner[203] through cycloaddition reaction of the readily accessible aceheptylene **318**[204] with dimethylacetylene dicarboxylate:

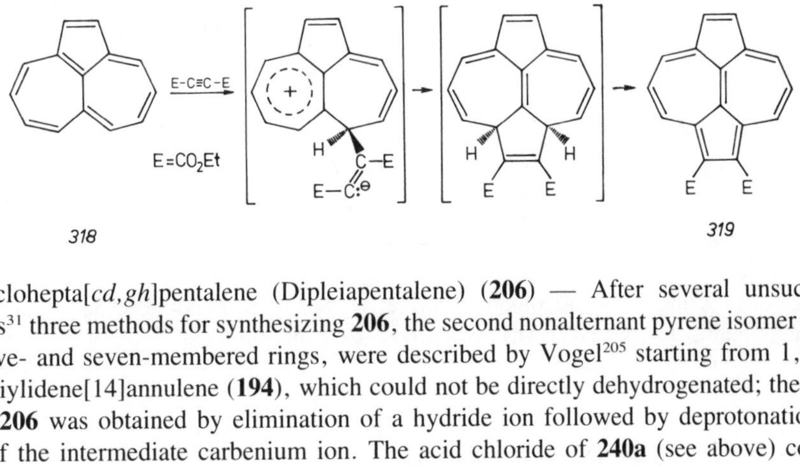

318   319

Dicyclohepta[*cd,gh*]pentalene (Dipleiapentalene) (**206**) — After several unsuccessful attempts[31] three methods for synthesizing **206**, the second nonalternant pyrene isomer formed from five- and seven-membered rings, were described by Vogel[205] starting from 1,6:8,13-ethanediylidene[14]annulene (**194**), which could not be directly dehydrogenated; the hydrocarbon **206** was obtained by elimination of a hydride ion followed by deprotonation with water of the intermediate carbenium ion. The acid chloride of **240a** (see above) could be converted to **206** with palladium chloride in decalin or with sodium azide followed by heating.[205]

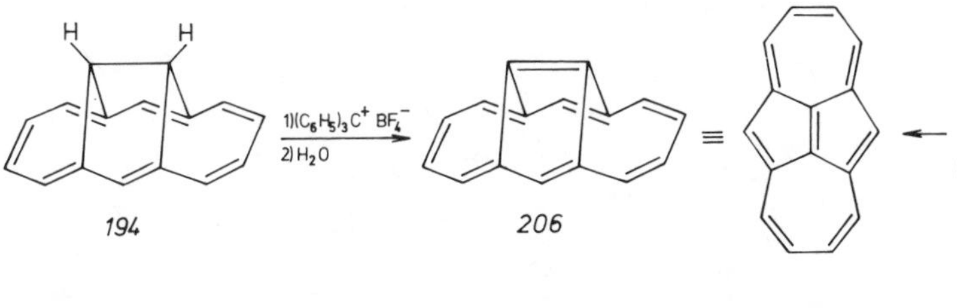

194   206   240 (R=COCl)

Recently, a facile synthesis of dicyclohepta[*cd,gh*] pentalene derivatives was announced.[206] Reaction of 5*H*-cyclohept[*a*]azulen-5-ones **320** with haloketenes R′CX=C=O afforded dipleiapentalene derivatives **321**. In some cases lactones **322** appeared (probably from intermediate cyclohept[*a*]azulenium ion **323**) and could be converted to **321** on treatment with triethylamine.

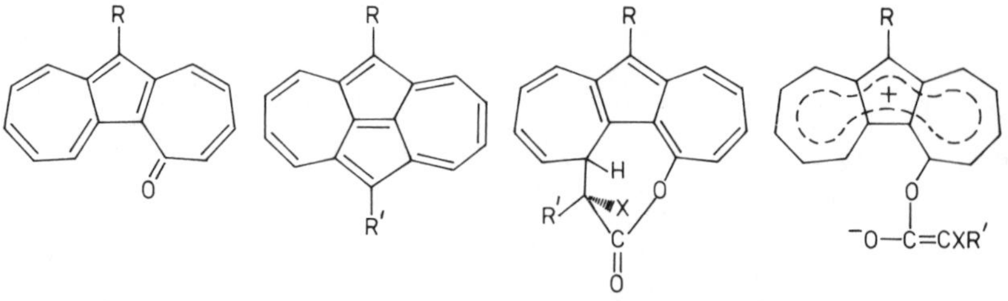

320 , R = H,COOMe   *321*   *322*   *323*

Pentaleno[216-*def*]heptalene (**207**), the third possible nonalternant isomer of pyrene formed from five- and seven-membered rings, is formally constituted from two azulenes or from a pentalene and a heptalene molecule. Two elegant synthetic approaches for **207** and derivatives, described by Hafner, involve either the formation of a new seven-membered ring on a cyclopent[*cd*]azulene molecule or of a new five-membered ring on an aceheptylene molecule. Thus, from 4,6-dimethyl-1,8-cyclopentenoazulene (**324**)[207] and *N*-methylanilinoacrolein, the immonium salt **325** resulted; heating with sodium methoxide yielded **326** (through intramolecular condensation accompanied by *N*-methylaniline loss) which afforded the methyl derivative **327** on treatment with chloranil.[208]

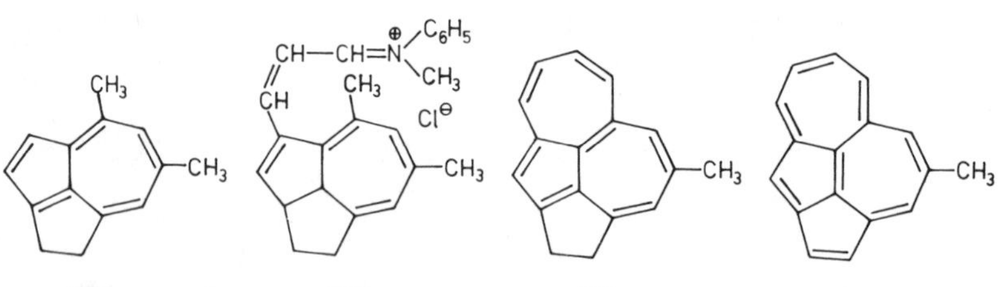

324   *325*   *326*   *327*

Similarly, starting from monomethyl derivative **328** corresponding to **324**, the unsubstituted pentalenoheptalene (**207**) was obtained in 1978 in two ways.[209]

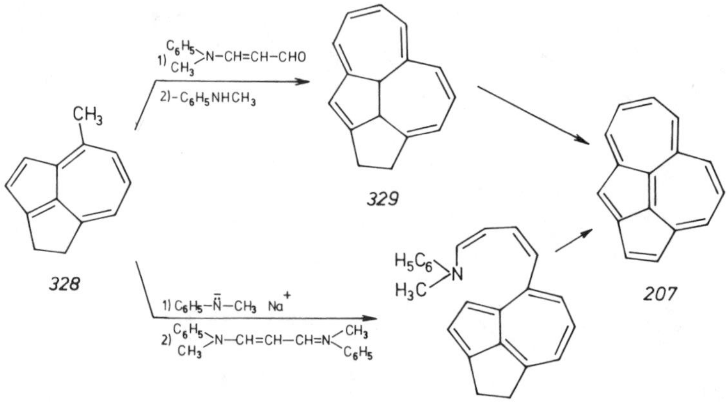

330

Another route to **207** uses 3-aceheptylenylacetic acid (**331**). The ketone **332** resulted on its cyclization is converted into dimethylaminoderivative **333** then into iminium salt **334** from which **207** is obtained through elimination of dimethylamine.[209]

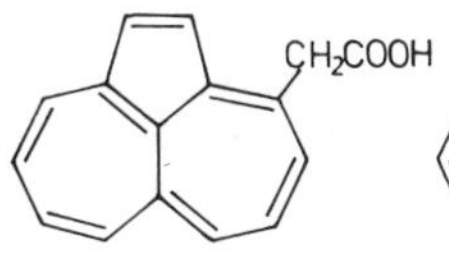

331 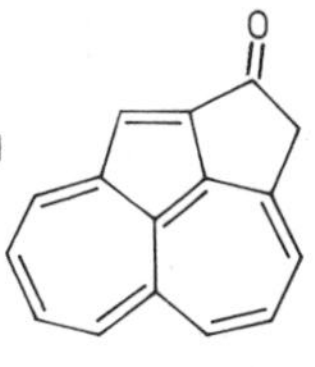   *332* 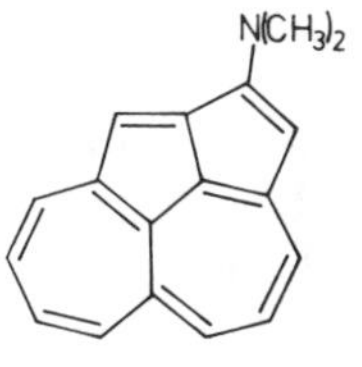   *333* 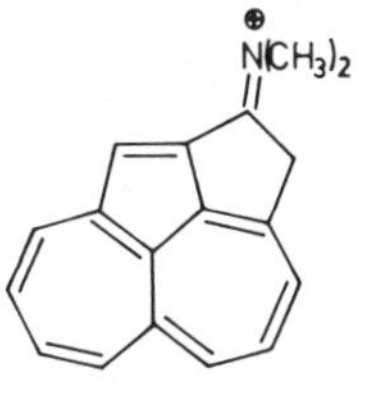   *334*

On the other hand, from aceheptylene derivative **335**[207,210] trimethyl-pentalenoheptalene **336** could be obtained.[204]

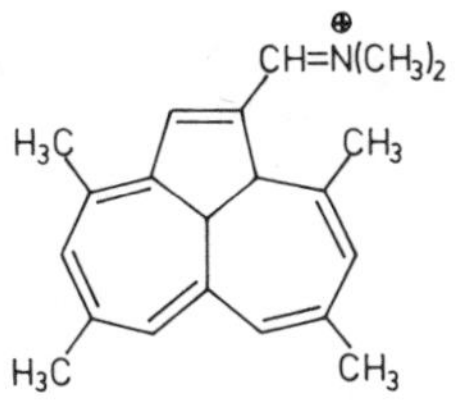

335 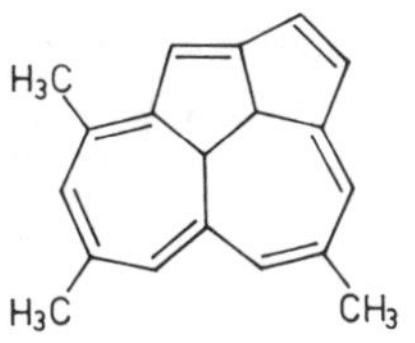   *336* 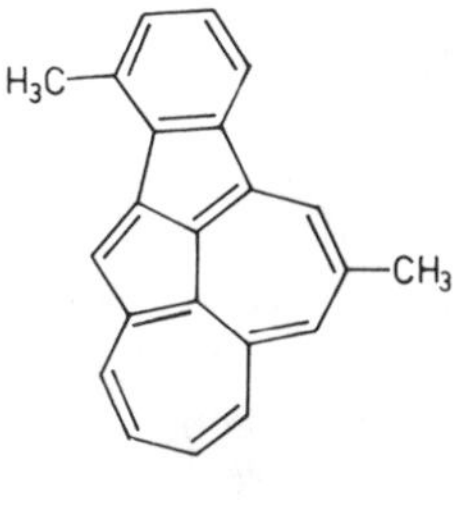   *337*

A dimethyl-benzoderivative of **207**, namely **337**, was obtained from the reaction of the corresponding benzocyclopentazulene with *N*-methyl-*N*[3(*N*-methylanilino)-2-propenylidene]anilinium perchlorate followed by thermal ring closure.[140]

Cyclohepta[*def*]fluorene (**208**) remains yet unsynthesized in spite of numerous attempts.

The addition of diazoacetic ester to 4*H*-cyclopenta[*def*]phenanthrene affords the esters **338a,b**[211,212] and not the ester **339**, as was initially suggested.[194]

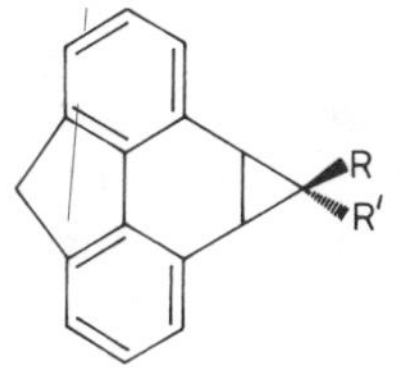

338 , *a* , R=H; R'=CO₂Et   *339* 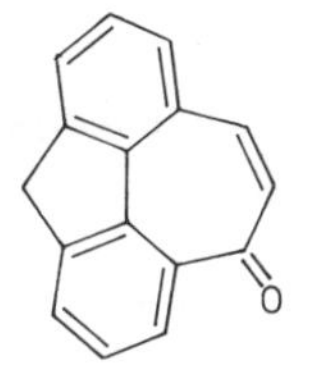   *340* 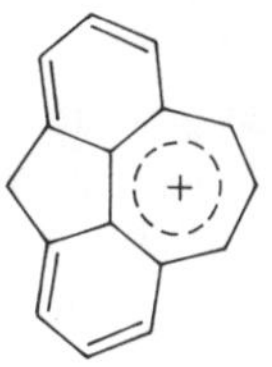   *341*
       b , R=CO₂Et ; R'=H

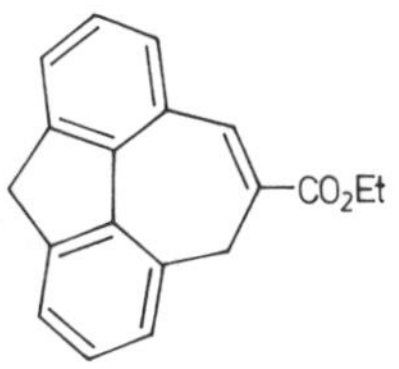

The ketone **340** obtained from the acid corresponding to **338a** does not show any sign of enolization.[211,212] Attempts to generate **208** by proton abstraction from the ion **341** obtained from **340** failed.[211,212] Attempts to tautomerize the methylenic hydrocarbon **342**,[212] to dehydrate the diol **343**,[212] to dehydrogenate **344**,[213] or to eliminate nitrogen from the 4-diazoderivative of **208**[213] also remained unsuccessful. A more recent attempt to synthesize a methylderivative of **208** through dehydrogenation of **345** also failed.[214]

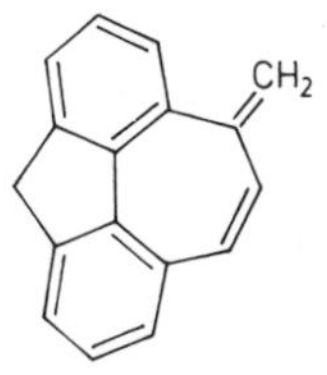

342 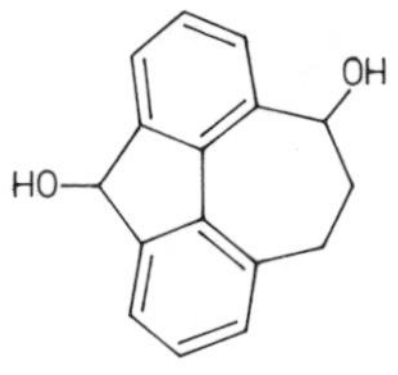   *343* 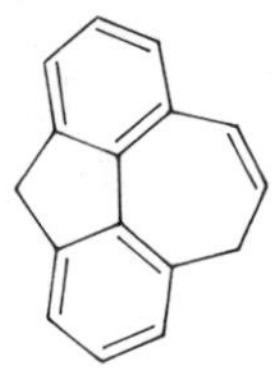   *344* 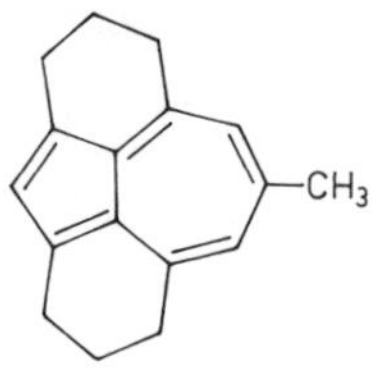   *345*

C. Structure and Physical Characterization

1. Tricyclic Bridged [14]Annulenes and Derivatives

Some physical data of carbon bridged [14]annulenes are collected in Table 4.

a. Syn-1,6:8,13-Bismethano[14]Annulene (188)

The large shielding of bridge protons in the ^{1}H-NMR spectrum as well as the important deshielding of vinylic protons (Table 4) are evidence for a significant diamagnetic ring current, thus proving the aromaticity of **188**.[156] The X-ray analysis revealed the structure presented in Figure 2,[215] in excellent agreement with the results of EHT calculations.[45] The distance of the inner hydrogen atoms is 1.78 Å, *one of the shortest distances of nonbonded hydrogens reported*.[11,11a] Force field calculations[43] also support the above structure and indicate that the contact between the inner hydrogens is partly relieved through an increased dihedral angle between the two bridges (calculated 26.4°; experimental 26.6°). Interestingly, the *syn*-isomer **188** is about 15 kcal · mol^{-1} more strained than the *anti*-isomer **189**.[43] Thus, the *syn*-hydrocarbon **188** is an aromatic molecule even if it contains a somewhat bent annulene ring; the aromatic character is obtained at the expense of the steric compression between the inner hydrogen atoms from the CH$_2$ bridges.

b. Anti-1,6:8,13-Bismethano[14]Annulene (189)

This compound is oxygen-sensitive and has a polyenic character. The ^{1}H-NMR spectrum at 20°C (CDCl$_3$; δ ppm): 1.88 and 2.48 (CH$_2$ protons) and 6.20 and 6.33 (s, vinylic H), changes considerably at lower temperatures (Table 4), indicating a dynamic process **189a** ⇄ **189b** (Figure 3). The calculated activation energy for double bond fluctuation in **189** is 7.1 kcal · mol^{-1} and the frequency factor $10^{12.2}$.[155] The resonance hybrid, **189c**, situated at a higher energy level than structures with localized double bonds, could represent the transition state of the valence tautomerism **189a** ⇄ **189b**. These unusual energy levels are probably due to strong deviation of the ring skeleton from planarity.

The X-ray analysis of **189**[216] proved the puckering of the annulene ring and indicated a torsional angle of 2p$_z$ orbitals at neighboring carbons of 70°. Force field calculations[43] indicated a C$_s$ symmetry of the molecule, the nonplanarity of the ring, as well as bond alternation throughout the perimeter.

In conclusion, **189** is a cyclic polyolefin with fluxional π-bonds.

c. Syn-1,6-Ethano-8,13-Methano[14]Annulene (190)

This compound is an oxygen-sensitive solid whose ^{1}H-NMR and electronic spectra (Table 4) reveal its olefinic nature. ^{1}H-NMR data, including a positive NOE effect, point to the *syn*-configuration.[158] Surprisingly, the X-ray analysis,[158,217] though indicating the *syn*-configuration, points to similar bond length for perimeter bonds. The 400-MHz ^{1}H-NMR analysis[217] proves the valence tautomerism **190a** ⇄ **190b** with an equilibrium constant of K = 2 and ΔG° = 2 kj · mol^{-1} in favor of **190b** at room temperature and rules out the valence tautomeric forms **190c** and **190d**.

Table 4
PHYSICAL CHARACTERIZATION OF SOME CARBON-BRIDGED [14]ANNULENES

Compound	Melting point (°C) (color)	^{1}H-NMR spectrum ^{13}C-NMR spectrum (δ ppm)	Electronic spectrum (λ nm)	Ref.
188	116 (Yellow-orange	−1.2 (d; H^a); 0.9(d; H^b); 7—8(m; 10H aryl) with 7.9(s,H-7,14) (in CDCl$_3$)	303; 320; 361; 469; 473; 479; 487; 493; 498	156
189	41—42 (Yellow)	1.47 and 2.39(AB; CH$_2$); 2.29 and 2.67(AB; CH$_2$); 6.33(s; H-7,14); 5.74(H-3,4) and 6.34(H-2,5) -AA′BB′; 6.07(H-9,12) and 6.57(H-10,11) -AA′BB′ (at −130°C, in CDCl$_3$)	217; 272; 350	155
190	Red	^{1}H-NMR: 1.53 and 4.28(H-17)-AX; 1.90—2.82(H-15,16) -AA′XX′; 5.98—6.55(m; 10 ring H)(in CCl$_4$); ^{13}C-NMR: 29.9; 32.1; 123.6; 126.4; 128.4; 132.0; 137.7; 133.9; 141.7 (in CD$_2$Cl$_2$)	287; 381	158
191	>250(Dec.) (Orange-red)	^{1}H-NMR: 7.81 and 8.11(H-2-5; 9-12) -AA′BB′; 8.53(s; H-7,14) (in d$_6$-DMSO); ^{13}C-NMR: 125.32(C-1,6,8,13); 129.53; 127.15(C-2-5; 9; 10-12); 129.22(C-7,14); 194.71 (C-15,16) (in d$_6$-DMSO)	292; 312; 365; 442; 509 (in dioxane)	
192	>138(Dec.) (carmine-red)	^{1}H-NMR: 6.85 and 7.18(H-2-5; 9-12) -AA′BB′; 7.78(s; H-7,14) (in d$_6$-DMSO); ^{13}C-NMR: 125.96; 128.29; 128.43; 144.89(C-1,6,8,13); 205.23 (C-15,16)(in d$_6$-acetone)	240; 318; 384; (in dioxane)	160
194	118—119 (Scarlet red)	−1.82(s; H-15,16); 8.0(s; H-7,14); 7.82(H-3,4,10,11) and 8.17(H-2,5,9,12) -AA′BB′ (in CDCl$_3$)	308; 339; 381; 398; 472; 551	162
196	180—181 (Orange)	^{1}H-NMR: -1.16(t; H-15,17); −0.61(t; H-16); 7.74 and 7.55(H-2,5; H9-12)-AA′ BB′; 7.88(s; H-7,14) (in CDCl$_3$);^{13}C-NMR: 18.7(C-16); 28.8(C-15,17); 111.9(C-1,6,8,13); 127.0(C-7,14); 127.7(C-2,5,9,12); 130.0(C-3,4,10,11) (in CCl$_4$)		163, 223, 226
197	175—176 (Yellow)	−0.96(H-15,18); 0.52(H-16,17); 7.12 and 7.57(H-2,5; H-9-12) -AA′BB′; 7.86(s; H-7,14) (in CCl$_4$)	302; 320; 360; 373; 467; 473; 480; 486; 492	164
203	156—162 (Red)	^{1}H-NMR: 6.89(H-5,6): 7.79(H-4,7) 7.83(H-12,13); 7.95(H-1,10); 8.33(H-2,9); ^{13}C-NMR: 125.8(C-1,10); 126.2 (C-12,13); 126.6(C-16); 126.9(C-4,7; 127.0(C-15); 127.4(C-2,9); 134.9(C-3,8); 137.0(C-5,6); 138.2(C-11,14)	190; 220; 255; 315; 330; 480; 500; 515; 550; 560 (in ethanol); values extracted from a figure	195

Table 4 (continued)
PHYSICAL CHARACTERIZATION OF SOME CARBON-BRIDGED [14]ANNULENES

Compound	Melting point (°C) (color)	¹H-NMR spectrum ¹³C-NMR spectrum (δ ppm)	Electronic spectrum (λ nm)	Ref.
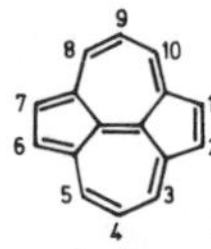 **205**	257—259 (Bronze)	7.34(t; H-4,9); 8.40(s; H-1,2,6,7) 8.68(d; H-3,5,8,10) (in CDCl₃)	252; 267; 285; 299; 308; 334; 343; 356; 409; 442; 452; 459; 470; 483	198, 200, 202
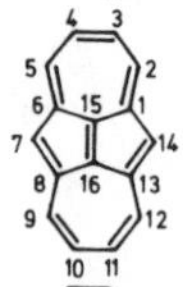 **206**	141—142 (Green-black)	8.06(s; H-7,14); 8.08(H-3,4, 10,11) and 8.66(H-2,5,9,12) - AA'BB' (in CS₂)	235; 284; 296; 310; 321; 363; 367; 374; 383; 387; 427; 486	205 220
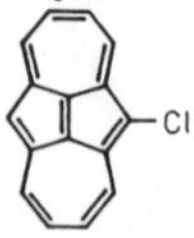 Cl	120—121 (Greenish-brown)	8.15(s; H-14); 8.07—8.43(M; H-3,4,10,11); 8.55—8.94(m; H-2,5,9,12) (in CDCl₃)	240; 284; 310; 372; 388; 394; 460; 490; 590 (in methanol)	206
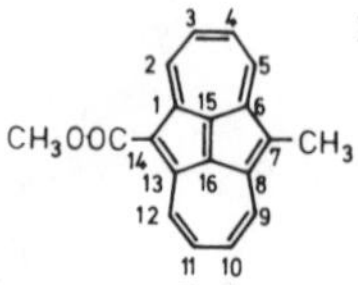 CH₃OOC / CH₃	168—169 (Green)	2.74(s; CH₃); 4.13(s; CH₃O); 7.62 − 8.50(m; H-3-5; 9-11); 9.26(dd; H-2,12) (in CDCl₃)	290; 302; 316; 350; 367; 386; 407; 447; 477; 640 (in methanol)	206
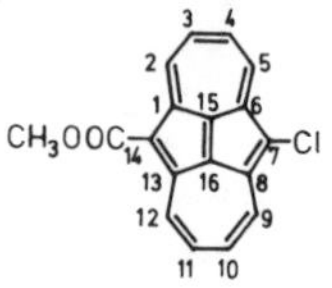 CH₃OOC / Cl	198—200 (Yellowish-green)	¹H-NMR: 4.61(s; OCH₃); 8.29 (m; H-3,4,10,11); 8.81(m; H-5,9); 9.46(m; H-2,12) (in CDCl₃); ¹³C-NMR: 112.8(s; C-14); 121.8(s; C-15,16); 113.5(s; C-7); 128.7(d; C-2,12); 131.0(d; C-3,11); 131.5(d; C-4,10); 134.2(d; C-5,9); 139.4(s; C-1,13); 145.6(s; C-6,8) (in CDCl₃)	289; 315; 368; 387; 409; 451; 480; 640 (in methanol)	206
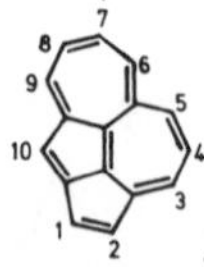 **207**	136—137 (Red-brown)	6.88—8.46(m; 8H arom); 7.49(d); 8.55(d)	287; 327; 376; 420; 448; 500; 527; 580; 725; (in n-hexane)	209
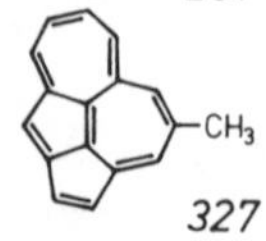 CH₃ **327**	109 (Black)	2.7(s; CH₃); 6.5—8.4(m; 9 ring H) (in CDCl₃)	255; 262; 284; 291; 317; 367; 382; 396; 405; 421; 449; 504; 539; 582; 710; 813; (in n-hexane)	204, 208, 272
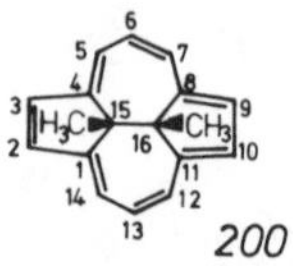 **200** (R=CH₃)	186 (Deep red)	¹H-NMR: − 4.53(CH₃); 8.04(H-6,13); 8.74(H-2,3,9,10); 8.77(H-5,7,12,14); ¹³C-NMR: 15.0(C-15a); 43.1(C-15); 119.2 (135.3)C-2; 135.3(119.2) C-5; 143.7(C-1); 152.6(C-6)	335; 346; 377; 397; 420; 440; 445; 507; 550; 575; 603	183
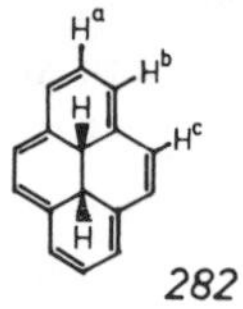 **282**	Unstable compound (green)	− 5.49(inner 2H); 7.82—8.02(m; Hᵃ); 8.5(d,Hᵇ); 8,58(s; Hᶜ)	262; 272; 305; 325; 355; 361; 394; 412; 432; 444; 497; 506; 516; 523.5; 553; 566; 571; 582; 595; 607; 623	175

Table 4 (continued)
PHYSICAL CHARACTERIZATION OF SOME CARBON-BRIDGED [14]ANNULENES

Compound	Melting point (°C) (color)	¹H-NMR spectrum ¹³C-NMR spectrum (δ ppm)	Electronic spectrum (λ nm)	Ref.
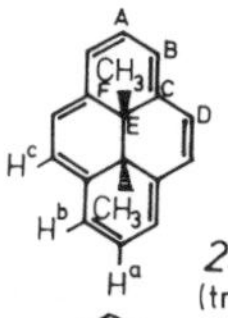 **254** (trans)	119—120 (Dark-green)	¹H-NMR: −4.25(s; CH₃); 7.98—8.23(m; Hᵃ); 8.57,8.67(d; Hᵇ); 8.67(s; Hᶜ); ¹³C-NMR: 14.0(F); 30.0(E); 122.8(A); 123.3 (B;D-): 136.6(C)	337.5; 377; 463; 528; 536; 586; 598; 611; 627; 634; 641	167, 168
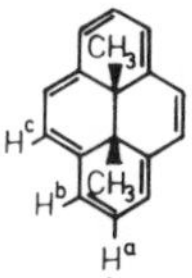 **283** (cis)	90—95 (Dark-green)	−2.06(s,CH₃); 7.50(t; Hᵃ); 8.24(d; Hᵇ); 8.74(s; Hᶜ)	310; 328; 335; 355; 396; 420; 439; 506; 564; 581; 602	175
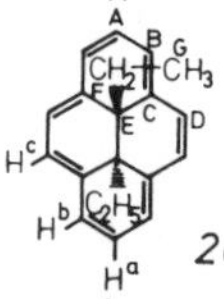 **268**	148—151 (Deep green)	¹H-NMR: −3.96(q; CH₂); −1.86(t; CH₃); 7.95(t; Ha); 8.64(s; Hᶜ); 8.67(d; Hᵇ); ¹³C-NMR; 5.1(G); 20.3(F); 35.1(E); 123.0(A); 123.8(D); 125.5(B); 134.5(C)	345; 349; 367; 386; 391; 428; 493; 545; 554; 606; 619; 633; 649; 655	171
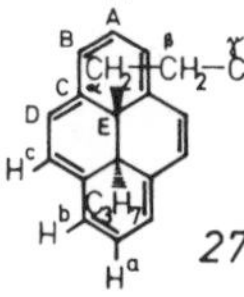 **271**	136.5—138 (Dark green)	¹H-NMR: −4.03(m; Hᵃ); −1.1 ÷ −2.2(m; Hᵇ); −0.65(t; Hᵞ); 7.95(t; Hᵃ); 8.67(d; Hᵇ); 8.67(s; Hᶜ); ¹³C-NMR: 13.3(γ); 13.9(β); 29.8(α); 34.9(E); 123.0(A); 123.9(D); 125.4(B); 135.0(C)	331; 345; 348; 366; 386; 391; 465; 544; 553; 605; 618; 632; 648; 664	172
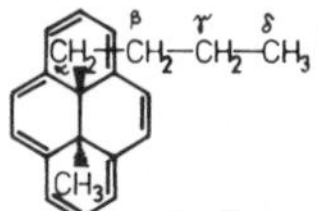 **285**	54—54.5 (Deep green)	−4.3(s,CH₃); −3.9 ÷ −4.14 (m; Hᵃ); −1.51 ÷ −1.91 (m; Hᵇ); −0.2 ÷ −0.62 (m; Hᵞ); −0.1(t; Hᵟ); 7.78 ÷ 8.11 (m; 2H aryl); 8.55(m: 8H aryl)	238; 278; 324; 340; 343; 358; 379; 383; 420; 441; 463; 478; 485; 533; 542; 575; 581; 592; 605; 618; 634; 647; 652	178
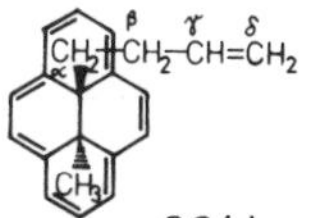 **284b**	42—43 (Dark green)	−4.25(s; CH₃); −4.21 ÷ −4.10(t; Hᵃ;10.29−0.42(m; Hᵇ); 3.39—3.61(m; Hᵞ,Hᵟ); 7.98—8.26(m; 2H aryl); 8.58—8.82(m; 8H aryl)	237; 275; 320; 341; 345; 358; 380; 384; 415; 439; 450; 475; 483; 534; 574; 590; 605; 617; 634; 651	177
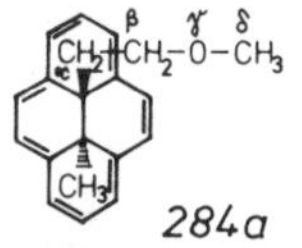 **284a**	50—51 (Dark green)	−4.30(s; CH₃); −3.43 ÷ −3.63 (t; Hᵃ); −0.40 ÷ −0.61 (t; Hᵇ); 2.40(s; CH₃O); 7.98—8.26 (m; 2H aryl); 8.58—8.82 (m; 8H aryl)	275; 324; 337; 346; 357; 380; 430; 481; 533; 569; 583; 635; 651	177
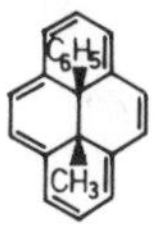 **286**	159—160 (Green)	−4.45(s; CH₃;2.55(d; H_ortho) 5.6(app.d; H_meta); 6.0(t; H_para 7.9—8.8(m; 10H aryl)		180
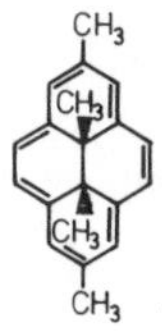 **358b**	194—195 (Deep green)	−4.09(s; inner CH₃); 3.03(s; outer CH₃); 8.39(s; 4H aryl); 8.49(s; 4H aryl)	340; 377.5; 470; 534.5; 600; 635; 648	275

Table 4 (continued)
PHYSICAL CHARACTERIZATION OF SOME CARBON-BRIDGED [14]ANNULENES

Compound	Melting point (°C) (color)	^{1}H-NMR spectrum ^{13}C-NMR spectrum (δ ppm)	Electronic spectrum (λ nm)	Ref.
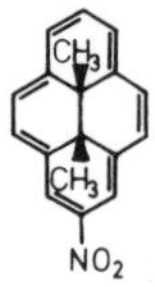 NO₂ (cis)	140—145 (Deep purple)	−1.89;s,CH₃); −1.98(s; CH₃); 7.77(t; 1H aryl); 8.39(d; 2H aryl); 8.10(d; 2H aryl); 9.10(d; 2H aryl); 9.28(s; 2H aryl)	288; 342; 378; 484; 562; 617	277
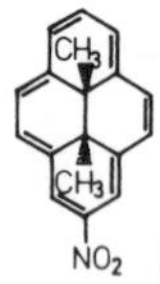 NO₂ *353a* (trans)	172—173 (Deep purple)	−4.03(CH₃); 8.25—8.50(m; 1H aryl); 8.72(d; 2H aryl); 8.87(d; 2H aryl); 9.07(d; 2H aryl); 9.63(s; 2H aryl)	348; 382; 406; 516; 598; 662	274
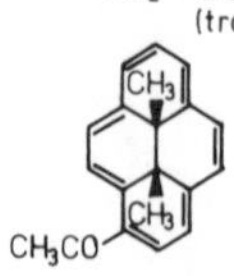 CH₃CO *360a* (cis)	Deep green crystals	−1.97(s; CH₃); −1.94(s; CH₃); 2.94(s; CH₃CO); 7.63(t; 1H aryl); 8.33(d; 2H aryl); 8.09 and 8.27(AB; 2H aryl); 8.83(s; 2H aryl); 8.87(d; 1H aryl) 9.65(d; 1H aryl)	358; 423; 442; 570; 616	277
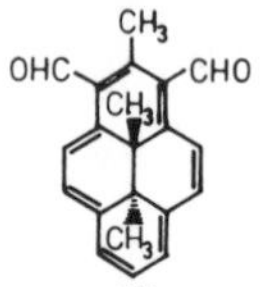 OHC CHO	190—192 (Deep green)	−3.78(s; CH₃); −3.61(s; CH₃); 3.46(s; CH₃); 8.08(1H aryl) and 8.64 (2H aryl)-A₂B; 8.78 and 9.40(AB; 4H aryl); 11.58(CHO)		314
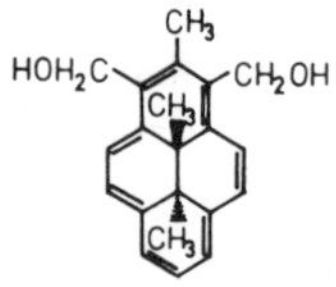 HOH₂C CH₂OH	210—212 (Green)	−4.08(s; CH₃); −3.99(s; CH₃); 3.27(s; CH₃); 5.70 and 5.84 (AB; CH₂O); 8.03(1H aryl) and 8.56 (2H aryl); A₂B system; 8.64 and 8.89(AB; 4H aryl)		314
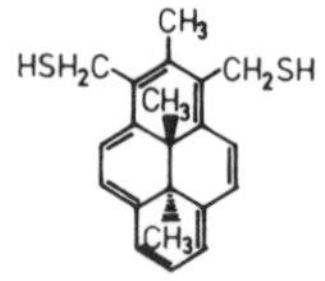 HSH₂C CH₂SH	103—105 (Green)	−4.07(s; CH₃) −3.97(s; CH₃); 3.14(s; CH₃); 1.96(t; SH); 4.66 and 4.92(CH₂S); 8.0(1H aryl) and 8.51 (2H aryl); A₂B; 8.59 and 8.63(AB; 4H aryl)		314

Valence Isomers

Compound	Melting point (°C) (color)	^{1}H-NMR spectrum ^{13}C-NMR spectrum (δ ppm)	Electronic spectrum (λ nm)	Ref.
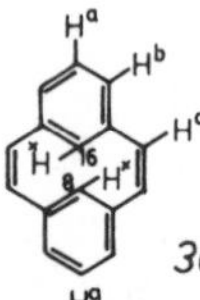 *362*	119—120 (Pale Yellow)	6.22(s; Hᶜ); 6.60(m; Hᵇ); 7.01(m; Hᵃ); 7.90(s; Hˣ)	280	174, 175
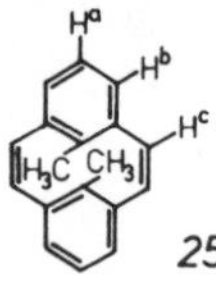 *253*		1.55(s; CH₃); 6.28(s; Hᶜ); 6.60(d; Hᵇ); 7.0(t; Hᵃ)	275	168, 175, 278

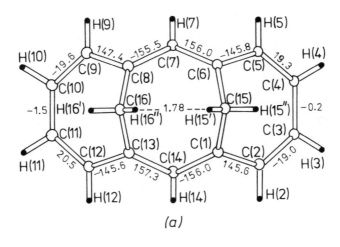

(a)

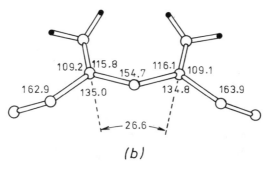

(b)

FIGURE 2. X-ray data for *syn* 1,6:8,13-bismethano[14]annulene (**188**): (a) numbering scheme, torsional angles along the annulene perimeter and the distance H-15'... H-16" Å; (b) dihedral angles. (From Destro, R., Pilati, T., and Simonetta, M., *Acta Cryst.*, B33, 940, 1977. With permission.)

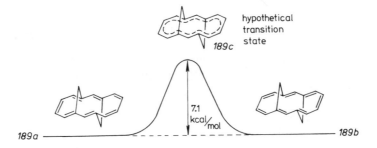

FIGURE 3. Double bond fluctuation in *anti*-1,6:8,13-bismethano [14]annulene (**189**). (From Vogel, E., Haberland, U., and Günther, H., *Angew. Chem. Int. Ed. Engl.*, 9, 513, 1970. With permission.)

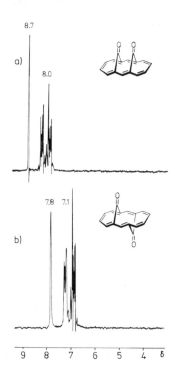

FIGURE 4.   ¹H-NMR spectra (90 MHz) of (a) 15,16-
dioxo-*syn*-1,6:8,13-bismethano[14]annulene (**191**) and
(b) 15,16-dioxo-*anti*-1,6:8,13-bismethano[14]annulene
(**192**); (D₆-DMSO; TMS). (From Vogel, E., Nitsche,
R., and Krieg, H.-U., *Angew. Chem. Int. Ed. Engl.*,
20, 811, 1981. With permission.)

In conclusion, **190** is an olefin which cannot attain aromaticity due to geometric reasons
(steric compressions between the bridges, thus nonplanarity of the perimeter) and which
valence tautomerizes by double bond fluctuation like an antiaromatic compound.

d. 15,16-Dioxo-Syn- and 15,16-Dioxo-Anti-1,6:8,13-Bismethano[14]Annulenes (191 and 192)

The π-systems of carbonyl groups and annulenic ring from *syn*-**191** are practically independent, as shown by models. The ¹H-NMR data[159] (see Table 4) as well as X-ray data[218]
evidence the aromaticity of **191** (C–C bonds of 1.376 to 1.413 Å throughout the perimeter).
The C–C–C bond angle of the bridge is 112.4° and the repulsion of oxygen atoms causes
their outer deviation from the corresponding plane of the bridge; consequently, the annulenic
ring in **191** is only slightly bent.[218]

The *anti*-derivative **192** rapidly polymerizes in air but shows a remarkable thermal stability.[160] The ¹H-NMR spectrum clearly indicates, especially in comparison with **191** (Figure
4), the olefinic character of **192**. The presence of only one AA′BB′ system suggests a rapid
π-bond shift whose activation energy is much smaller than in the case of the corresponding
hydrocarbon **189** because the ¹H-NMR spectrum remains unchanged up to − 130°C.

Whereas the electronic spectrum of **192** denotes a conjugative interaction of CO and annulenic moieties the, νCO of 1695 cm^{-1} rules out such an interaction.[160] The X-ray analysis[219] showed no bond alternation but it was suggested that this reported geometry of **192** could be due to disorder in the crystals.[160,219]

In conclusion, the pair of isomeric dioxobismethano[14]annulenes includes an aromatic compound (**191**) and an olefinic one (**192**). The great difference between **191** and **192** arises, as in the case of the corresponding hydrocarbons **188** and **189**, from the steric arrangements of the bridges.

e. Syn-1,6:7,12-Bismethano[14]Annulene (193)

This compound shows after preliminary physical data, a diamagnetic ring current,[109] indicating aromatic character.

2. Tetracyclic-Bridged [14]Annulenes and Derivatives
a. 1,6:8,13-Ethanediylidene[14]Annulene (194)

This compound presents, according to its ^{1}H-NMR data (Table 4), a diamagnetic ring current, which is stronger than in higher homologues **196** or **197**.[162] The UV spectrum of **194** is quite similar to those of **196** and **197**, but indicates a bathochromic and hyperchromic shift for the longest-wavelength band.[162]

b. 15,16-Methylene-1,6:8,13-Ethanediylidene[14]Annulene (195)

This compound is a stable aromatic molecule. The ^{1}H-NMR spectrum (δ ppm): 7.95 (H-3,4,10,11); 8.09 (H-7,14); 8.48 (H-2,5,9,12)[220] resembles that of dicyclohepta[*cd,gh*]pentalene (**206**). From the ^{13}C–H coupling constant for the CH$_2$ group (160 Hz), it appears that **195** is not contaminated with its valence isomer **195a** which would be devoid of an annulene perimeter. This situation is entirely reversed in the isomeric system **346** where only the benzenoid form exists.[221] The X-ray analysis of **195**[222] shows the similarity with **194**.

195   195a   346a   346

c. 1,6:8,13-Propanediylidene[14]Annulene (196)

This is an aromatic compound, as the ^{1}H-NMR and UV spectra show[155,223] (Table 4). The X-ray examination of **196** indicated[224,225] the nearly planar C$_{14}$ perimeter with C–C bonds characteristic for an arene. A comparative ^{13}C-NMR study of **196** and related derivatives gave[226] positive evidence for the existence of a diamagnetic ring current effect on the ^{13}C-resonances; however, the utility of ring current effects in ^{13}C-NMR spectroscopy seems to be limited due to other factors (ring strain, angle deformation) which more strongly affect the ^{13}C-resonances.

d. 1,6:8,13-Butanediylidene[14]Annulene (197)

This compound presents UV and ^{1}H-NMR data comparable to the lower homologue **196**.[164] From an X-ray investigation[227] it resulted that the perimeter bond lengths are between 1.371 and 1.422 Å (thus in a narrower interval than in anthracene). It seems that **197** is an aromatic compound with a nearly planar perimeter, the deformations appearing mostly at the bridge; introduction of a new carbon atom in the bridge (as compared to **196**) influences the bond angles much more than the bond lengths.

e. 15,16-Dialkyl-15,16-Dihydropyrenes (Boekelheide's Type Annulenes)

The *trans*-15,16-dialkyl-15,16-dihydropyrenes are typical aromatic compounds. Some physical characteristics of these hydrocarbons as well as of their valence tautomers are collected in Table 4.

^{1}H-NMR spectra are particularly relevant for the demonstration of the aromaticity of *trans*-15,16-dialkyldihydropyrenes. Thus, e.g., in dimethylderivative **254** the important shielding of CH$_3$ groups within the cavity of π-electron cloud (δ = −4.25 ppm) as well as the low field position of peripheric protons signals (δ = 7.98 to 8.67 ppm) clearly demonstrate the existence of a diamagnetic ring current.[168] The magnitude of the ring current remains practically the same in different 15,16-dialkyl derivatives (unsubstituted on the periphery). A steady falloff of the shielding effect is observed when the internal protons become more distant from the center of the molecule. Thus, e.g., in *trans*-15,16-diethyldihydropyrene (**268**) the usual sequence of CH$_3$- and CH$_2$-signals from ethyl groups is reversed due to ring current effects (δ CH$_2$ = −3.96 ppm; δ CH$_3$ = −1.86 ppm.[171]) An interesting comparison of the ring current effects on the various internal protons depending on their distance from the center of the molecule was published[177] (Table 5).

An interesting feature is manifest in the ^{1}H-NMR spectrum of *trans*-15-phenyl-16-methyl-dihydropyrene (**286**) where the signal of the *ortho* aromatic protons at δ = 2.55 ppm makes these protons the most shielded aryl protons yet known[180] (4.8 ppm shielding in comparison with usual phenyl protons, so that their signal appears in the range of saturated protons).

The ^{13}C-NMR spectra of some 15,16-dialkyl-dihydropyrenes **198** were analyzed[228] (see also Table 4) and compared with those of dihydroderivatives **347** (lacking ring current).

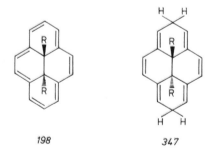

198   347

A mapping of ring current effects can be made for **198**.[228] The magnitude of ring current effects[229,230] for ^{13}C- and ^{1}H-resonances is essentially the same, and is in good agreement with calculations by Johnson and Bovey.[229]

The electronic spectra (Table 4) of practically all *trans*-15,16-dialkyldihydropyrenes (unsubstituted on the perimeter) are similar and agree well with an aromatic [14]annulenic structure. All these compounds are green solids. The electronic structures of derivatives **198** were discussed[231] and connections with Vogel's bridged [14]annulenes were made.[232]

The photoelectron spectrum of **254** was studied and was found to agree well with the calculated one.[233]

The ESR spectra of radical ions (cations and anions) of **254** were analyzed;[234] the observed spin densities for both radical ion types were in good agreement with MO calculations considering an unrestrained delocalization of the π-electron system.[234]

The mass spectra of 15,16-dialkyldihydropyrenes indicate as important fragmentation the loss of substituents from 15- and 16-positions and the appearance of the molecular ion of

Table 5
EFFECTS OF RING CURRENTS IN SOME 15,16-
DIALKYL-15,16-DIHYDROPYRENES (δ ppm)

Compound	Observed chemical shift	Reference value	Upfield shift due to ring current
One Atom Removed			
R=CH_2–CH_2–CH=CH_2	−4.14	1.33	5.47
CH_3	−4.25	0.91	5.15
R=CH_2–CH_2–O–CH_3	−3.52	1.33	4.85
CH_3	−4.30	0.91	5.21
R=CH_2–$(CH_2)_2$–CH_3	−4.02	1.33	5.35
CH_3	−4.30	0.91	5.21
Two Atoms Removed			
R=CH_2–CH_2–CH=CH_2	0.36	2.00	1.70
R=CH_2–CH_2–O–CH_3	0.51	3.40	2.89
R=CH_2–CH_2–CH_2–CH_3	−1.71	1.33	3.04
Three Atoms Removed			
R=CH_2–CH_2–CH=CH_2	3.86	5.72	1.86
R=CH_2–CH_2–CH_2–CH_3	−0.41	1.33	1.74
Four Atoms Removed			
R=$(CH_2)_2$–CH=CH_2	3.50	5.28	1.78
R=$(CH_2)_2$–O–CH_3	2.40	3.30	0.90
R= $(CH_2)_3$–CH_3	−0.10	0.91	1.01

From Mao, Yuh-lin and Boekelheide, V., *J. Org. Chem.*, 45, 2746, 1980. With permission.)

pyrene as the most intense peak;[175] when the substituents from 15- and 16-positions are different, the fragmentation first occurs at the larger group.[178]

The X-ray analysis of some derivatives from this class indicated peripheral bond lengths similar to benzene (e.g., 1.386 to 1.401 Å in the diacetate **260**[235] or 1.392 to 1.401 Å in **268**,[171] as compared to 1.397 Å for benzene.

The unsubstituted parent compound, *trans*-15,16-dihydropyrene **282**, though unstable and difficult to handle, indicates ^{1}H-NMR- and electronic spectra characteristic for aromatic compounds[175] (Table 4).

The *cis*-15,16-disubstituted-15,16-dihydropyrenes also indicate, according to physical (and chemical) evidence and aromatic character (Table 4). The molecular model of dimethyl-derivative **283** indicates an overall shallow saucer form with a perfectly planar perimeter and with methyl groups shifted out from the center of the plane as compared with *trans*-isomer **254**. In agreement with these observations the δ value for CH_3 groups in **283** is −2.06 ppm as compared to −4.25 ppm for **254**, whereas the resonances of perimeter protons are similar in both compounds.[175]

The valence isomeric forms (e.g., **280**), are colorless compounds, difficult to isolate and purify, exhibiting all physical characteristics of strained olefinic *meta*-cyclophanedienes (Table 4).

The spectral data of derivative **199** compare favorably with those of *cis*-15,16-dimethyl-dihydropyrene (**283**).[11]

Hydrocarbon **200**, an isomer of Boekelheide's *trans*-15,16-dimethyldihydropyrene (**154**), is an aromatic compound according to its spectral data (Table 4) and X-ray analysis.[183] Also, hydrocarbon **287** is essentially diatropic.[184]

f. Pyrene and Isomers

Numerous calculations concerning charge densities, π-electronic structures, energetic characteristics, electronic transitions for pyrene and isomers, **201** to **208** are available.[33,34,151,236-243] Different indices, e.g., the Kekulé index,[244] stability index,[33] aromaticity index,[245] and reactivity index,[33] were calculated for pyrene and isomers permitting a comparative overview of their structures and reactivity.

Numerous calculations were devoted to resonance energies of these hydrocarbons. Some comparative data are collected in Table 6.[33,34,238,239]

From REPE values a more or less aromatic character results for **201** to **207** whereas **208** is polyenic in nature.[239] A classification was proposed[239] using the Dewar and de Llano method:[246] (1) benzenoids (0.114 < REPE < 0.138), (2) semibenzenoids (0.120 < REPE < 0.055), and (3) nonbenzenoids (0.011 < REPE < 0.057). Using the Hess and Schaad method[34] the above categories present REPE values which permit a sharper differentiation: (1) 0.051 to 0.055, (2) 0.030 to 0.039, and (3) 0.009 to 0.023, respectively. However, the Dewar and de Llano method affords results which are the most consistent with the chemical properties.

The RE/bond values exhibit the same trend as REPE values, decreasing in the order: (1) benzenoid (0.32 to 0.33), (2) semibenzenoid (0.251 to 0.30), (3) nonbenzenoid (0.22 to 0.249).

The more recently proposed empirical calculation of stabilities of polycyclic nonalternant hydrocarbons[245] using stability and aromaticity indices results in good agreement with experimental data.

Dipole moments,[247] magnetic susceptibilities,[248] and electronic spectra[152] were also calculated for pyrene and isomers.

Pyrene (**201**) is a colorless sublimable solid with mp 156°C and bp 392°C. The X-ray analysis[249] confirmed the planar skeleton and indicated C–C bonds of 1.39 to 1.45 Å. The strong aromatic character of pyrene is confirmed by its ^{1}H- and ^{13}C-NMR data,[250] the δ values for hydrogen atoms being 8.11 to 8.31 ppm. Theoretically the aromaticity of pyrene was explained by two approaches: (1) the peripheral model[30,251] considering a Hückel-type 14π perimeter perturbed by an inner double bond as seen in formula **201a** and (2) by the partial contributions of the components[35,36] as indicated by formula **201b**.

201a   201b

Cyclohepta[*bc*]acenaphthylene (**202**) is a red solid with mp 143.5 to 144.5°C. The electronic spectrum presents a pattern common with azulene.[194] Calculations indicate about 63% from the aromaticity of pyrene (Table 6).

Cyclohepta[*fg*]acenaphthylene (acepleiadylene) (**203**) — The first calculations of acepleiadylene appeared[252] before its synthesis in 1956.[195] The acepleiadylene molecule seems to be the most studied pyrene isomer (theoretically as well as experimentally; see also Table 4).

Table 6

RESONANCE ENERGIES CALCULATED FOR PYRENE AND ITS ISOMERS

Compound	R_E[34] (β units)	REPE[34] (β units)	R_E[239] (β units)[34]	REPE[239] (β units)[34]	% aromaticity[239]	DE[33] (β units)	P[33] (stability index)
Pyrene (201)	0.508	0.032	0.81	0.051	100	6.51	1.000
Cyclohepta[bc]acenaphthylene (202)	0.554	0.035	0.51	0.032	62.7	6.21	0.101
Acepleiadylene (203)	0.522	0.033	0.55	0.034	66.7	6.25	0.442
Naphth[cde]azulene (204)	0.353	0.022	0.52	0.033	64.7	6.22	0.182
Azupyrene (205)	0.332	0.021	0.35	0.022	43.2	6.05	0.276
Dicyclohepta[cd, gh]pentalene (206)	0.251	0.016	0.33	0.021	40.7	6.03	0.147
Pentaleno[216-def]heptalene (207)	0.293	0.018	0.25	0.016	31.4	5.95	0.096
Cyclohepta[def]fluorene (208)			0.29	0.018		5.99	0.000

Compound	RE[238] (eV)[417]	RE[238] (eV)	RE/bond[238] (eV)[417]	RE[239,418] (eV)[246]	REPE[239,246]	% Aromaticity[239]	RE/bond[34,239a]
Pyrene (201)	6.489	6.004	0.3415	1.822	0.114	100	0.328
Cyclohepta[bc]acenaphthylene (202)	5.438	5.277	0.286	1.141	0.071	62.6	0.274
	4.896	5.229	0.258				
Acepleiadylene (203)	5.815	5.479	0.306	1.517	0.095	83.3	0.278
Naphth[cde]azulene (204)	4.559	5.253	0.240	0.995	0.062	54.7	0.285
Azupyrene (205)	5.292	5.631	0.279	0.563	0.035	30.9	0.249
Dicyclohepta[cd, gh]pentalene (206)	5.440	5.277	0.286	0.369	0.023	20.3	0.250
Pentaleno[216-def]heptalene (207)	4.898	4.638	0.258	0.600	0.038	32.9	0.234
Cyclohepta[def]fluorene (208)	3.765	4.482	0.198	−0.122	−0.008		0.220

a   Average of values from different methods.

Acepleiadylene (**203**) is a red solid with an electronic spectrum resembling that of ace-pleiadiene but hyperchromically shifted.[195] The experimentally determined electronic transitions correspond fairly well with the calculated ones.[253] The experimental dipole moment (0.49 D)[254] does not agree with the calculated values.[252] The calculated and experimental ionization potentials are 7.05 and 7.13 eV, respectively, in comparison with 7.62 and 7.41 eV for pyrene.[255] The calculated diamagnetic susceptibility (in 10^6 $cm^3 \cdot mol^{-1}$) of -368 (as compared to -324 for pyrene) and the anisotropy agree well with experimental data.[256,257] The exaltation of diamagnetic susceptibility is $53 \cdot 10^{-6}$ $cm^3 \cdot mol^{-1}$ in comparison with $57 \cdot 10^{-6}$ $cm^3 \cdot mol^{-1}$ for pyrene.[258] Such exaltations were proposed by Dauben[259] as a measure of aromaticity. X-ray analyses of **203**[260,261] proved its planar structure.

The ^{1}H-NMR data[258,262,263] show a linear correlation of δ values with local π-electron densities on the carbon atom to which the hydrogen is bonded. The calculated ring current intensities[147] show an unambiguous diatropism, comparable to that of pyrene, The ^{13}C-NMR data[258,264] indicated a uniform π-electron distribution throughout the periphery and a marked upfield shift of inner C-15 and C-16 atoms in comparison with the remaining quaternary carbons. Calculated ^{13}C-NMR data resulting from additivities of chemical shift contributions agree well with experimental data.[264]

The ESR study of radical anions and radical cations of **203** reveals spin distributions in agreement with the HMO calculated ones.[265]

Naphth[*cde*]azulene (**204**) is a black solid with mp 197 to 200°C, showing a gray-purple color in solution and a characteristic electronic spectrum.[197]

Dicyclopenta[*ef,kl*]heptalene (**205**) is a thermally stable aromatic hydrocarbon with a higher mp (259°C) than that of pyrene (156°C), acepleiadylene (156 to 162°C), or other related compounds. Its trivial name "azupyrene" was suggested[198] in recognition of the double relationship with azulene and pyrene. In agreement with experimental results, the applications of Craig's rule[266,267] predicts a fully symmetric valence bond ground state, and thus a normal aromatic stability.[198] The aromatic character is proved by NMR and UV data (Table 4)[198,200] as well as by the diamagnetic susceptibility (Faraday method): Δ/Δ_{bz} of 3.9 $\pm$ 0.3 is comparable with that of pyrene (4.2 $\pm$ 0.1).[198,200] The ESR spectrum of radical ions was found to be in agreement with calculated data.[198,200]

Dicyclohepta[*cd,gh*]pentalene (**206**) — The high thermodynamic stability of **206** was suggested a long time before its synthesis.[31] The consideration of **206** as a [14]annulene perturbed by a central double bond[268] finds support in semiempirical calculations[33,149] as well as in ESR data of radical anions.[220] The electronic and ^{1}H-NMR spectra (Table 4) are in agreement with an aromatic bridged annulene structure.[205,220]

Pentaleno[216-*def*]heptalene (**207**) was the object of extensive theoretical investigations.[32,33,36,41,151,152,238,269,270] The results of most calculations agree with the description of **207** as being formed from two azulene moieties; all four rings should possess an induced diamagnetic ring current and the molecule could be classified as aromatic. The X-ray analysis[271] indicated slight deviations from planarity towards pyramidal geometry about the two central atoms and azulene-like bond orders in two of its rings.

The spectral data of the thermally very stable **207** as well as of its methylderivative **327**[204,209] (Table 4) indicate diamagnetic ring current and aromaticity. The electronic spectrum of **327** is in agreement with the calculated one,[208,272] whereas its dipole moment (1.5 D) is incompatible with charge localization in the ground state.[204]

Cyclohepta[*def*]fluorene (**208**) — In agreement with the low or negative calculated RE, all attempts to synthesize **208** failed. It seems very probable that **208** should have a diradical ground state or a low-lying, thermally excitable triplet state.[268]

D. Chemical Properties

1. Vogel's Bridged [14]Annulenes

In contrast to physical evidence proving the aromaticity of *syn*-1,6:8,13-bis-methano[14]annulene (**188**), the bromination reaction affords the addition product **348** which shows no tendency to eliminate hydrogen bromide.[11]

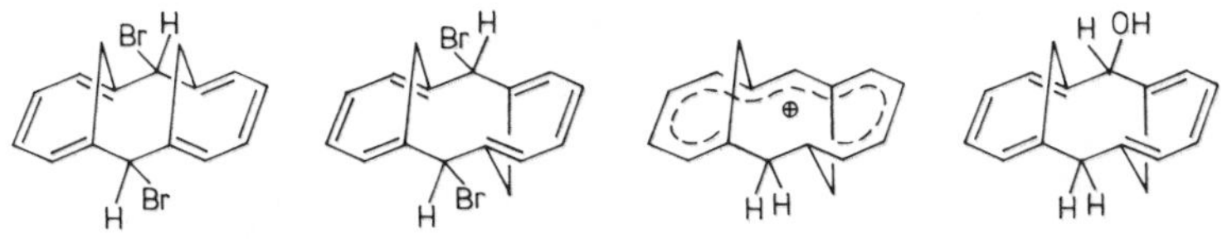

348   349   350   351

This unusual course of bromination is governed by the tendency to relieve the steric interactions between the two bridges (the molecule **348** is folded like 9,10-dihydroanthracene).

The *anti*-isomer **189** manifests a pronounced olefinic behavior resembling vitamin A and carotenoids. Thus, it affords on bromination the dibromoderivative **349** showing no tendency to eliminate HBr.[155] Protonation of **189** occurs selectively at C-7 position, and the stable carbonium ion **350** thus formed gives rise easily to the alcohol **351**.[155]

In striking opposition to **188** and **189**, 1,6:8,13-propanediylidene[14]annulene (**196**) affords substitution products, e.g., **352a** or **352b**.

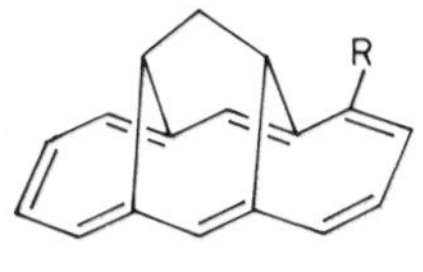

352, *a* , R = Br
b , R = D

The most interesting chemical behavior of 1,6:8,13-ethanediylidene[14]annulene (**194**) is its stability towards dehydrogenation (e.g., on treatment with Pd/BaSO$_4$ in boiling decalin[162] it does not undergo dehydrogenation to **206**).

2. 15,16-Dialkyl-15,16-Dihydropyrenes (Boekelheide's Annulenes)

The chemistry of 15,16-dihydropyrenes is dominated by two essential reaction types, namely aromatic substitution and valence tautomerism.

a. Aromatic Substitutions

Numerous substitution products were obtained from 15,16-dialkyldihydropyrenes (see also Table 4). Thus, from *trans*-15,16-dimethylderivative **254** compounds **353a** to **e** were obtained through nitration, bromination, and Friedel-Crafts acylation; **353f** was obtained through reaction with dichloromethyl *n*-butyl ether.[274]

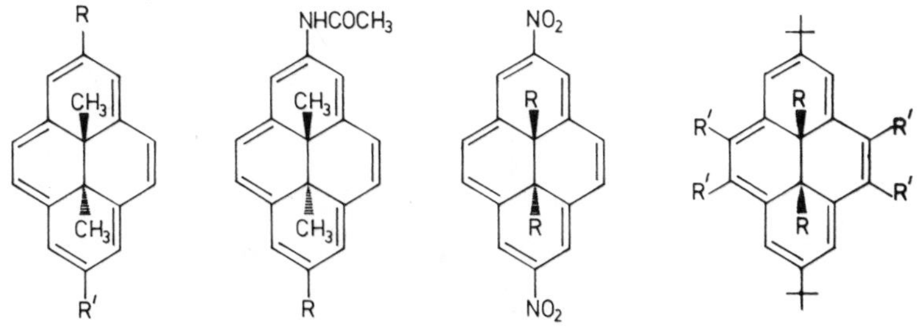

353

a, R=NO$_2$; R'=H
b, R=R'=Br
c, R=COCH$_3$; R'=H
d, R=COC$_6$H$_5$; R'=H
e, R=R'=COC$_6$H$_5$
f, R=CHO; R'=H

354

a, R=H
b, R=CHO

355

a, R=C$_2$H$_5$
b, R=n-C$_3$H$_7$

356

R=CH$_3$, C$_2$H$_5$
R'=Cl, Br

Reduction of **353a** with zinc in acetic anhydride affords **354a**[274] (a behavior resembling that of 1-nitroazulene) which could be easily converted to **354b**, a dihydropyrene derivative containing an elctron-withdrawing and an electron-donating substituent.[274] Nitration of *trans*-15,16-diethyldihydropyrene (**268**) or of its di-*n*-propyl homologue **271** affords **355a** and **355b**, respectively.[171,172] Tetrabromoderivative **356** (R = CH$_3$; R' = Br) was obtained on treatment of 8,6-dimethyl-*anti*-5,13-di-*t*-butyl[2.2]metacyclophane with bromine.[181] Tetrahaloderivatives **356** were obtained on SO$_2$Cl$_2$ or *N*-bromosuccinimide reaction of 2,7-di-*t*-butyl derivative of *trans*-15,16-dialkyldihydropyrenes.[182]

Numerous substituted derivatives with 15,16-dialkyldihydropyrene skeleton were obtained through indirect methods. Thus, formyl derivatives **357** were obtained on treatment of the quinone **259** (see above) with dimethylsulfonium dimethylide.[275]

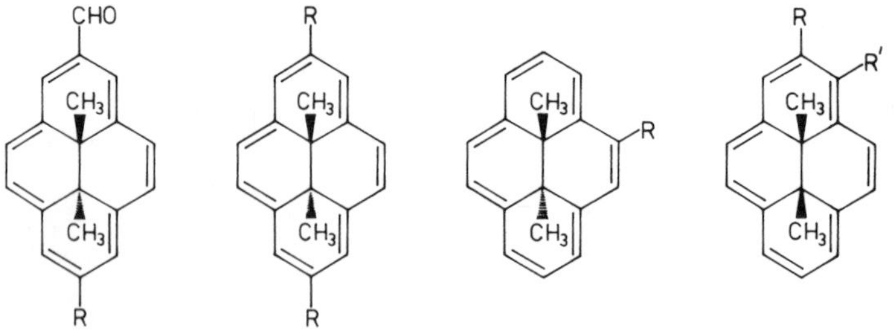

357
a, R=CHO
b, R=CH$_2$OH

358
a, R=COOCH$_3$
b, R=CH$_3$

359
a, R=Br
b, R='COOH
c, R=COOCH$_3$
d, R=CH$_3$

360
a, R=H; R'=COCH$_3$
b, R=COCH$_3$; R'=H

From **357a** the derivative **358** could be obtained by the oxidative cyanide procedure whereas from **357b** the tetramethylderivative **358b** is obtained on reduction.[275] Starting from the same quinone **259** the bromoderivative **359a** was obtained in three usual steps and from **359a** the derivatives **359b** to **d** become easily accessible.[276]

Whereas the *trans*-15,16-dialkyldihydropyrenes undergo initial substitution only at the 2-position, the *cis*-isomers (e.g., **283**) give mixtures of 1- and 2-substituted products, e.g., a 2:1 mixture of 1-acetyl and 2-acetyl derivatives **360a,b**.[277] However, nitration of **283** gives the 2-nitroderivative.[277] The reactivity towards oxygen of *cis*-**283** is in contrast with the stability of the *trans*-isomer **254**.[277]

b. Valence Tautomerism and Related Reactions
The interconversion of compounds **198** and **280** through a valence tautomerism was suggested even from the beginning of the research in this field.[165]

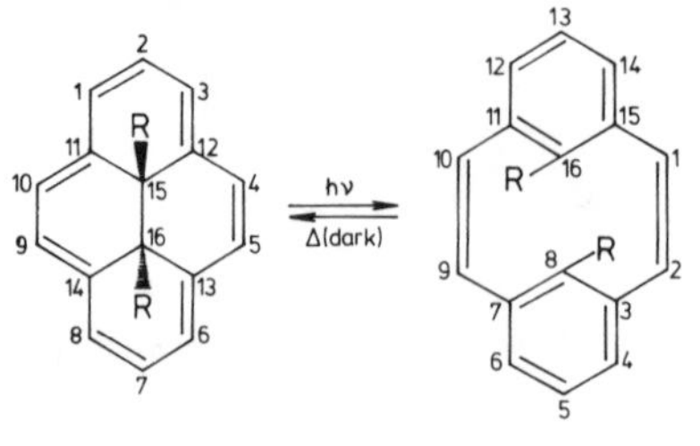

198
(254; R=CH$_3$)

280
(253; R=CH$_3$)

Unfortunately, the initial attempt to synthesize **198** through a spontaneous isomerization of **280** failed due to the impossibility of obtaining **280** by introduction of double side chain unsaturation in the corresponding *m*-cyclophanes.[165] After the synthesis of **254**,[167,168] it was found immediately that its irradiation afforded **253**,[278] whereas on standing in the dark **253** reverts easily to **254**.[168,278]

Later, when a suitable method for the synthesis of **253** became available (Hofmann elimination) it was found that this compound isomerizes into **254** also during column chromatography;[175] thus the initial assumption concerning the synthesis of 15,16-dihydropyrenes through valence tautomerism was confirmed.

The valence tautomerism of the above type (a forward photochemical reaction excited by visible light, here with $\lambda \geq 365$ nm, and a backward thermal dark reaction) was later established for numerous substituted derivatives.[176] In a detailed investigation,[176] kinetic and thermodynamic data for the first-order dark reactions were determined; some selected examples are collected in Table 7.

Experimental data have shown that the photochemical isomerization **198** → **280** proceeds only via singlet states;[176] the quantum yields of this isomerization parallel the dark reaction rate constants (Table 7).

Interestingly, the derivatives **280** revert to **198** also by irradiation with light of suitable wavelength ($\lambda \leq 313$ nm); the quantum yields of these reactions approaches 1 in numerous cases.[176] This reversible photochemical rearrangement **198** ⇌ **280** is a case of the more general *cis*-stylbene ⇌ 4a,4b-dihydrophenanthrene rearrangement.[279] For steric reasons it clearly follows that the above reversible photochemical reactions can occur in a concerted manner only if they are conrotatory; according to the Woodward-Hoffmann selection rules such a process is allowed only in excited electronic states.[280] On the other hand, if the thermal reaction **280** → **198** occurs synchronously, it would be a symmetry-forbidden reaction.[280] However, it has been argued that some symmetry-forbidden concerted electrocyclic processes may take place if the activation energies are sufficiently low.[280]

Interestingly, the conversion **198** → **280** can also occur thermally on heating to about 70°C,[176] similar to the known thermal conversion of 4a,4b-dihydrophenanthrene into *cis*-stilbene.[281] The calculated free enthalpy of activation for the reaction **198** $\overset{\Delta}{\to}$ **280** is about 27 kcal · mol^{-1}, irrespective of the substitution pattern;[176] hence the higher the ΔG value, the faster the dark reaction **280** $\overset{\Delta}{\to}$ **198**.

Heating *trans*-15,16-dimethyldihydropyrene (**254**) at about 200°C produces an irreversible thermal rearrangement to *trans*-13,15-dimethyldihydropyrene (**361**),[276] probably proceeding through a [1.5]sigmatropic shift.[172] A similar reaction was also found for the diethylderivative **268**.[172]

No experimental evidence for a valence tautomerism of *cis*-15,16-dimethyldihydropyrene (**283**) is available[175] (a thermal isomerization to a *meta*-cyclophanediene, if concerted, would be an allowed process according to the Woodward-Hoffmann rules).

361   *362*   *282*

In the case of the unsubstituted *trans*-15,16-dihydropyrene (**282**) the valence tautomeric form **362** is more stable,[174,175] permitting its isolation and characterization (Table 4). Heating

Table 7

KINETIC AND THERMODYNAMIC DATA FOR THE DARK REACTION 280 → 198 AND QUANTUM
YEILDS (φ) OF PHOTOCHEMICAL REACTION 198 → 280[a]

Compound	$k^{30°}$	$k^{50°}$	Ea	log A	$\Delta G^{\ddagger}$	$\Delta H^{\ddagger}$	$\Delta S^{\ddagger}$	φ (25°)
trans-15,16-Dimethyldihydropyrene (254)	0.001	0.01	23.0	13.6	24.4	22.4	−7	0.02
2-Formyl-254	0.052	0.44	20.5	13.5	21.9	19.9	−7	0.26 (15°)
2-Nitro-254	0.069	0.51	19.7	13.1	21.8	19.1	−9	0.37 (15°)
4-Bromo-254	0.00058	0.0081	25.2	15.0	25.2	24.6	−2	0.011
4-Formyl-254	0.0008	0.0061	21.6	12.5	24.8	21.0	−12	0.007
trans-15,16-Diethyldihydropyrene (268)	0.0061	0.056	21.7	13.4	23.4	21.1	−7	0.012
1,3,6,8-*trans*-15,16-Hexamethyldihydropyrene (265)	0.0009	0.009	24.5	14.5	24.6	23.9	−2	0.0085

[a] k in min^{-1}; $\Delta S^{\ddagger}$ in cal. grd. $^{-1}$mol^{-1}; φ%; other data in kcal · mol^{-1}.

From Blattmann, M. R. and Schmidt, W., *Tetrahedron*, 26, 5885, 1970. With permission.

to 120°C or irradiation of **362** easily produce **282**.[175] The study of **282** is difficult owing to its easy conversion into pyrene (oxidatively or photochemically).[175]

c. *Pyrene and Isomers*

As a "classical" aromatic hydrocarbon, pyrene (**201**) undergoes different aromatic substitution reactions affecting the 3,5,8- and 10-positions, which are the most reactive.[186,282] Thus, 3-chloro, 3-bromo, 3-nitro, 3-sulfonic acid, 3-acyl derivatives are easily obtained then 3,8- and 3,10-disubstitution products, and finally 3,5,8,10-tetrasubstituted derivatives can be prepared.

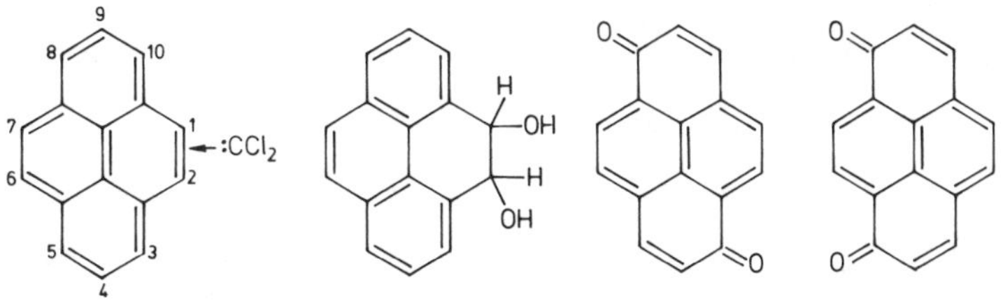

201	363	364	365

Addition reactions take place at the 1,2-bond, e.g., ozonization, hydrogenation, etc.[186,282] Treatment with OsO_4/pyridine followed by hydrolysis affords **363**[283] and oxidation with chromic acid in acetic acid yields quinones **364** and **365**. Dichlorocarbene adds[284] at the 1,2- and 6,7-double bonds and not at the central double bond. Pyrene forms colored addition compounds with alkaline metals, e.g., red with lithium, dark-blue with sodium, etc.[282] Concerning the automerization of pyrene, see Chapter 7.

Cyclohepta[*bc*]acenaphthylene (**202**) is a hydrocarbon with a high basicity, comparable to azulene.[194] Though calculations suggested easy reactions with electrophiles in the 10-position and with nucleophiles in the 1-position,[33] no experimental chemistry of **202** seems to have been developed.

Acepleiadylene (**203**) is an aromatic hydrocarbon which affords a red mononitroderivative on treatment with cupric nitrate/acetic anhydride and which does not react with maleic anhydride.[195] Calculations indicate that electrophilic reagents should attack **203** in the 5- or 7-position and the nucleophilic reagents in the 1-position.[285] On hydrogenation of **203** the derivatives **366** and **367** are formed.[195,286] From thermodynamic data it was appreciated that the conjugation energy of the three nonnaphthalenic double bonds from **203** is at least 18.9 kcal · mol^{-1}.[286] The brown-colored complex of **203** with tetracyanoethene is utilized in paper chromatography.[287]

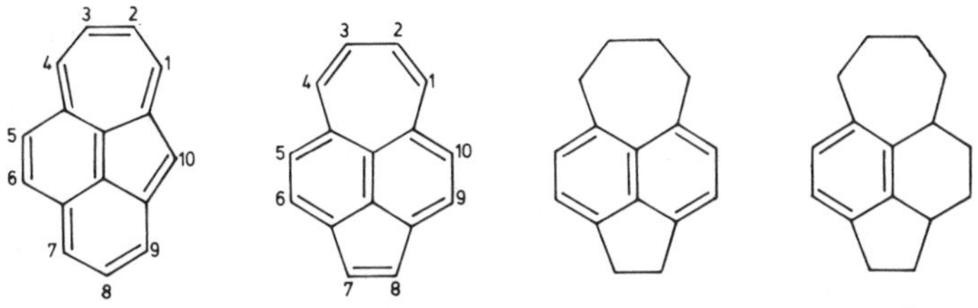

202	203	366	367

Naphth[*cde*]azulene (**204**) was studied mostly theoretically; the attack of electrophiles in the 1-position, of nucleophiles in the 8-position, and a basicity comparable to azulene were suggested.[33]

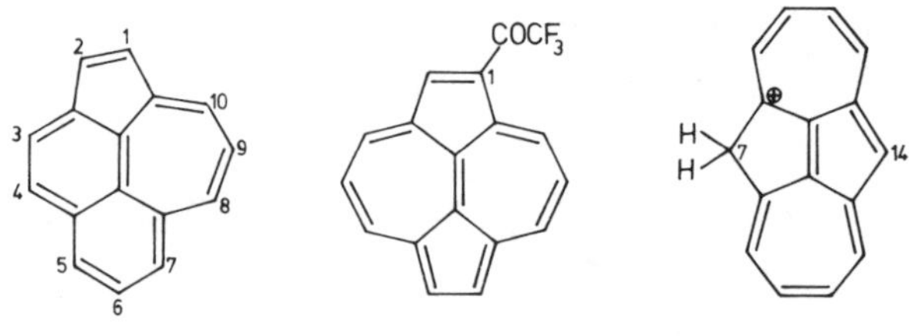

204   368   369

Azupyrene (**205**) reacts with trifluoroacetic anhydride affording **368** in agreement with the results of simple HMO calculations.[288] Nitration with AgNO$_2$ affords the 3-nitroderivative whereas thermolysis at 500°C yields pyrene.[289] Formation of **200**[183] and **287**[184] starting from **205** was mentioned before.

Dicyclohepta[*cd,gh*]pentalene (**206**) can be protonated in the 7-(or 14-)position, affording the stabilized carbonium ion **369** which regenerates **206** on treatment with bases.[205] On catalytic hydrogenation an azulene derivative is formed.[205]

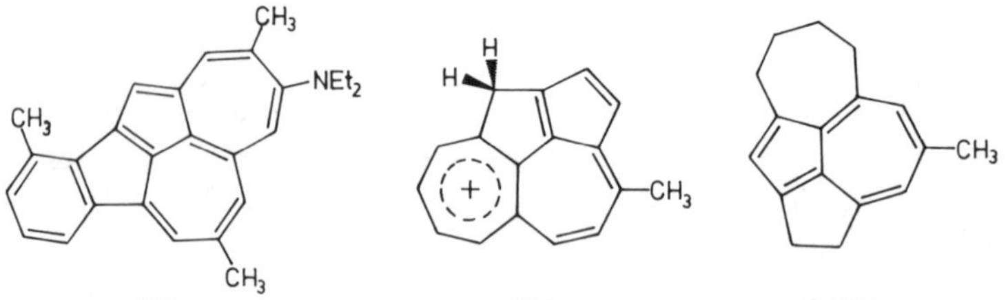

370   371   372

Pentaleno[216-*def*]heptalene (**207**) undergoes no cycloadditions with acetylenes.[140] However, the methylderivative **327** reacts with *N,N*-diethyl-1-propylylamine, affording the benzopentalenoheptalene **370**.[140] Protonation of **327** yields the cation **371** which regenerates the starting hydrocarbon with bases.[208,272] The methylderivative **327** is inert towards electrophilic reagents[208] but can be hydrogenated to **372**.[208]

E. Conclusions

This class of bridged [14]annulenes enriched the chemistry of aromatic hydrocarbons with numerous and various interesting representatives. The factors which decide whether the structure is propellanic or bridged [14]annulenic (also for [10]annulenic systems) were exhaustively discussed by Ginsburg.[72,290] In these reviews,[72,290] the chemistry of the propellanic valence isomers is thoroughly investigated.

The large difference shown between the aromatic *syn*-bis-methano[14]annulene (**188**) and the olefinic *anti*-bis-methano[14]annulene (**189**) "provide the most impressive demonstration yet available that the molecule geometry is of crucial importance for the occurrence of aromaticity, . . . , in accord with the steric criterion of the Hückel rule."[156] On the other hand, the comparison between the olefinic *anti*-bis-methano[14]annulene (**189**) and the aromatic propanediylidene[14]annulene (**196**) — a compound similar to **188** which eliminates, however, the unfavorable H–H interactions through a new CH$_2$ group — "amounts to a textbook demonstration of the importance of a planar π-electron system as a geometrical prerequisite for aromaticity."[38]

A very interesting comparison can be made between hydrocarbons **206, 195, 194, 196, 197**, and **373**,[11] which may be arranged in a series of progressive bending of the skeleton. This series also includes the announced compound **373**,[11,291,292] the first derivative in which a conformational mobility of the bridge appears (bridge inversion similar to that found in [3,3]*para*-cyclophanes. The comparative ^{1}H-NMR study of **194, 196**, and **197**[291] shows no indication of a significant reduction of diamagnetic ring current. In this connection it has been shown that H–H coupling constants seem to be a better tool in the evaluation of π-electron delocalization for these [14]annulenes than the chemical shifts.[291]

progressive bending

The results of calculations[43] agree well with the bond lengths and angles determined by X-ray structure determinations of **195, 194, 196, 197,** and **373.** The steric energy increases progressively from left to right in the above-mentioned series and this trend is determined especially by the twisting and nonbonded energy contributions.[43]

By means of PE spectroscopy[293] the following order of planarity was determined: **195 > 194 > 196 > 197.**

A detailed comparative study of light absorptions, polarized emissions, and dichroism in the above series was reported.[294] It has been shown that this series is a good illustration of a gradual transition from a hypothetical planar monocyclic [14]annulene to anthracene by introduction of cross-links.

The ESR spectra of radical anions of the above hydrocarbons were studied[295] and the coupling constants of the ring protons in these radicals were proposed as a criterion for planarity of the perimeter. On the basis of this criterion the following order of planarity was established:[295] **195 > 194 > 196 > 197 > 189.**

According to spectral data, hydrocarbon **195** seems to be the "as yet best model for the fictitious unbridged planar [14]annulene with anthracene perimeter."[11]

Calculations of different characteristic indices[45] led to the following order of aromaticity: **195 > 194 > 196 > 197 > 189,** which agrees with the order derived from qualitative ring puckering and steric strain as well as from spectral data.

Going to the right hand of the above series, it can be stated that the progressive bending could not significantly diminuate the π-electron delocalization. Even the hydrocarbon **373,** in which the torsional angle of $2p_z$ orbitals at adjacent carbons is about 40° (compare with 70° in **189**) and the side-view outline of the annulene skeleton is approximately a semicircle, still behaves as an aromatic molecule. It seems that the energy gain due to the tendency of 14π-electron systems from the above series to achieve maximum p-overlap overcomes much steric strain apparent from stereomodels, yet this gain is not sufficient in the case of the *anti*-bismethano[14]annulene (**188**). Thus, the steric conditions for aromaticity in the Hückel rule are much less rigid than it was assumed earlier.

Another interesting parallel may be made between bridged annulenes having an anthracene perimeter (Vogel's) and with pyrene perimeter (Boekelheide's). In both series the *cis*-isomers possess a slightly bent perimeter ("saucer shaped") whereas *trans*-isomers indicate a planar carbon perimeter.

The recently synthesized **199** and **200** along with **283** form a triad of two-carbon bridged [14]annulenes which all present aromatic character.

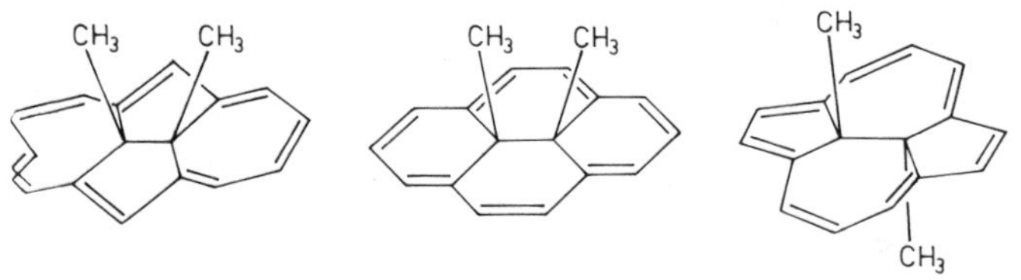

199 (R=CH₃)   283   200 (R=CH₃)

The hydrocarbon **196** may also be considered as a member of a new series: **3, 196, 374** (last member yet unknown) which allows comparison since these systems are expected to possess similar symmetry, rigidity, and planarity.

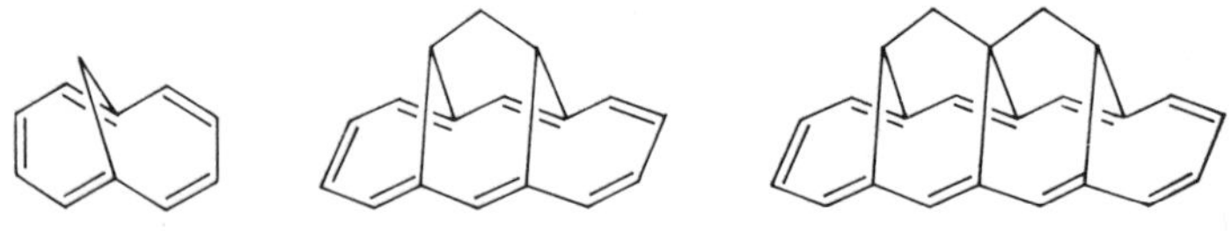

3 (10π)   196 (14π)   374 (18π)

An analogous comparison between the corresponding monocyclic [10]-, [14]-, and [18]annulenes cannot unfortunately be made owing to large geometric differences (in co-planarity, conformations, etc.).

Boekelheide's 15,16-dialkyl-15,16-dihydropyrenes constitute a unitary class of aromatic compounds. Their improved syntheses make readily available significant amounts of such compounds which may be thus thoroughly studied (by physical and especially chemical means). Interesting observations concerning ring currents were made on the basis of their ^{1}H-NMR spectra, whereas the chemical studies reveal, along with the aromatic character-istics, their interesting valence tautomerism with *meta*-cyclophanedienes.

The class of pyrene and its isomers, which was thoroughly investigated theoretically, includes, beside well-known aromatic compounds such as pyrene (**201**) and acepleiadylene (**203**), a compound such as azupyrene (**205**) whose aromatic reactivity was only recently demonstrated, compounds such as cyclohepta[*bc*]acenaphthylene (**202**) and naphth[*cde*]azulene (**204**) which, in spite of their earlier syntheses are not yet chemically investigated in detail, difficultly accessible compounds such as dipleiapentalene (**206**) and pentaleno[216-*def*]-heptalene (**207**) which show a limited chemical reactivity and finally, a compound, cyclo-hepta[*def*]fluorene (**208**), which has so far resisted all attempts for its synthesis, and which is predicted to be less stable than the remaining compounds although it does not possess a significant bond angle strain.

Vogel and co-workers[406] announced a new synthesis of dipleiapentalene (**206**) from **444** via a dioxetane, in a reaction sequence resembling methathesis.

Macrocyclic biphenyls containing the dihydropyrene nucleus were synthesized by Mitchell et al.[407]

Gerson et al.[408] published an ESR study of radical cations of 1,6,8,13-bridged[14]annulenes.

The previously reported data[288,289] concerning electrophilic substitution of azupyrene were supplemented by Anderson et al.[409]

V. CARBON-BRIDGED [16]ANNULENES

A. Tricyclic-Bridged [16]Annulenes

1,6:9,14-Bismethano[16]annulene (**375**) was recently obtained by Vogel[296] in a one-step synthesis through reductive coupling of cycloheptatriene-1,6-dialdehyde (**10**) using titanium tetrachloride and zinc in the tetrahydrofuran (McMurry reaction).

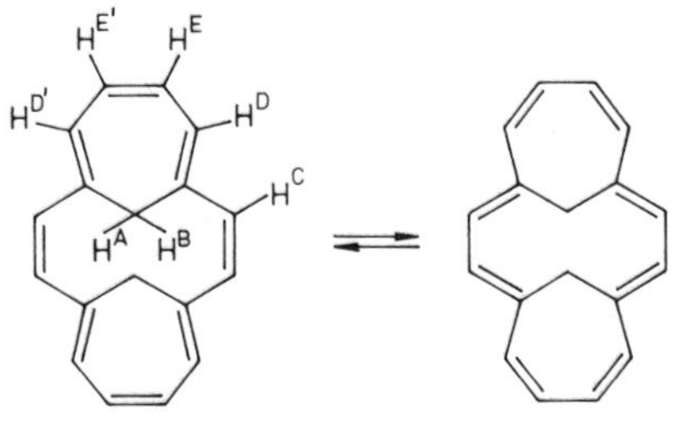

375a   375b

The ^{1}H-NMR spectrum of the golden-brown compound **375** (CDCl$_3$; δ ppm): 4.80 (s, H^C); 5.10 (AA'BB' system; H$^{D,D',E,E'}$); 5.58 and 8.20 (AX system; H^{A,B}) proves its paratropic character. Very probably a mixture of both double bonds isomers **375a** and **375b** exists with the former one being favored.[296] The electrolytic reduction of **375** to the radical anion and dianion proceeds similarly as for [16]annulene.[296]*

A recent thorough investigation by Vogel et al.[410] of the ^{1}H-HMR spectra of 1,6:9,14-bis-methano[16]annulene (**375**) indicated the existence of only one isomer, **375b.** The X-ray analysis showed, in agreement with molecular mechanic calculations, the *anti*-arrangement of the CH$_2$-bridges.

B. Tetracyclic-Bridged [16]Annulenes
1. Ethane-Bridged Systems

The two carbon-bridged [16]annulene **378** was obtained in 1970 by Boekelheide[297] from the dianion of *trans*-15,16-dimethyldihydropyrene (**376**). A usual six-step conversion of **376** into ditosylate **377** was followed by the solvolytic ring enlargement to **378** (as a mixture of two positional isomers).

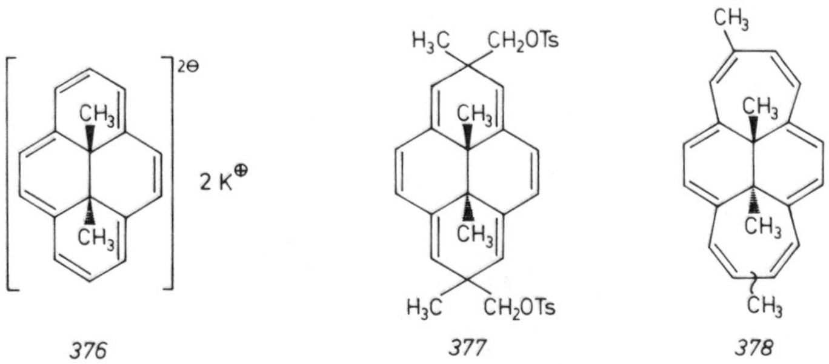

376   377   378

The very complex ^{1}H-NMR spectrum of **378** (due to the nonplanar perimeter and to the conformational mobility) suggests a small paramagnetic ring current characteristic for a [4n]annulene: δ = 4.81 ppm and 0.59 to 0.60 ppm for inner H and outer CH$_3$ protons, respectively, and δ ~ 4 ppm for vinylic protons.[297]

2. Ethene-Bridged Systems

Five formulas of tetracyclic ethene-bridged [16]-annulenes containing five-, six-, and seven-membered rings **379** to **383** may be written. (Inclusion of eight-membered rings raises the number of such isomers to 16.) In spite of numerous theoretical works concerning these

---

* Also by using the McMurry reaction, a doubly-bridged bismethano[24]annulene with two tropylidene rings joined by two hexatrienic chains was prepared by Yamamoto, K., Shibutani, M., Kuroda, S., Ejiri, E., and Ojima, J. (*Tetrahedron Lett.*, 27, 975, 1986). It presents stronger paratropicity than the analogous bismethano[16]annulene **375**.

compounds, only one of them, azuleno[8,8a,1,2-*def*]heptalene (**379**) has been obtained so far.

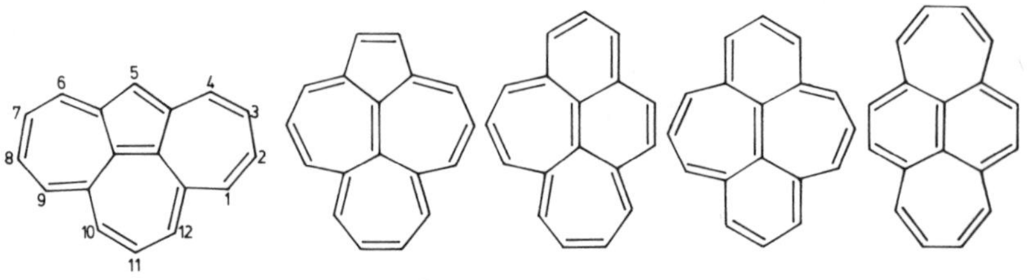

379   *380*   *381*   *382*   *383*

3. Syntheses

Very ingenious syntheses of azulenoheptalene **379**[204] as well as of its 11-methyl- and 2,4,11-trimethyl derivatives[298] starting with 3-methylaceheptylene or from its polymethyl-derivatives were described by Hafner. Partial reduction of **384** affords **385** which, being an azulene derivative, is exclusively substituted at C-2 by the electrophilic reagent 3(N-methylanilino)acrolein/POCl$_3$. Dehydrogenation of the resulting **386** with chloranil followed by treatment with N-methylaniline/perchloric acid afforded the aldiminium perchlorate **387** which, on treatment with alkali, undergoes intramolecular ring closure to **379**. This indirect route was used because **387** could not be obtained directly by Vilsmeyer reaction due to the preferential attack of **384** by electrophiles in 1-, 4- or 9-, and 6- or 7-positions.

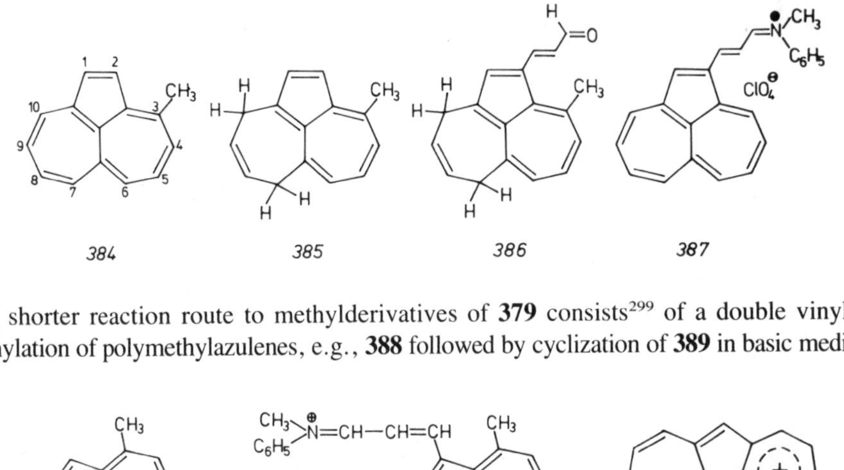

384   *385*   *386*   *387*

A shorter reaction route to methylderivatives of **379** consists[299] of a double vinylogous formylation of polymethylazulenes, e.g., **388** followed by cyclization of **389** in basic medium.

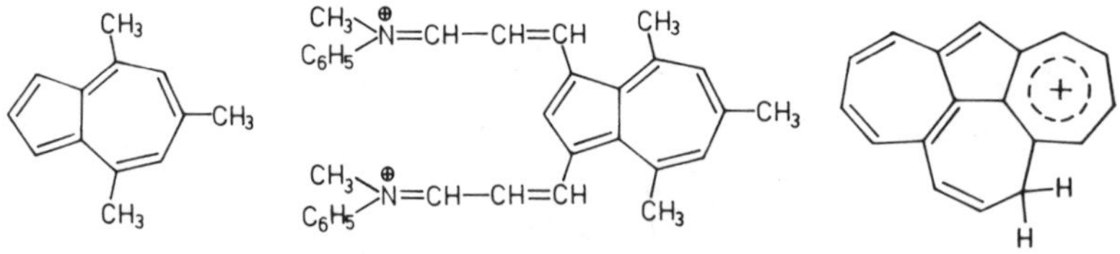

388   *389*   *390*

4. Structure and Physicochemical Characterization

Numerous theoretical papers treated the azulenoheptalene molecule (**379**). Thus, from the calculated geometry[41,242] it results that aromatic bonds should coexist in **379** with nonaromatic ones. The electron density is maximal in the 5-position (in the five-membered ring) and is also high in the 10- and 12-positions.[32] The calculated dipole moment is in the range of 0.43 to 1.32 D,[242] the resonance energy per bond is 0.239 eV,[239,300] the resonance energy per π-electron is 0.011 eV or 0.009 β,[239,300] and the resonance energy is 0.195 eV.[36]

Though from the above data some degree of aromaticity of **379** could be inferred, the physicochemical properties do not confirm those predictions.

Azulenoheptalene (**379**)[204,298] is a brownish-yellow thermally unstable compound whereas its methyl-substituted derivatives are stable. In the electronic spectrum the longest absorption extends to the near infrared range (up to 1700 nm). In the ^{1}H-NMR spectra of **379** and of its 11-methylderivative the proton of the five-membered ring resonates at δ ~ 4.5 ppm (upfield relatively to azulene) demonstrating a small or inexistent diamagnetic ring current whereas the signals of seven-membered ring protons appearing at δ = 1.6 to 4.4 ppm indicate a strong paramagnetic ring current. These results are in good agreement with the calculated ring currents for each ring in **379**.[32] The strong basicity of **379** was proved by the formation of the cation **390** with a suggested pK_a = 2 on protonation with sulfuric acid.[298] This fact explains why azulenoheptalene hardly undergoes electrophilic substitutions (e.g., Vilsmeyer reaction[299]).

In conclusion, as Hafner says, "azulenoheptalene has an extraordinary position in the series of the polycyclic conjugated nonbenzenoid systems known so far."[204]

From results of calculations for isomeric-bridged [16]annulenes **380** to **383** one can mention:

1. For **380** a RE of -0.005 eV[36] was obtained; excepting the five-membered ring with a small diamagnetic ring current in all the seven-membered rings paramagnetic ring currents were calculated.[32]
2. For **381** a RE of 0.835 eV was calculated.[36]
3. For **382** a triplet ground state with no appreciable bond fixation in the peripheral carbon skeleton was suggested. The antiaromatic character results from its $4n$ circuits.[36]
4. For "dipleiadiene" (**383**) (also named "dipleiadadiene"[239]) a RE of 0.902 eV,[36] a RE per bond equal to 0.295 eV,[239] a RE per π-electron of 0.035 eV or 0.017β,[239] as well as a predominating paramagnetic ring current[147] were calculated.

C. Conclusions

Paratropicities were demonstrated for bridged [16]annulenes **375** and **378** along with the special position of azulenoheptalene (**379**).

VI. CARBON-BRIDGED [18]- AND [22]ANNULENES

A. Syntheses

1. Bridged [18]Annulenes

a. Syn,syn-1,6:8,17:10,15-Trismethano[18]Annulene (394)

The idea of constructing higher "homologues" of bismethano[14]annulene comes from the finding that the pentaenedialdehyde **211c** (see Section II.B.1) exists as a *syn*-isomer.[156] If the homologous dialdehyde **391** also exists as a *syn*-form, then the synthesis of trismethano[18]annulene (**394**) could be attempted by the "building block approach".

211c   391   216

Indeed, the synthesis of **391** was effected starting from **211c** in a manner analogous to the lower "homologue"[8] using the same Wittig-Horner reagent **216**.[8,301] The *syn*-geometry of **391** was proved by X-ray analysis.[302] The conversion of **391** into trismethano[18]annulene **394** was performed in three ways:[301] (1) through reaction with thiodimethylenebis(triphenylphosphoniumbromide) *via* **392a** and **393a** (as in the case of lower homologue **188**); (2) and (3) through double olefination to **392b, 392c** followed by an 18 π electrocyclic process accompanied by elimination of hydrogen chloride (reflux in DMF) or of hydrogen, respectively (thermolysis with DDQ).[301]

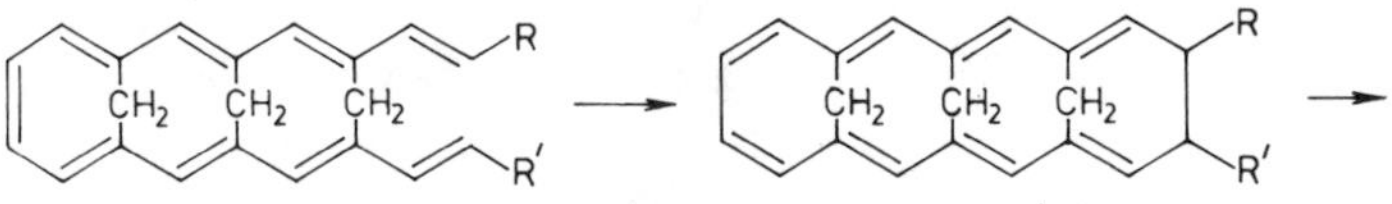

392, *a*, R—R'= $>$S
     b, R=Cl ; R'=H
     c, R=R'=H

393, *a*, R—R'= $>$S
     b, R=Cl ; R'=H
     c, R=R'=H

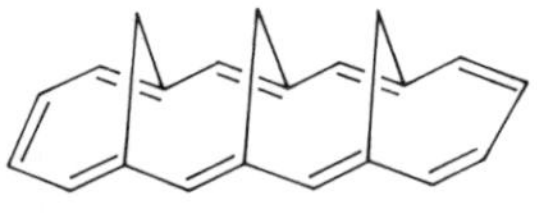

394

b. Tetracyclic Ethene-Bridged [18]Annulenes

No such system seems to be yet known. From the 12 possible ethene-bridged [18]annulenes containing five- to eight-membered rings, 11 include at least one eight-membered ring. The single compound without eight-membered rings, the symmetric hydrocarbon **395**, was the object of early calculations which indicated a resonance energy of -0.325 eV[36] and a resonance energy per π-electron of -0.011β,[35] thus suggesting an antiaromatic and very reactive character. However, other calculations suggested diamagnetic ring currents in all rings of **395**.[32]

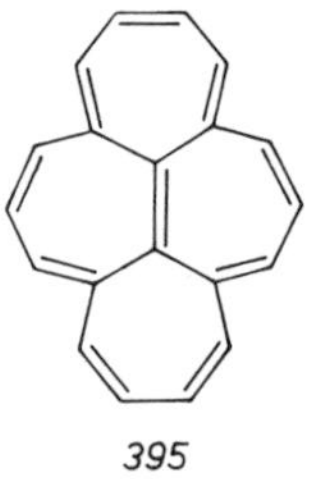

395 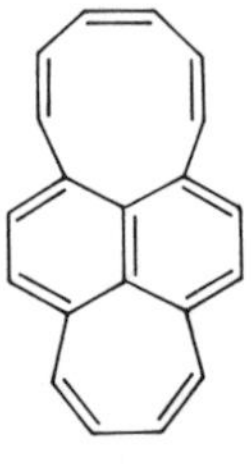   396

Interestingly, some calculations[303] suggested aromatic character similar to acepleiadylene for cyclonona[*cd*]pleiadene (**396**), a tetracyclic ethene-bridged [18]annulene containing a nine-membered ring.

c. Pentacyclic [18]Annulenes

Such systems are known so far with an interannular bond and an ethene bridge. Azuleno[5,6,7-*cd*]phenalene (**400**) was obtained in two steps from the phenalene anion (**397**) via (**399**):[304,305]

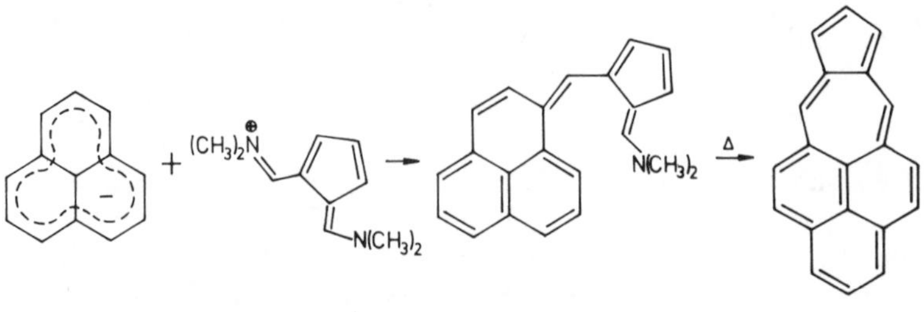

397   398   399   400

The isomeric azuleno[1,2,3-*cd*]phenalene (**403**) was produced in a Ziegler-Hafner azulene-type synthesis from dihydro-cyclopenta[*cd*]phenalene **401** and **402**.[306]

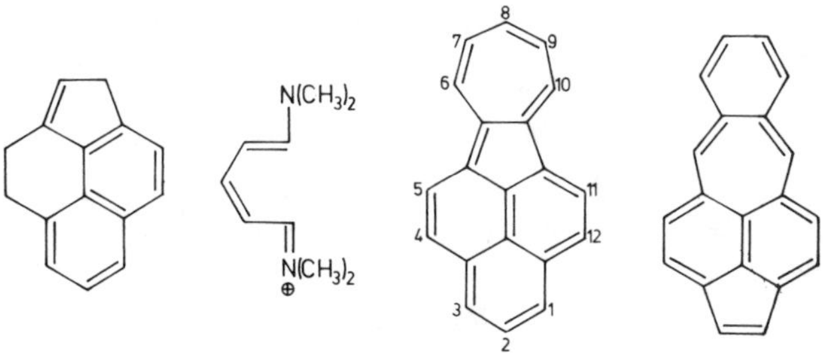

401   *402*   *403*   *404*

On treatment with acids, 5,10-dihydroacepleiadene-5,10-diol gives, through a vinylogous elimination, the unstable blue acepleiadylene derivative **404** with a periphery of [18]annulene.[307,308]

d. Hexa- and Heptacyclic [18]Annulenes

The 12c,12d,12e,12f-tetrahydrobenzo[*ghi*]perylene (**406a**) and derivatives **406b** to **f** were obtained by Boekelheide[309-311] via **405a** to **f** following the general method utilized in the series of 15,16-dialkyldihydropyrenes (including Stevens rearrangement and Hofmann elimination; see Section II.B.2).

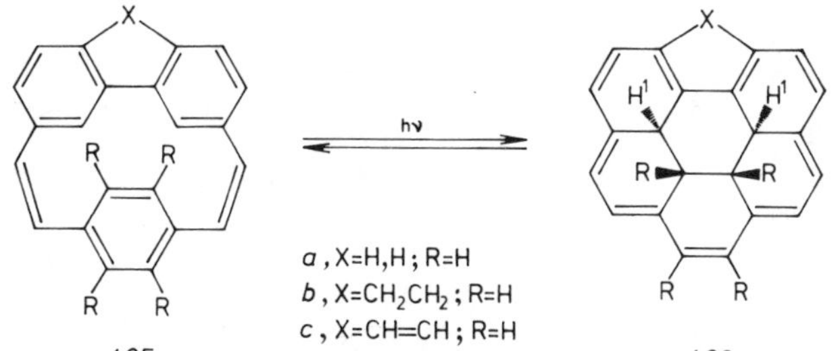

a, X=H,H; R=H
b, X=CH₂CH₂; R=H
c, X=CH=CH; R=H
d, X=CH=CH; R=CH₃
e, X=S; R=H
f, X=S; R=CH₃

405   *406*

The valence tautomerization **405** → **406** occurs only photochemically, though the Woodward-Hoffmann rules predict a thermally allowed process for the concerted reaction.

The synthesis of 12b,12c,12d,12e,12f,12g-hexahydrocoronene (**410**), a multibridged [18]annulene, unsuccessfully attempted by Cram in 1959,[312] was perfected in 1978 by Boekelheide[313] through photochemical cyclization of [2.2.2]*para*-cyclophanetriene (**408**) at −60°C (instead of room temperature as in Cram's attempt). The intermediate (**409**) is converted to hexahydrocoronene in the presence of oxygen.[313] Through an analogous photoisomerization, benzo- and dibenzoderivatives of hexahydrocoronene **410** were obtained.[313]

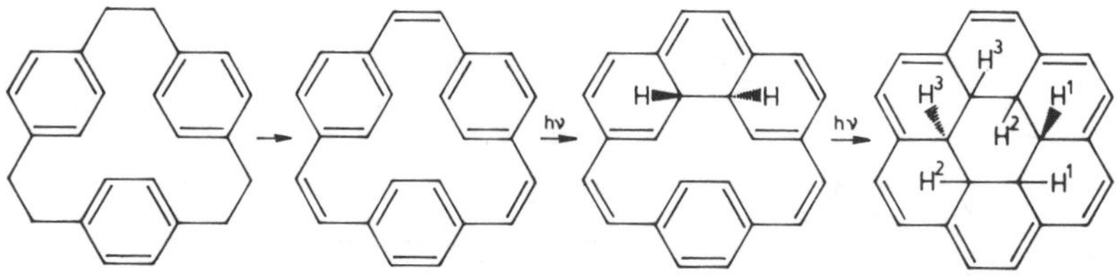

407   *408*   *409*   *410*

2. Bridged [22]Annulenes

Syn,syn,syn-1,6:8,21:10,19:12,17-Tetrakismethano[22]annulene (**413**) was synthesized after the "building block approach", including homologation of **391** to **411**[8,9] (*all-syn* configuration proved by X-rays[8]) double olefination to **412** followed by thermal cyclodehydrochlorination.[9]

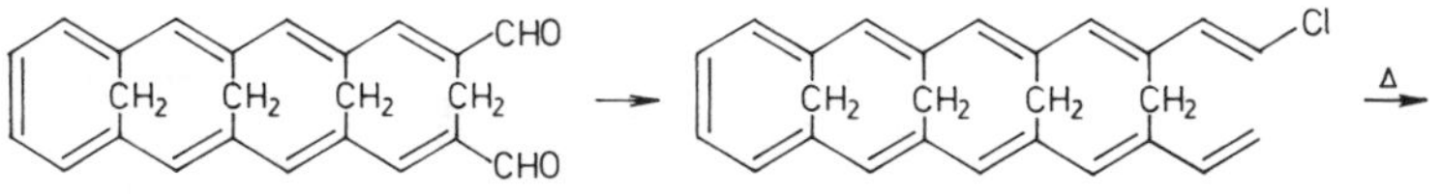

411

412

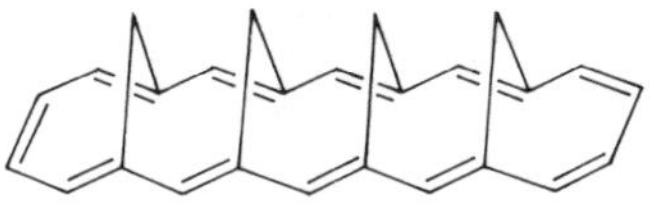

413

The attempted synthesis of the multibridged [22]annulene **414** failed.[314,315] Thus, ring contraction and sulfur elimination from **415** was unsuccessful,[314] attempts to remove sulfur for introducing double bonds in **416** failed,[314] and the quinone **417** could not be converted to **414**.[315]

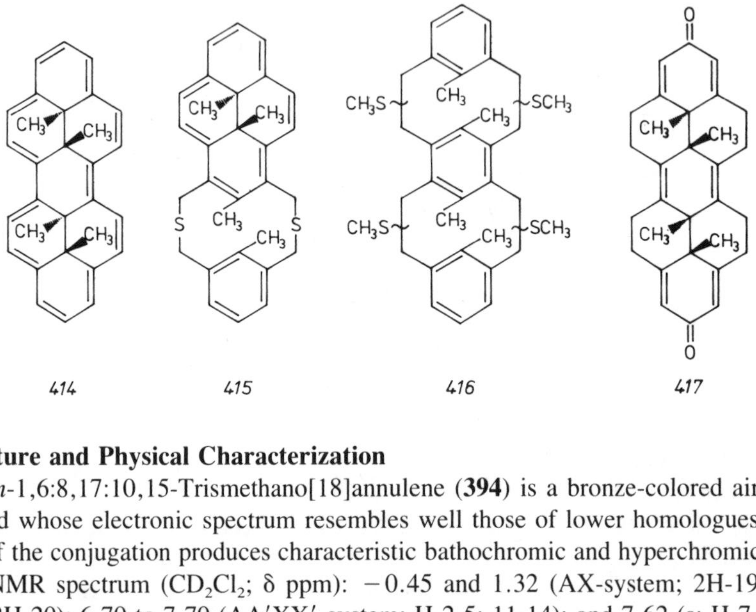

414

415

416

417

B. Structure and Physical Characterization

Syn,syn-1,6:8,17:10,15-Trismethano[18]annulene (**394**) is a bronze-colored air-sensitive compound whose electronic spectrum resembles well those of lower homologues, The extension of the conjugation produces characteristic bathochromic and hyperchromic shifts,[301] The ^{1}H-NMR spectrum (CD$_2$Cl$_2$; δ ppm): -0.45 and 1.32 (AX-system; 2H-19; 2H-21); 0.53 (s; 2H-20); 6.70 to 7.70 (AA′XX′-system; H-2-5; 11-14); and 7.62 (s; H-7, 9,16,18) clearly proves the diatropicity of the molecule and the existence of only one isomer, namely, the *syn,syn* one with C$_{2v}$ symmetry.[301]

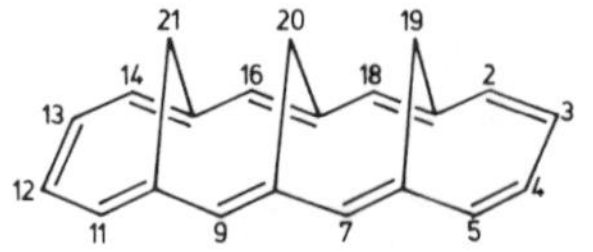

394

Azuleno[5,6,7-*cd*]phenalene (**400**) is a green solid with a dipole moment of 0.7 ± 0.7 D (much smaller than the calculated one) and with a complex electronic spectrum (in agreement to the calculated one).[305] Physical data (as well as chemical evidence, see later) show that **400** may be classified as a slightly aromatic compound.[305]

The aromaticity of azuleno[1,2,3-*cd*]phenalene (**403**) was evidenced by ^{1}H-NMR signals at δ values (CDCl$_3$; ppm): 7.50 (t; H-7,9); 7.8 to 8.1 (m; H-1,2,3,8); 7.85, and 8.36 (AB-system; H-4,5,11,12), and 9.14 (d; H-6,10).[306] The unusual electronic spectrum of **403** (with the longest-wavelength absorption maximum at 1010 nm; lg ε = 1.99) is, however, in agreement with calculations.[306] The PE spectrum and the electronic structure of **403** were investigated in comparison with **400**.[316]

The tetrahydrobenzo[*ghi*]perylenes (**406**) show the characteristics of aromatic-bridged annulenes possessing strong diamagnetic ring currents.[311] Thus, the H$_1$ resonances appear between δ = −1.20 (in **406d**) and −4.96 ppm (in **406f**) whereas the CH$_3$ group signals appear at −1.94 (in **406d**) and −3.74 ppm (in **406f**),[311] i.e., the thiophene derivatives **406e,f** show a larger diamagnetic ring current than **406a** to **c**.

The hexahydrocoronene (**410**), a red compound, could not be obtained as a pure sample. In the visible spectrum, absorptions resembling those of [18]annulene appear. The ^{1}H-NMR spectrum at −97°C indicates a multiplet at 9.30 to 9.55 ppm for the 12 peripheral outer protons and an AA'BB'XX' system at δ = −6.54 to −7.96 ppm for inner protons, being thus in agreement with a strongly diatropic character of the aromatic **410** which possesses a C$_2$ symmetry.[313]

It can be mentioned in this connection that Boekelheide and co-workers[313] extended to polycyclic annulenes, e.g., **410**, Haddon's approach[317] for quantitative assessment of chemical shifts of simple annulenes (the degree of aromaticity is the ratio between the observed and calculated ring currents) utilizing, when X-ray determinations are absent, the geometry inferred from shadowgrams of Dreiding stereomodels.

The ^{1}H-NMR spectrum of *syn,syn,syn*-tetrakismethano[22]annulene (**413**), a brownish-black compound, indicates diamagnetic ring current: annulene protons resonate (δ values) between 6.6 and 7.7 ppm whereas bridge protons between 0.24 and 2.03 ppm.[109] The same spectrum indicates the C$_{2v}$ symmetry of the molecule with delocalized or rapidly fluctuating double bonds.[109] The electronic spectrum is shifted to longer wavelengths[109] (as compared to **394**).

C. Chemical Properties

The trismethano[18]annulene (**394**) as well as the tetrakismethano[22]annulene (**413**) show a great tendency to polymerization.[9,109,301] The olefinic reactivity was attributed,[301] in the case of **394**, to Baeyer- and Pitzer strain rather to the loss of resonance, for according to physical data, the hydrocarbon is clearly aromatic. On attempting protonation with fluorosulfonic acid a 16 π-dication is formed by oxidation (see Section VIII.E).

The azuleno[567-*cd*]phenalene (**400**) does not react with maleic anhydride but gives a 1-substituted formiminium salt on treatment with POCl$_3$/dimethylformamide.[305] With acids it affords a deep-red carbonium ion.[305]

Some tetrahydrobenzo[*ghi*]perylenes, e.g., **406d,f** may be easily dehydrogenated on gentle heating affording compounds **418** with phenanthrene or dibenzothiophene moieties.[311] The driving force of this concerted elimination is the formation of benzenoid aromatic rings as well as the relief of steric strain.

418, a, X= CH=CH
b, X= S

When the bridged [18]annulene lacks the internal methyl groups oxidation occurs more easily; thus from **406e** thiacoronene is obtained quantitatively.[311]

D. Conclusions

The series of aromatic bridged annulenes which fulfill Hückel's rule, starting with **3** and **188** can be extended to [18]annulene **394** and to [22]annulene **413** using Vogel's "building block approach" synthesis.

3   *188*   *394*   *413*

In this new strategy the C_9 synthon cycloheptatriene-1,6-dicarbaldehyde is homologated with one, two, or three C_5 units affording the corresponding α,ω-dialdehydes with *syn*-stereochemistry. The bridged annulenes **188, 394,** or **413** could be obtained by adding a final C_2 unit in a three-sequence path: (1) Wittig olefination, (2) a $[4n + 2]\pi$-electrocyclic process, and (3) elimination. It has been also shown that excluding the C_5-units the above approach could produce the 1,6-methano[10]annulene (**3**) itself.

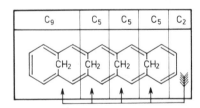

The comparison of the electronic spectra within the above series of bridged annulenes reveals the stepwise bathochromic and hyperchromic shifts (Figure 5).

Comparison of the ^{1}H-NMR spectra (Figure 6) indicates a reduction of diamagnetic ring current passing from **188** to **394** and further to **413**. However, comparison of the chemical shifts of bridge protons (e.g., in **188** and **394**) indicates that the steric compression of the inner bridge protons (and consequently, the H–H distances) are very similar.

Interestingly, though the side-view profile of the tetrakismethano-bridged compound **411** shows, according to X-ray data, a significant curvature (dihedral angle of 84.9° between terminal rings), the conjugation in the polyenic system seems to be largely unaffected.[8]

Consistent with a stepwise reduction of the diamagnetic ring current towards the right-hand end of the above series of bridged annulenes, a gradual replacement of aromatic reactivity by an olefinic one is noted.

Also interesting are the aromatic characteristics of tetrahydrobenzoperylenes **406** as well as their easy dehydrogenation.

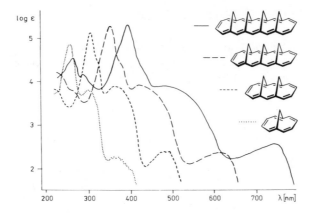

FIGURE 5. Electronic spectra of bridged annulenes **3, 188, 394,** and **413** (in cyclohexane). (From Vogel, E., *Current Trends in Organic Synthesis,* Nozaki, H., Ed. Permagon Press, New York, 1983, 393.With permission.)

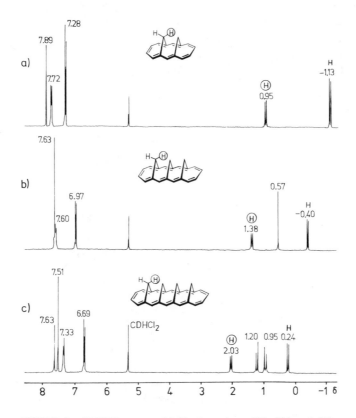

FIGURE 6. ¹H-NMR spectra of bridged annulenes **188, 394,** and **413** (300 MHz-FT; CS_2/CD_2Cl_2, lock CD_2Cl_2). (From Vogel, E., *Current Trends in Organic Synthesis,* Nozaki, H., Ed., Pergamon Press, New York, 1983, 394. With permission.)

VII. HETEROATOM-CONTAINING BRIDGED ANNULENES

The heteroatom-containing bridged annulenes can be divided in two fundamentally different classes according to the position of heteroatom(s): (1) heteroatom-bridged annulenes **420** and (2) bridged heteroannulenes **421**.

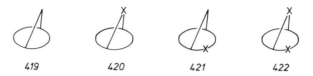

$$419 \qquad 420 \qquad 421 \qquad 422$$

The formula **419** is a bridged annulene; **420** to **422** are heteroatom-containing bridged annulenes.

Representatives of heteroatom-bridged heteroannulenes **422** have yet to be synthesized.

A. Heteroatom-Bridged Annulenes
1. Heteroatom-Bridged [10]Annulenes
a. Syntheses

The synthetic strategy utilized for 1,6-methano[10]annulene was extended for the generation of heteroatom bridged [10]annulenes.

1,6-Oxido[10]annulene (**425**) was independently synthesized in 1964 by Sondheimer[318,319] and Vogel[320] using the same principle.

$$5 \qquad 423 \qquad 424 \qquad 425$$

1,6-Imino[10]annulene (**427**) and its *N*-acetyl derivative **429** were obtained from the aziridine **426** synthesized from **5** via the 9,10-nitrosochloride. Bromination of **426** followed by HBr-elimination afforded **427**.[321] Acetylation of aziridine **426** to **428** followed by the same aromatization procedure yielded **429**.[320]

$$426 \qquad 427 \qquad 428 \qquad 429$$

Interestingly, the *N*-acetyl compound **429** could not be hydrolyzed to **427** whereas **427** is easily acetylated to **429**.[321]

b. Structure and Properties

The 1,6-oxido[10]annulene (**425**) is a pale yellow compound whose ^{1}H-NMR spectrum contains an A_4B_4 system centered at $\delta = 7.44$ ppm[320] and whose electronic spectrum presents maxima at λ_{max} 256, 299, and 394 nm.[47] These spectra are similar to those of 1,6-methano[10]annulene. The dipole moment of **425** is 1.94 D.[59]

The 1,6-imino[10]annulene (**427**) is a yellow compound with mp 16°C. The ^{1}H-NMR spectrum proves the diamagnetic ring current through its signals at $\delta = -1.1$ ppm (NH) and $\delta = 7.2$ ppm (peripheral protons).[321] The UV spectrum: λ_{max} 273; 312 and 394[47] presents slight deviation from that of 1,6-methano[10]annulene. The dipole moment of **427** is 1.42 D.[59] The ESR spectrum of the radical anions of **427**[322] reflects the reduced symmetry (C_s) in comparison with 1,6-methano[10]annulene (C_{2v}). The stabilities of radical anions of 1,6-X-bridged[10]annulenes show the following order:[322] X=CH$_2$ >> NH > O > NCH$_3$.

The magnetic circular dichroism of **427** (as well as of **425**) suggested some degree of interaction of the nonbonding heteroatom electrons with the perimeter 10 π-electron system.[50] The enthalpies of formation are $\Delta H_f^{298} = 47.8 \pm 1.2$ kcal $\cdot$ mol^{-1} for **425** and 87.8 ± 0.7 kcal $\cdot$ mol^{-1} for **427**.[60] The spectral data of **429** agree with the bridged annulene structure.[320]

Among the chemical reactions of **425**, the conversion to α-naphthol on acidic[319,320] or basic treatment[323] and the isomerization to 1-benzoxepin (**430**) deserve mention.

430   431   432   433,a,R=H
                                            b,R=Br

The isomerization to α-naphthol does not proceed through **430** as it was initially proposed[318,319] because **430** subjected to acid catalysis does not afford α-naphthol.[69] Bromination of **425** with 1 mol of bromine at $-75°C$ affords **431** which at 0°C takes up another mole of bromine, giving **432**. Treatment of **431** and **432** with potassium *t*-butoxide causes the return to annulenic structures since bromoderivatives **433a** and **433b**, respectively, are formed.[73] Starting with **433a** different 1-substituted 1,6-oxido[10]annulenes were obtained[73] analogously to 1,6-methano[10]annulene. The notable difference between **425** and 1,6-methano[10]annulene on bromination at low temperature (initial formation of tricyclic vs. bicyclic products) was ascribed to the difference in the equilibrium position of the related systems: oxepin $\rightleftarrows$ benzeneoxide and cycloheptatriene $\rightleftarrows$ norcaradiene (ring strains in ethylene-oxide and cyclopropane are 13 and 27 kcal $\cdot$ mol^{-1}, respectively. This larger ring strain of norcaradiene than of benzeneoxide is discussed in more detail in the chapter on heteroannulenes).

On the other hand, nitration of **425** affords a mixture of isomeric 2- and 3-nitroderivatives,[318,319] whereas the reaction with 4-phenyl- or 4-methyl-1,2,4-triazolin-3,5-dione[71] gives rise to bis adducts similar to those of 1,6-methano[10]annulene.

The 1,6-imino[10]annulene (**427**) is a base affording a hydrochloride, an *N*-methylderivative,[321] and an *endo*-peroxide **434**.[9] Starting from **434**, by standard procedures, interesting derivatives, e.g., naphthalene-*endo*-peroxide **435** (a source of singlet oxygen), the naphthalenedioxide **436** and the naphthalene pentoxide **437** were obtained.[9]

434   435   436   437

On the other hand, thermolysis of N-substituted 1,6-imino[10]annulenes led, probably via benzeneimine valence tautomers **438**, either to 1-H-benzazepines **439** (through Berson-Willcott rearrangement) or to fission to two C–N bonds leading to nitrenes (:NR) along with naphthalene; the nitrenes were trapped with cyclohexene as **440**.[9] Diene synthesis of **427** with 4-substituted-1,2,4-triazoline-3,5-diones proceed similarly to those of 1,6-methano[10]annulene.[71]

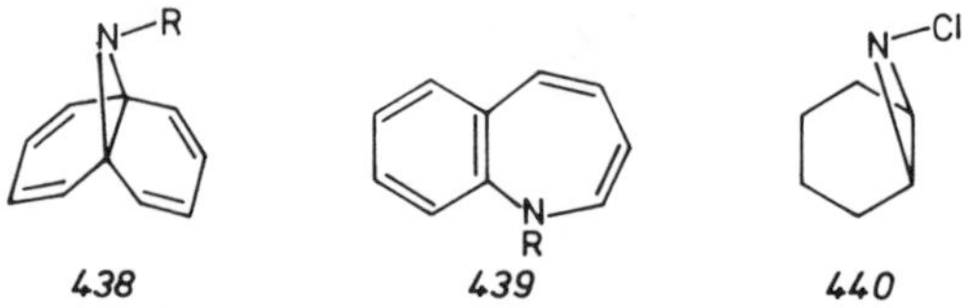

Unlike methylene-bridged [10]annulenes, in O- or NR-bridged [10]annulene, there is no evidence for a valence isomerism with a heteropropellanic structure such as **438** at room temperature; however, such structures may be involved in high-temperature reactions.

2. Heteroatom-Bridged [14]Annulenes
a. Syntheses

Syn-1,6:8,13-bisoxido[14]annulene (**444**) was obtained in 1966 by Vogel[324] (see the structure below). Epoxidation of hexahydroanthracene **218** afforded a mixture of *syn*- and *anti*-diepoxyderivatives whose bromination gave tetrabromoderivatives **441** and **442**.

Scheme 5

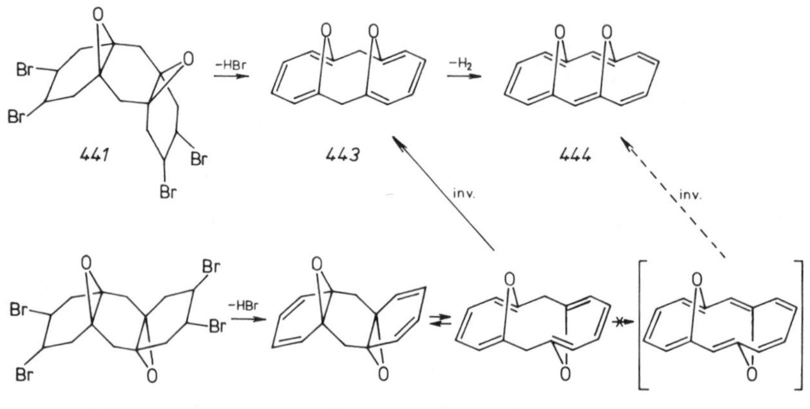

The treatment of **441** with potassium *t*-butoxide easily afforded **444** via **443** (which could be isolated in inert atmosphere).[38,324] The similar treatment of **442** leads only to **445** in equilibrium with a small concentration of *anti*-bis-oxepin **446**.[325] Attempts to dehydrogenate the mixture **445** ⇄ **446** did not afford *anti*-bisoxido[14]annulene (**447**), but the same *syn*-isomer **444**.[38] The bridge flip involved in this reaction occurs most probably at the stage of bis-oxepines **446** → **443**, because a sample of **445** was isomerized thermally at 100°C to **443**.[38,325] It also seems possible that the elusive *anti*-bisoxido[14]annulene **447** could undergo an easy oxygen bridge inversion to the aromatic, hence more stable, **444** (an analogous inversion of carbon-bridged [14]annulenes cannot occur owing to the bulkier CH$_2$ groups).

Syn-1,6-methano-8,13-oxido[14]annulene (**450**) was obtained[325] in a manner similar to **444**. Both *syn*- and *anti*-epoxides obtained on epoxidation of **448** (resulted from **218** in a Simmons-Smith reaction) afford the same *syn*-product **449** on bromination-dehydrobromination (via bridge flip); the final aromatization of **449** to **450** succeeded again with DDQ.

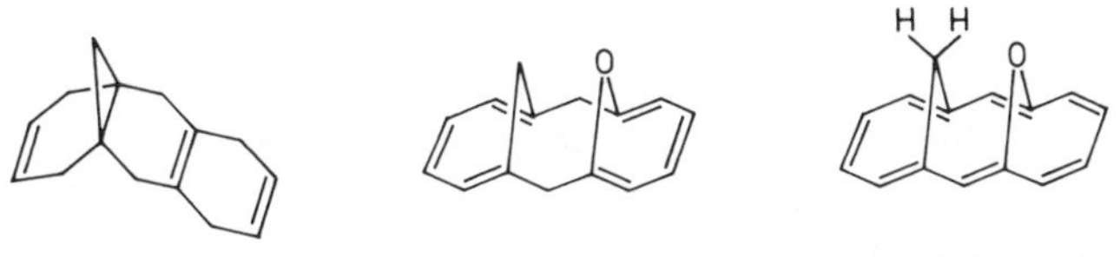

448   449   450

Syn-1,6-imino-8,13-methano[14]annulene (**454**) was prepared[326] from **448** via **451** (obtained on treatment with sodium azide and *N*-bromosuccinimide) and **452** (obtained on reduction with LiAlH₄). The subsequent bromination-dehydrobromination to **453** followed by dehydrogenation with oxygen and potassium *t*-butoxide afforded **454**.

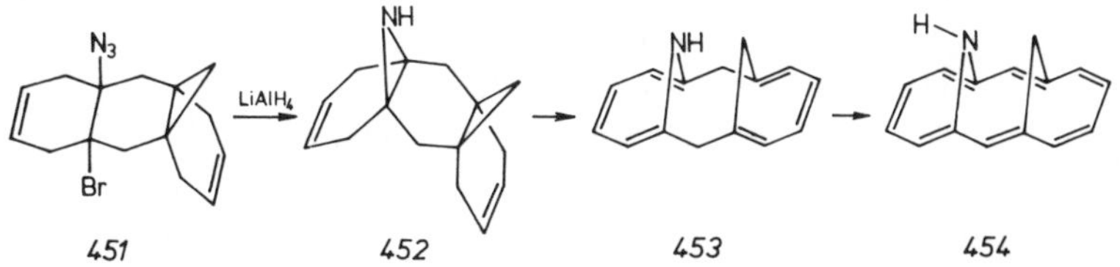

451   452   453   454

Syn-1,6:8,13-bisimino[14]annulene (**461**)* was synthesized quite recently by Vogel[109,327] in an original variation of his general method. The nucleophilic opening of the epoxide rings in hexahydroanthracene bisepoxide by azide ions afforded **456**, whose reaction with SO₃/H₂SO₄ afforded **457** (all other attempts to functionalize the strongly hydrogen bonded OH groups failed). Reduction of **457** to **458** was followed by reaction with thionyl chloride affording **459**, by aromatization to **460**, and by final acid hydrolysis to **461**.[109,327] Other conversions of **458** into **461** were not conclusive.

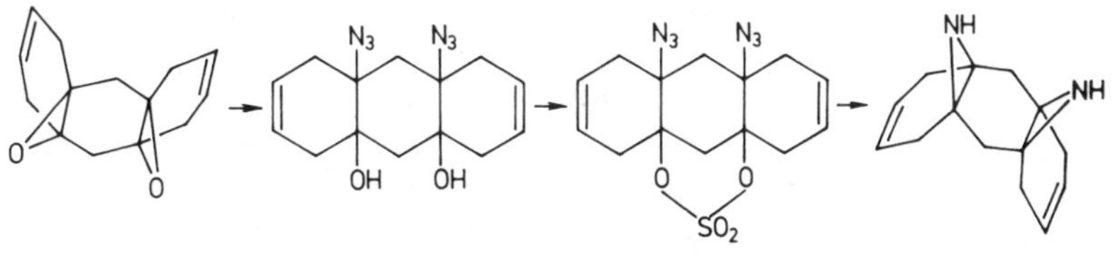

455   456   457   458

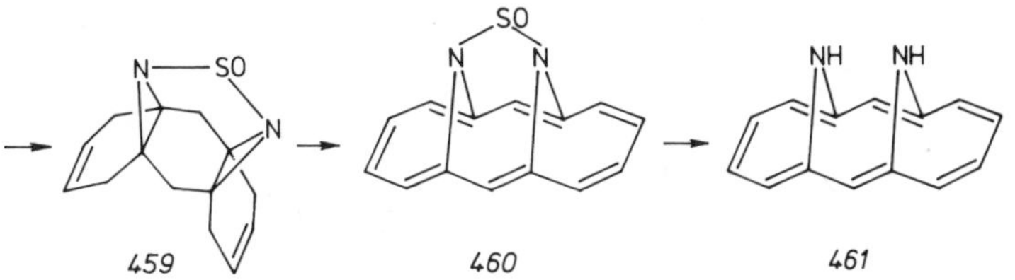

459   460   461

b. Structure and Properties

Syn-bisoxido[14]annulene is a bright red solid with a ¹H-NMR spectrum similar to that of 1,6-oxido[10]annulene: δ = 7.67 ppm (eight outer perimeter protons) and 7.94 (two central perimeter protons,[114] and an electronic spectrum in agreement with the structure **444**: λ_max 306; 345; 382; 555 nm.[324] The dipole moment is 3.2 D.[81] On the basis of coupling constants in the ESR spectrum of radical anions of **444** an almost equal planarity of *syn*-bisoxido[14]annulene and methano-oxido[14]annulene (**450**) was inferred.[295] The X-ray anal-

---

*   1,4:8,11-Bisimino[14]annulene, a diatropic molecule, was obtained[411] from *N*-(pyrrolo)pyrrole derivatives in three steps.

ysis of **444**[328] confirmed the geometry with a C–O–C angle of 108°, with O⋯O distance of 2.55 Å, probably one of the shortest distances recorded for two nonbonded oxygens, and with a nearly equal aromatic C–C distances throughout the perimeter.

Compound **444** is stable on heating or towards an oxidation and easily affords 2-substitution products (bromo, nitro, acetyl).[81] Treatment with chromium hexacarbonyl affords anthracene by deoxygenation.

Syn-methano-oxido[14]annulene (**450**), an orange-red solid, presents in the ¹H-NMR spectrum the following signals: δ = 0.92 and −1.4 ppm (bridge protons); 7.75 (2H, s) for the 7- and 14-protons, and 7.3 to 7.8 (8H, m) characteristic for an aromatic bridged annulene. The 2.3 ppm difference between the signals of CH_2 bridge protons indicates an important interaction of the two bridges whereas the coupling constant of CH_2 protons, with 3.7 Hz greater than in 1,6-methano[10]annulene, proves the sprawling of the **450** perimeter.[291] The UV spectrum of **450** resembles that of **444**.[325]

Compound **450** undergoes aromatic substitutions (e.g., acylation).[325]

Syn-iminomethano[14]annulene (**454**) is a brick-red solid whose ¹H-NMR spectrum proves its aromaticity; δ = −2.07 (NH); −1.52 and 2.08 (CH_2), and 7.45 (s, H-7,14) and 7.23 to 7.84 (eight outer perimeters protons).[326] The *exo*-position of NH proton, resulted from its shielded ¹H-NMR signal as well as from X-ray determination,[329] is of great interest in view of the much debated question of relative sizes of the unshared electron pair and a bonded hydrogen.[330] The electronic spectrum of **454** presents evidence for interactions between the N-electron pair and the π-electrons.[326]

No important chemical reaction of **454** was described; an acetyl-derivative was obtained but not directly.[326]

Syn-bisimino[14]annulene (**461**) is a scarlet-red stable compound, mp 158°C, with a pK_a = 5.74.[327] The ¹H-NMR spectrum at room temperature: 1.47(NH); 7.53(H-3,4,10,11)-AA′BB′ system; 7.80(H-2,5,9,12) and 7.89(s; H-7,14) is changed at −100°C into two AA′BB′ systems and two NH signals (δ = 4.22 H*endo*; δ = −1.58 H*exo*). The ¹³C-NMR spectrum at room temperature: δ = 114.21(C-1,6,8,13); 128.42(C-2,5,9,12); 131.90(C-3,4,10,11); 123.55(C-7,14) changes into a spectrum of seven signals at −133°C.[327] NMR-Data point to a dynamic process (intramolecular, because the rate constants are concentration independent) of configurational inversion of N-atoms ("windshield wiper") **461a** ⇄ **461b** with $\Delta H^{\ddagger}$ = 26.1 kJ · mol⁻¹ and $\Delta S^{\ddagger}$ = −76.5 J · mol⁻¹K⁻¹.[327]

Crystal and molecular structures of *syn*-1,6:8,13-diimino[14]annulene (**461**) and of its derivatives were investigated by Destro et al.[419]

461a   461b

In conclusion, the O- or NH-heteroatom-bridged ([14]annulenes present no evidence for reversible valence tautomerism (involving three-membered ring compounds such as **438**); their ¹H-NMR spectra indicate a diamagnetic ring current involving the planar aromatic 14 π-electron system in the *syn*-configuration **444, 450, 454, 461**; similarly, oxepines with 2,7-electronegative substitutents, e.g., oxepin-2,7-dialdehyde, presents only the monocyclic valence isomer.[331] Recalling that the carbon-bridged [14]annulene in the *anti*-configuration is not aromatic, it is easy to understand why the heteroanalogs were isolated only in the *syn*-configuration; indeed, inversion of heteroatom configurations occurs much more readily than inversion involving carbon.

B. Bridged Heteroannulenes

Bridged heteroannulenes represent a relatively new research area which was developed as an extension of that of bridged annulenes; the important differences in conformation and mobility of monocyclic annulenes are avoided. On the other hand, interactions with other research areas, such as aza [4n + 2]annulenes and porphyrins, appear possible.

1. Syntheses

1-Aza-2,7-methano[10]annulene (**465**) a 10 π-analog of pyridine, was synthesized in 1978 by Vogel[332] starting with cycloheptatriene-1,6-dicarbaldehyde (**9**). The acid **462a** (obtained by Wittig olefination followed by oxidation was converted into the corresponding azide **462b** which was subjected to Curtius rearrangement. The resulted isocyanate **462c** was thermally isomerized (through a 10 π-electrocyclic process or by successive cycloheptatriene-norcaradiene and Cope rearrangements) into the unstable aza-enone **463** which affords the tautomeric lactam **464** by prototropic migration. The O-tosyl derivative of **464** was reduced with LiAlH₄, then aromatized with DDQ to **465**.

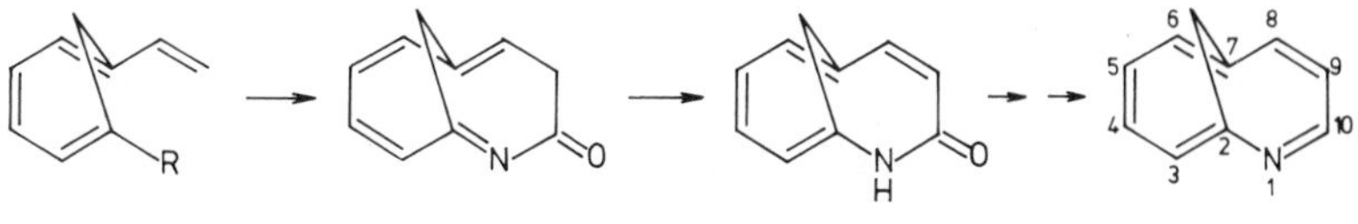

462,a,R=COOH
   b,R=CON₃
   c,R= N=C=O   463   464   465

Methyl derivatives of **465** — A similar reaction sequence as above led to 8-methyl derivative **466**.[9] However, this method could not be applied for the 9-methyl derivative **467** because the corresponding isocyanate intermediate (analogous to **462c**) did not possess the driving force to undergo the 10 π-electrocyclic process at low temperature in order to compete with side reactions.

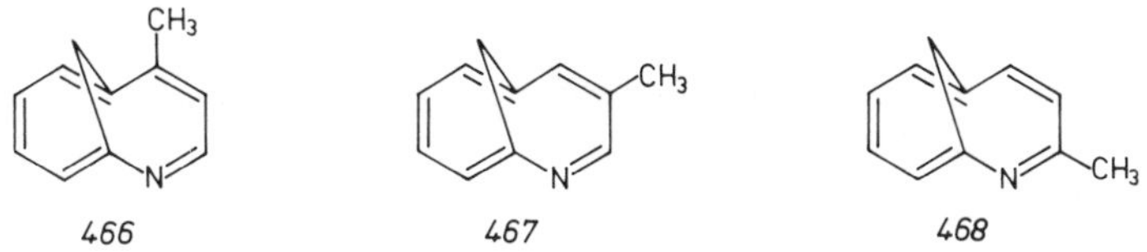

466   467   468

The synthesis of **467** (as well as of the 9-cyano derivative) was, however, achieved[109] via the substituted cis-2-vinylcyclopropylisocyanate **469b** obtained from the bicyclic ester **469a** by usual steps. The Cope reaction of **469b** affords the azatrienone **470** which isomerizes further by sigmatropic migration yielding the lactam **471**. The O-tosyl derivative **472** of **471** is converted then by the same mild steps as above into **467**.

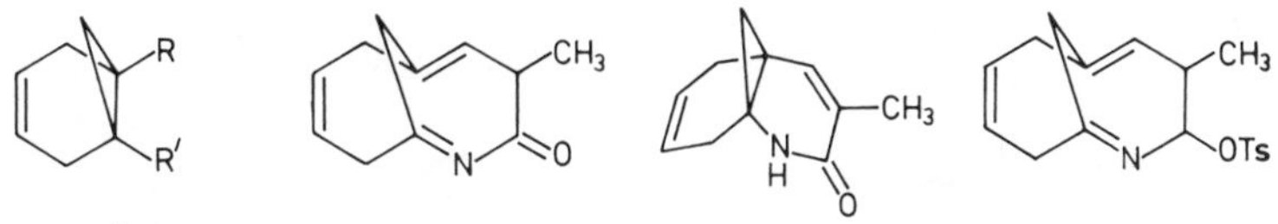

469,a,R=R'=COOMe
   b,R= CH=CH—CH₃
   R'= N=C=O   470   471   472

The 10-methyl isomer, **468,** was obtained using the same method[109] from isocyanate **473** via **474** (directly formed through a Wittig reaction followed by Cope rearrangement and isomerization).

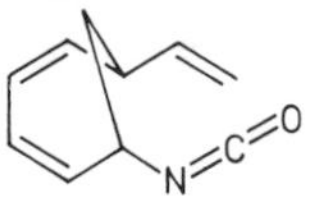

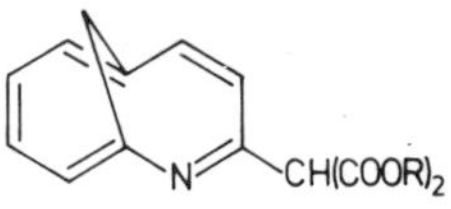

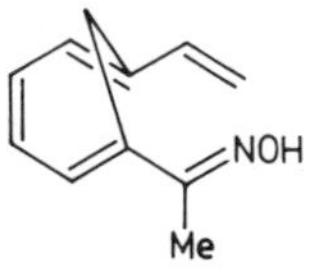

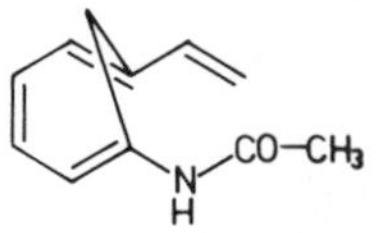

473   474   475   476

An alternative synthesis of **468** was found[109] by treating the oxime of 1-acetyl-6-vinyl-cycloheptatriene (**475**) with *p*-toluenesulfonyl chloride and trimethyl amine (a Beckmann rearrangement to **476** accompanied by neighboring group participation of the vinyl double bond).

The 10-methoxy derivative **479** of **465** was prepared from the oxime **477** via the lactam **478**.[333]

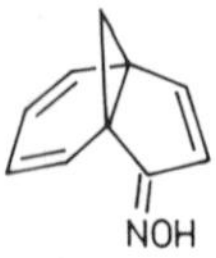

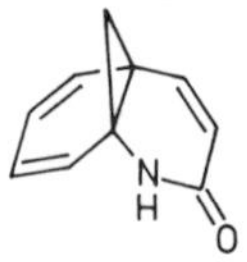

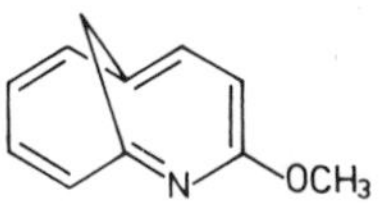

477   478   479

1-Aza-10-methoxy-3,8-methano[10]annulene (**484**), the first derivative of the yet unknown 1-aza-3,8-methano[10]annulene, was obtained in 1978 independently by Vogel[332] and Helmkamp.[333] The oxime **480** affords on Beckmann rearrangement the lactam **481** which is converted to **482** (bromination-dehydrobromination), then with trimethyloxonium tetrafluoroborate to **483** which is dehydrogenated to **484**.[332]

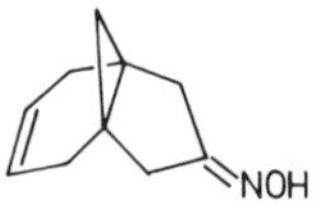

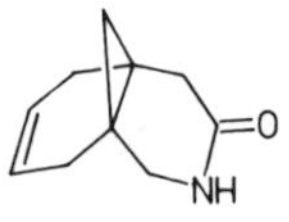

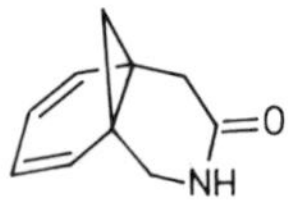

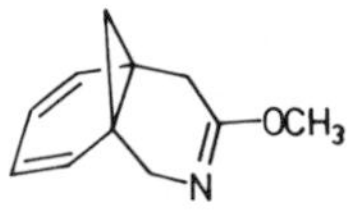

480   481   482   483

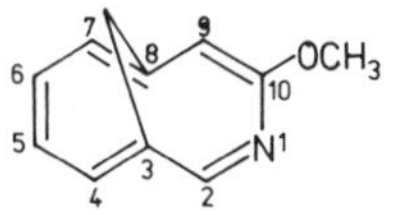

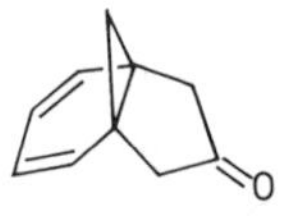

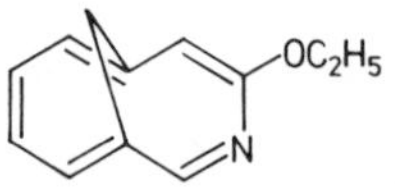

484   485   486

In the second synthesis[333] the lactam which resulted from Beckmann rearrangement of the oxime of **485** was aromatized by addition of a methyl-thio group which was oxidized to a sulfone, followed by *O*-methylation of the amide and finally by methanesulfinic acid elimination.

The 10-ethoxyderivative **486** was independently obtained in 1978 by Muchowsky et al.[334] using practically the same scheme as Vogel.[332]

Substituted derivatives **488** of 1-aza-3,8-methano[10]annulene resulted on condensation of 1,2,4-triazines **487** with benzocyclopropene:[335]

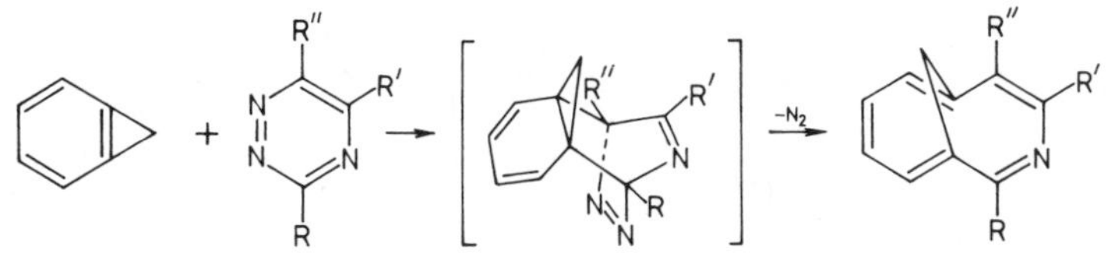

487   488

The 1-thia-4,9-methano[11]annulene (**490a**) and 1-oxa-4,9-methano[11]annulene (**490b**) were obtained[336] through Wittig reaction of cycloheptatriene-1,6-dialdehyde (**9**) with phosphonium salts **489**

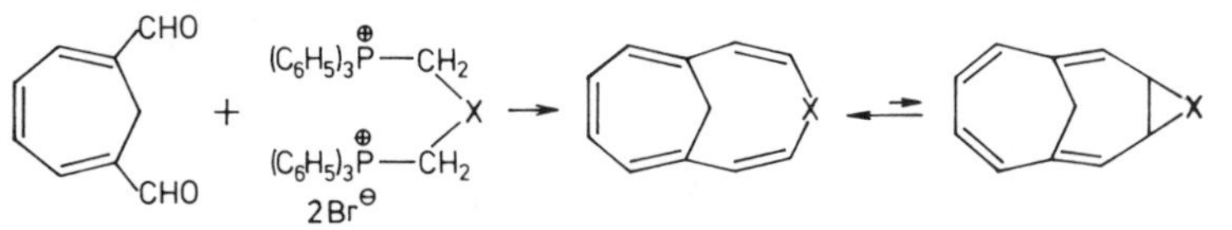

489, a, X=S   490, a, X=S   491 a, X=S
      b, X=O   b, X=O
                          c, X=CH$_2$

1-Aza-2-methoxy-5,10-methano[12]annulene (**494**) was described in 1975 by Paquette.[82] Its synthesis starts from the known ketone **492**[337] whose oxime afforded the lactam **493** on Beckmann rearrangement; treatment of **493** with methanol produced **494**. An attempted synthesis of the isomeric compound **495** failed at the step of Beckmann rearrangement.[82]

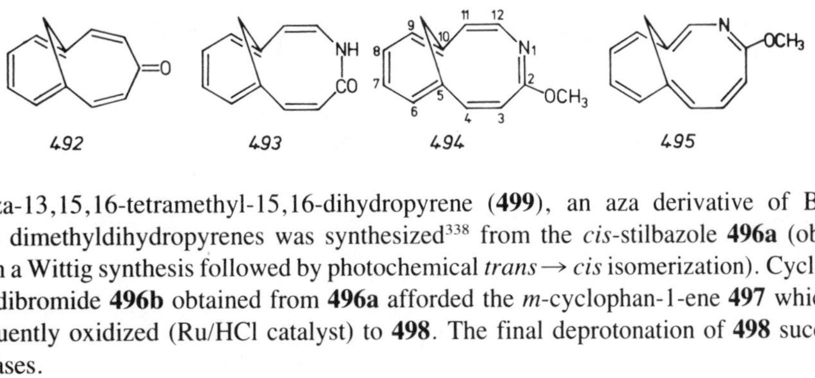

492   493   494   495

2-Aza-13,15,16-tetramethyl-15,16-dihydropyrene (**499**), an aza derivative of Boekelheide's dimethyldihydropyrenes was synthesized[338] from the *cis*-stilbazole **496a** (obtained through a Wittig synthesis followed by photochemical *trans* → *cis* isomerization). Cyclization of the dibromide **496b** obtained from **496a** afforded the *m*-cyclophan-1-ene **497** which was subsequently oxidized (Ru/HCl catalyst) to **498**. The final deprotonation of **498** succeeded with bases.

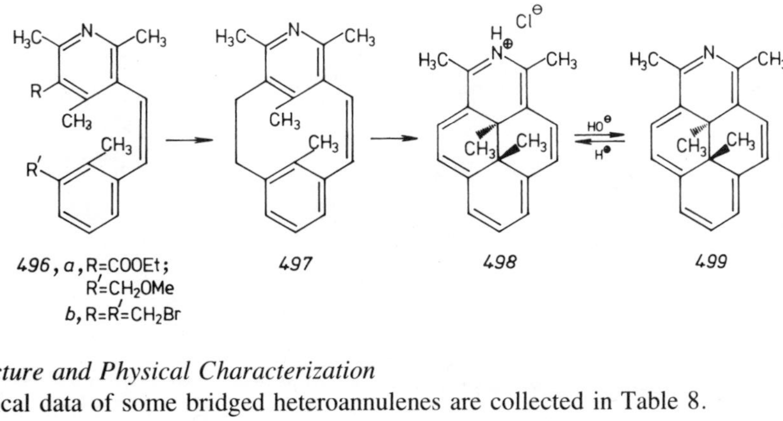

496, a, R=COOEt;   497   498   499
      R′=CH$_2$OMe
      b, R=R′=CH$_2$Br

2. Structure and Physical Characterization
Physical data of some bridged heteroannulenes are collected in Table 8.

Table 8

PHYSICAL CHARACTERIZATION OF SOME BRIDGED HETEROANNULENES

Compound	Melting point or boiling point (°C)/ pressure (Torr) (color)	¹H-NMR spectrum; ¹³C-NMR spectrum (δ ppm)	Electronic spectrum (λ nm)	Ref.
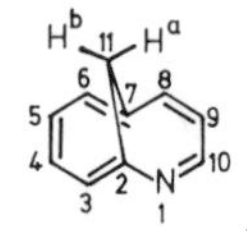 **465**	B.p. = 63/1 (yellow)	−0.4 and 0.65(H-11a,b) AX system(J = 8.4 Hz); 6.5(dd;H-9);6.6—7.2(m;H-4-6); 7.38(two superposing d,H-3,8); 8.23(d;H-10;J = 6.3 Hz) (in CCl₄)	240; 258; 297; 364 (in cyclohexane)	332
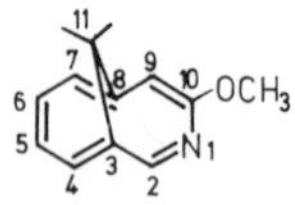 **484**	B.p.62—3/0.1 (yellow)	−0.14 and 0.37(H-11) AX system; (J = 9.5 Hz); 3.86(s;CH₃O); 5.90(s,H-9); 6.66 and 7.28(m;H-4-7); 7.46(s;H-2)(in CCl₄)		332, 333
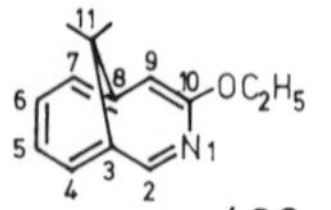 **486**	B.p.70/0.1	¹H-NMR: 0.0 and 0.54(H-11) AX system; 1.40(t;CH₃); 4.37(q;CH₂O); 5.94(s;H-9); 6.55 and 7.20(m;H-4-7); 7.44(s;H-2); ¹³C-NMR: 14.79(CH₃); 33.84(C-11); 61.35(OCH₂); 100.62(C-9); 123.96(C-5); 124.84(C-3); 125.29(C-8); 127.99(C-6); 130.33(C-7); 131.27(C-2); 157.02 (C10)	258; 315 (in methanol)	334
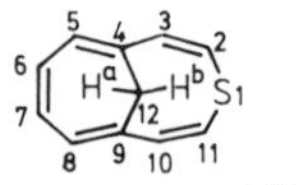 **490a**	27—28 (Red)	1.24(H-12a) and 6.28(H-12b) AX system (J = 11.5 Hz); 4.78(H-2,11) and 6.45(H-3,10) AX system (J = 9.6 Hz); 6.09(H-5,8) and 6.54(H-6,7)- AA′XX′ system (in CCl₄)	233; 284; 395(in cyclohexane)	336
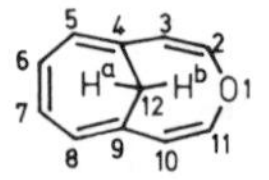 **490b**	B.p. ∼70/0.1 (yellow)	0.63(H-12a) and 4.52(H-12b) AX system (J = 10.4 Hz); 5.48(H-2,11) and 5.71(H-3,10) AX system (J = 6.2 Hz); 6.15(H-5,8) and 6.67(H-6,7)- AA′XX′ system (in CCl₄)	248; 328 (in cyclohexane)	336
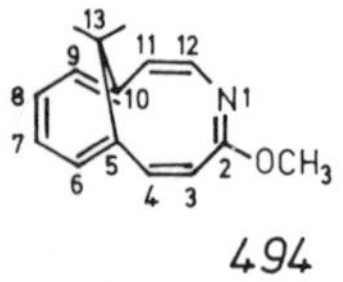 **494**	68—70 (Bright orange)	1.64(d;J = 12 Hz;H-13); 3.76(s;CH₃O); 5.08(d;J = 10 Hz;H-11); 5.34(d;J = 12 Hz;H-13); 5.85—6.55(m;7 ring H) (in CDCl₃)		82
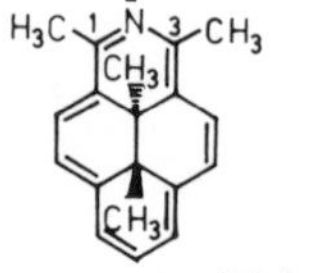 **499**	105—106 (Black)	−3.75 and −3.80(s;internal CH₃; 3.43(s;external CH₃); 8.1—9(m;7 ring H)(in CDCl₃); ¹H-NMR of hydrochloride: −3.80 and −3.82(s;internal CH₃; 3.87(s;external CH₃); 8.7—9.5(m;7 ring H) (in CDCl₃)	199; 233; 349; 393; 472; 540; 593; 602; 655 (in cyclohexane)	338

1-Aza-2,7-methano[10]annulene (**465**) presents an electronic spectrum similar to that of 1,6-methano[10]annulene and an ^{1}H-NMR spectrum which recalls both 1,6-methano[10]annulene and quinoline.[332] The X-ray investigation of the crystalline 3-bromoderivative revealed:[339] (1) the equivalence of both C–N bond lengths, (2) a C–N–C bond angle of 119.5° (larger than in pyridine), (3) important distorsions in the ring caused by the nitrogen atom, and (4) a shorter C-2 to C-7 distance (2.202 Å) than in 1,6-methano[10]annulene, indicative of stronger interaction of the respective atoms.

1-Aza-10-(*m*)ethoxy-3,8-methano[10]annulene (**484** and **486**) present related ^{1}H-NMR spectra (see Table 8) indicative of diamagnetic ring currents.[332,334]

Recently, aza-methano[10]annulenes were separated in optical isomers.[340]

As expected, 1-thia-4,9-methano[11]annulene (**490a**) does not exhibit in the ^{1}H-NMR spectrum characteristics of a significant ring current. The large difference between CH_2 bridge protons (5 ppm in ^{1}H-NMR spectrum) is in agreement with: (1) a steric compression between the sulfur atom and the bridge proton H-12b situated in the *syn*-position, (2) the diamagnetic anisotropy of C–S bonds, and (3) the field effect of sulfur.[336] The presence of an episulfidic valence isomer **491a** is ruled out by the lack of temperature dependence of ^{1}H-NMR spectra around normal temperatures. The UV spectrum of **490a** (Table 8) is different from that of the related hydrocarbon **490c** and this suggests interactions between the sulfur atom and the π-system in spite of the nonplanar conformation of thia-annulene.[336]

490a (X=S)

1-Oxa-4,9-methano[11]annulene (**490b**), is an air-sensitive liquid whose ^{1}H-NMR spectrum occupies an intermediate position between **490a** and **490c**. The identity of electronic spectra of **490b** and the hydrocarbon congener **490c**[336] indicates that both compounds exist in a similar conformation.

1-Aza-2-methoxy-5,10-methano[12]annulene (**494**) has, according to the ^{1}H-NMR data (Table 8) a nonplanar molecule with localized π-bonds. Analogously to 1,6-methano[12]annulene (**153**), the structure **494** with a bridged cycloheptatrienic unit has a larger contribution to the real structure of the molecule than the resonant formula with shifted π-bonds. No indication for the presence of a norcaradienic valence isomer could be obtained.[82]

2-Aza-1,3,15,16-tetramethyl-15,16-dihydropyrene (**499**) presents an ^{1}H-NMR spectrum characteristic for an aromatic compound. Interestingly the upfield shift of the internal CH_3 protons signal observed on conversion of the free base **499** into the corresponding conjugate acid **498** (instead of a normal downfield shift) points to a stronger diamagnetic ring current in **498** than in **499**.[338]

3. Chemical Properties

1-Aza-2,7-methano[10]annulene (**465**) with $pK_a = 3.20$ (20°C)[332] is a weaker base than pyridine or quinoline. It affords a stable hydrochloride,[332] can be N-alkylated, and gives rise to an *N*-oxide analogous to pyridine oxide.[9] The oxidation of methylderivatives **466** to **468** parallels the behavior of the corresponding methylpyridines; whereas **466** and **468** could be oxidized with SeO_2 and Ag_2O to the acids **500** and **501**, the methylderivative **467** remains unchanged.[109]

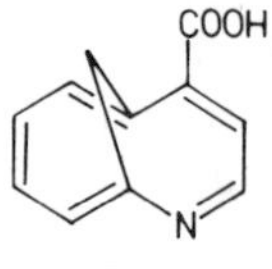

500 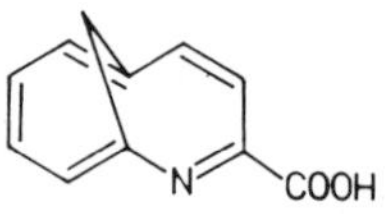   501 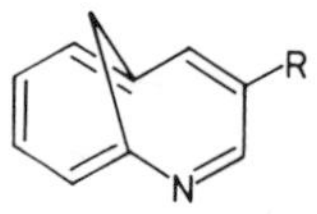   502, a, R=CN
                               b, R=CONH₂

The amide **502b**, a 10-π analog of nicotinamide, could be, however, obtained on hydrolysis (95% H_2SO_4!) of the nitrile **502a**.[109] Bromination of **465** parallels the corresponding reaction of 1,6-methano[10]annulene affording initially the labile *cis-syn*-dibromoadduct **503** (structure proved by X-ray analysis), then, on elimination, the substitution product **504**.[109]

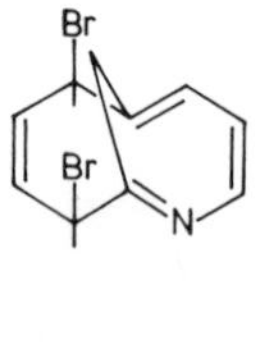

503 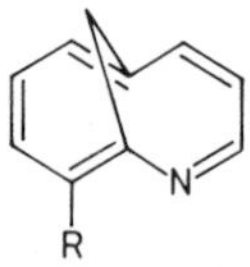   504, a, R=Br   505, a, R=H   506
            b, R=COCH₃   b, R=CH₃
            c, R=NH₂
            d, R=OH

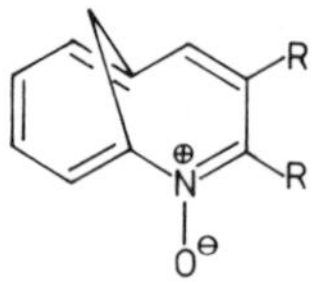

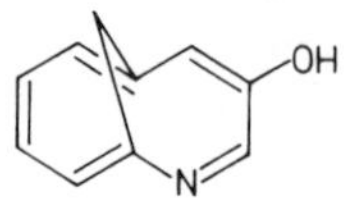

On the other hand, acetylation yields **504b** which could be further converted to **504c,d**. Attempts to convert **504a** directly into **504d** failed.[109]

Interesting photochemical conversions of aza-annulene oxides **505a,b** to **506** and **507**, respectively, were reported[109] and epoxides **508** were proposed as intermediates.

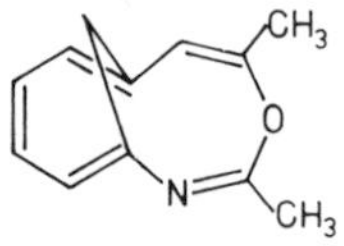

507 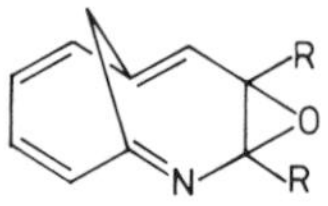   508, R=H or CH₃   509

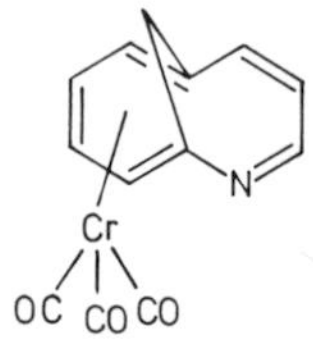

A chromium tricarbonyl π-complex **509**, maintaining the 10-π electron delocalization similar to the complex **95** of 1,6-methano[10]annulene could be obtained from **465**;[341] the X-ray analysis showed an unsymmetric bonding of the $Cr(CO)_3$ group to the heteroatom-free part of the molecule, in *anti*-orientation to the CH_2 bridge.[341]

The unique reaction mentioned for 1-thia-4,9-methano[11]annulene (**490a**) is the easy thermal decomposition into 1,6-methano[10]annulene and sulfur, proceeding probably through the episulfide valence isomer **491a**.[336]

1-Aza-2-methoxy-5,10-methano[12]annulene (**494**) could not be converted into diene adducts.[82] Polarographic reduction affords a dianion (possibly **510**) which could not be further investigated owing to its instability. Alkali metal reduction afforded two dihydroderivatives of **494** but no evidence for traces of **510** could be obtained.[82]

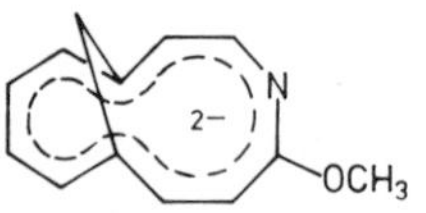

510 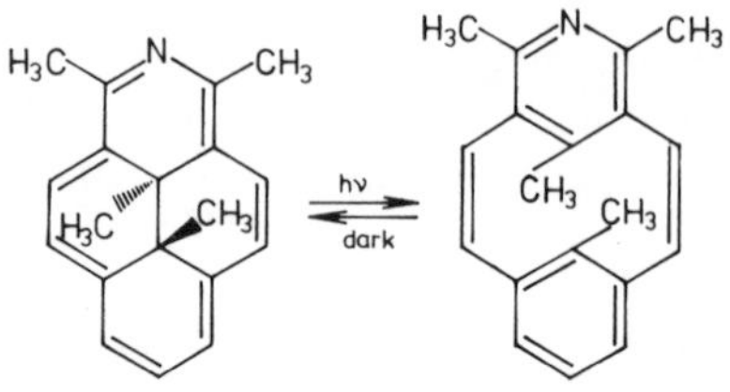   499   511

2-Aza-1,3,15,16-tetramethyl-15,16-dihydropyrene (**499**) presents a reversible valence isomerization to the dienic *m*-cyclophane **511** similarly to 15,16-dialkylhydropyrenes.[338]

C. Conclusions

The heteroatom-bridged [10]- and [14]annulenes are stable aromatic compounds. Whereas the chemistry of the O-bridged **425** is that of an aromatic compound, some reactions of the NH-bridged **427** form a theoretical link between bridged annulenes and epoxides of aromatic hydrocarbons.

Among the bridged [14]annulenes, two interesting series are now available for comparison, namely **444**, **450**, **188** and **461**, **454**, **188** (stepwise substitution of oxygen or NH-groups by CH_2 groups). Whereas **444** and **450** are quite similar compounds (thus the first CH_2 group does not sterically impair the delocalization of the annulene perimeter) the introduction of the second CH_2 group conducts to the sterically congested hydrocarbon **188**, which is aromatic in physical behavior but less aromatic chemically.

444 ; X=O ; Y=O   *461* ; X=NH ; Y=NH
450 ; X=CH_2 ; Y=O   *454* ; X=CH_2 ; Y=NH

188 (X=Y=CH_2)

A complete comparison with the members of the second series still awaits the chemical characterization of **461**. According to X-ray data,[329] the trend of increasing steric hindrance between the bridges is **444** < **454** < **188** (namely, distances between the bridge atoms, and the dihedral angles between the planes determined by one bridge and two bridge-head atoms).

Correlation of aromaticity with geometric parameters allowed to infer similar aromaticities for **454**, **444**, and **188**.[215]

A "building block approach" synthesis starting with oxepin-2,7-dialdehyde (obtained by Vogel in 1976[331]) could lead to oxygen bridged annulenes in an expanded series.

The new area of bridged heteroannulenes has expanded in recent years to new dimensions. The synthesis of aza-methano[10]annulenes and the exploration of their chemistry allowed chemists to assess their aromaticity and to compare these compounds with carbocyclic methano[10]annulenes and, on the other hand, with aromatic heterocycles, such as pyridine, quinoline, isoquinoline, etc. From the above description some similarities between the three classes of compounds as well as limits of such a parallelism resulted.

The methano-bridged heteroatom (O,S)[11]annulenes **490a,b** are nonplanar *syn*-compounds presenting paramagnetic ring current; the extrusion of sulfur atom from **490a** may proceed through an episulfidic valence isomer **491a**.

The methanoaza[12]annulenes **494** are stable compounds with localized π-bonds favoring cycloheptatrienic structures. The possibility of electron delocalization of the neutral homoaromatic type was discussed[82] but deserves further investigation.

The tetracyclic aza[14]annulene derivative **499** of Boekelheide's type is a typical aromatic compound prone to valence isomerism.

Bridged 10-π-electron analogs of pyridinium systems have been prepared from **465**, but so far no analog of pyrylium cation was reported; such systems and physicochemical behavior are interesting since in the case of [14]annulene analogs, the pyridinium analog **498** was inferred to have more pronounced delocalization than the neutral congener **499**.

Okazaki et al.[412] described the 13-step synthesis of 2,7-methanothia[9]annulene (starting from 1,6-diiodocycloheptatriene); the former exists (16%) in equilibrium with its norcara-diene valence isomer (84%).

VIII. BRIDGED ANNULENE IONS

The area of bridged annulenes was developed from the very beginning in close connection with that of related ionic species. Such a connection seems to be a natural one, keeping in mind that one of the most spectacular achievements of the Hückel theory was the discovery of ionic aromatic species.

As it will be shown in this chapter, numerous bridged annulene ions were obtained and their behavior brought important confirmations of the theory of aromaticity.

A. Cations of 1,6-Methano[10]Annulenes

Protonation of 1,6-methano[10]annulene (**3**) with $FSO_3H/ClSO_2F$ at $-78°C$ affords the 8π monocation **512** (orange solution) which lacks any paramagnetic ring current.[342] The seven olefinic protons from **512** indicate an average chemical shift of $\delta = 7.79$ ppm (a mere 0.12 ppm higher than olefinic protons of monohomotropylium ion) and an average of $\delta = 1.90$ ppm for bridge protons. The C-1 to C-6 distance may be smaller than in the starting hydrocarbon **3**.[342]

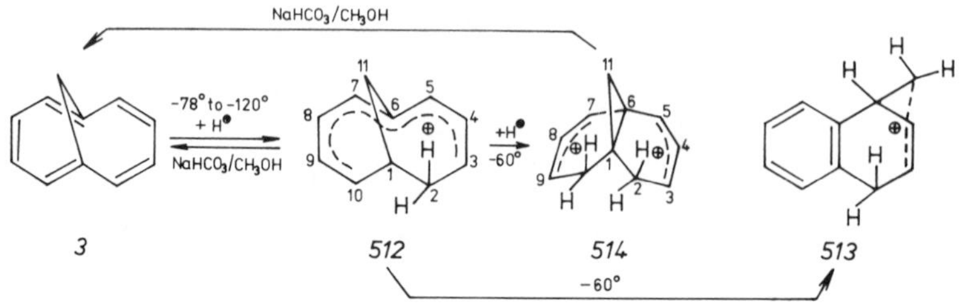

The cation **512** rearranges slowly at temperatures higher than $-60°C$ to the cyclopropyl-carbinyl cation **513** whose structure was demonstrated by ¹H- and ¹³C-NMR spectra.[343] The conversion **512** → **513** was rationalized through a ring closure between C-1 and C-6 followed by wandering of bridge methano group over the ''naphthalenium skeleton'' probably via successive [1,2]- and [1,4]- or via two subsequent [1,3] sigmatropic shifts.[343] Quenching of **513** solutions with $NaHCO_3$-buffered methanol afforded **516** (probably via the unstable cation **515**) and **517** (by proton loss).[343]

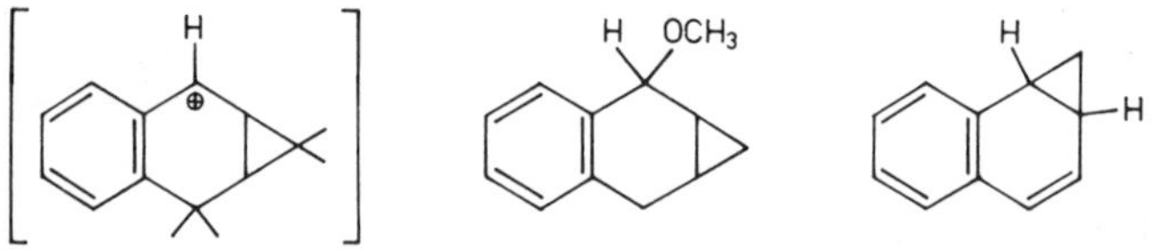

On the other hand, the cation **512** generated from **3** with $FSO_3H/ClSO_2F$ at $-120°C$ is further protonated on raising the temperature to $-60°C$ to the unusual dark red stable propellanic species **514** with two allylic cations bounded to the same cyclopropanic carbon atom[344,345] (probably the first stable dication without heteroatom substituents). The structure **514** was inferred on the basis of ¹H- and ¹³C-NMR data.

Interestingly, 11-methylene-1,6-methano[10]annulene **102** is protonated with $FSO_3H/ClSO_2F$ at $-80°C$ to the 8π-cation **518** similar to **512**, and not to bridge protonated ion **519**.[346]

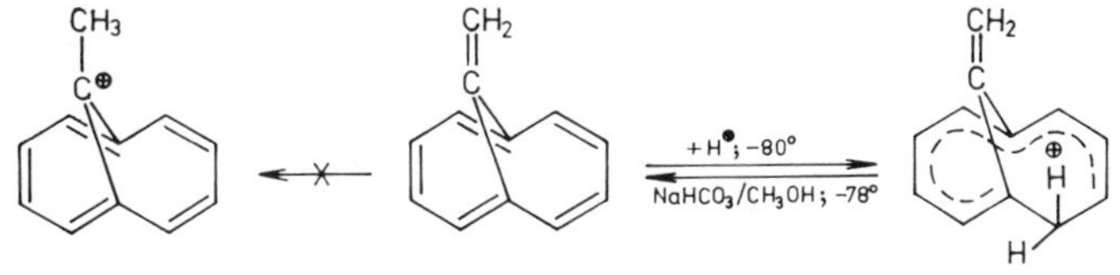

519   102   518

Quenching of **518** regenerates the starting hydrocarbon. It seems that the annulene moiety has a higher basicity than the double bond. The ^{1}H- and ^{13}C-NMR spectra prove the delocalization of charge on the annulenic skeleton without enhancement of 1,6 interaction. Calculations confirm that **518** is favored over **519** by 13.7 kcal $\cdot$ mol^{-1} (MINDO/3) or by 28.4 kcal $\cdot$ mol^{-1} (MNDO).[346]

B. Bridged 10π-Annulene Ions

The bicyclo[5.4.1]dodecapentaenylium cation **521** was obtained,[347] similar to tropylium cation, on treatment of olefins **520** and **490c** with triphenylmethyl fluoroborate which abstracts a hydride ion and forms triphenylmethane. The first olefin resulted on treatment of 1,6-methano[10]annulene with diazomethane, whereas **490c** was obtained via a double Wittig reaction of cycloheptatriene-1,6-dialdehyde with trimethylene-1,3-bis(triphenylphosphonium)-bromide.[6,347]

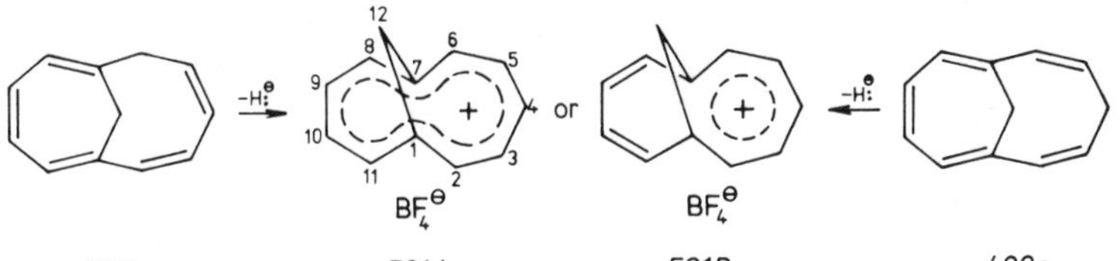

520   521A   521B   490c

The ^{1}H-NMR spectrum of the yellow-orange stable salt **521** (m.p. 190 to 195°C) indicates a diamagnetic ring current. (CD$_2$HCN, δ ppm): −0.3 and −1.8(H-12) and 8.3 to 9.6(9H).[6,347] The UV spectrum of **521** (60% H$_2$SO$_4$): λ$_{max}$ 272; 302; 320; 385 and 423 nm[347,348] is in agreement with results of calculations;[348] this system can be considered as a perturbed [11]annulenium cation. The same conclusion of a perturbed [11]annulenium ion results from the X-ray analysis of the hexafluorophosphate of **521** presenting a C-1 to C-6 distance of 2.30 Å (equal to that in the neutral methanobridged[10]annulenes).[349,350] However, the ^{13}C-NMR data (CD$_3$CN, δ ppm): 32.6(C-12); 139.3(C-8,11 or C-3,5); 140.1(C-3,5 or C-8,11); 141.2(C-9,10); 144.3(C-4); 149.6(C-1,7); and 158.0(C-2,6), suggest[351] that the substituted homotropylium cation structure **521B** is a better representation than **521A**.

The bridged 10π-electron carbanion **525** with a 9C-perimeter was synthesized[69,352,353] from the propellanic hydrocarbon **524** (on treatment with CH$_3$SOCH$_2$Na). This hydrocarbon could be obtained either from dichlorocarbene adduct **522** of reduced indanone ethylene ketal via **523**[69,352] or from **526** via **527** and **528**:[353]

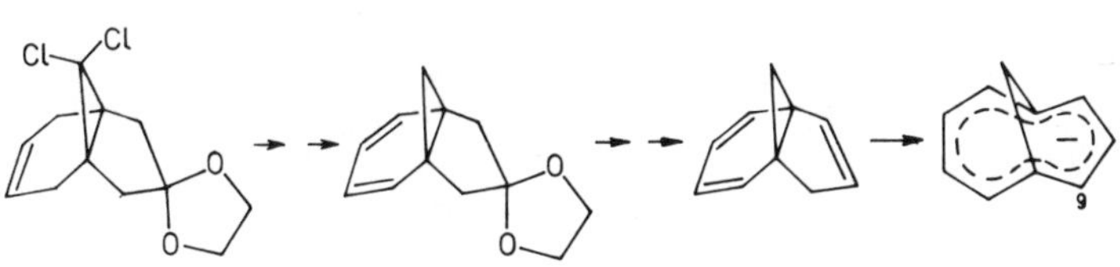

522   523   524   525

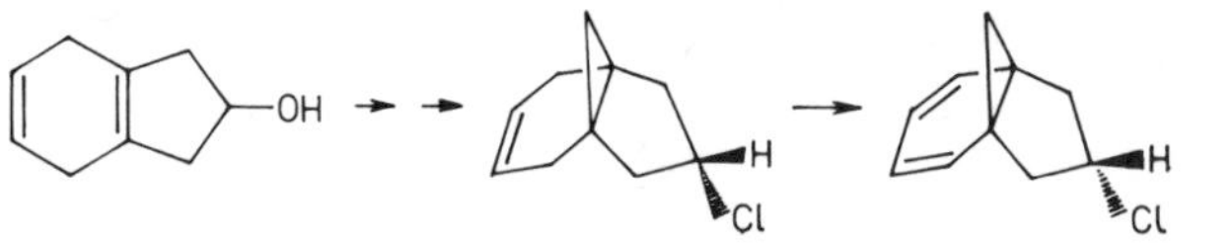

526   527   528

The ^{1}H-NMR spectrum of **525** indicates the ring opening of the three-membered ring simultaneously to the appearance of a diamagnetic ring current (bridge protons resonating at $\delta = -0.7$ and -1.2 ppm[69] or at -0.45 and -0.95 ppm[353]). Treatment of **525** with D_2O afford the 9-*endo*-deuterio compound;[354a] however, the exclusive 9-*exo* stereoselectivity was observed at the 9-carboxymethyl derivative of **525**,[354b,c] then for **525** itself.[354d]

C. Bridged 12π-Annulene Ions

1. Bicyclic Systems

The bicyclo[5.4.1]dodecapentaenyl anion (**529**) was obtained in 1973[355] by KND_2 treatment of the above-mentioned alkene **520**. The ^{1}H-NMR spectrum suggests the paratropic character (ND_3; δ ppm): 1.21(H-4); 2.31(H-2,6); 2.99(H-3,5); 3.16(H-8,11); 3.92(H-9,10); 10.31(H-12a); 14.19(H-12b), taking into account the downfield positions of bridge protons and the upfield positions of perimeter protons. The value of $J_{8,9} = J_{10,11} = 4.7$ Hz, "the smallest value of a J_{vic} between olefinic protons yet reported for a cycloheptatriene"[355] suggests that the anion **529** is twisted about the C-8 to C-9 and C-10 to C-11 bonds. It was suggested that the folding of the molecule is a mechanism for diminishing the antiaromaticity due to 12π-electrons and a possibility for relieving the angle strain.[355] Thus, in the more planar diatropic 10π-electron analogous cation **521** the $J_{8,9} = J_{9,10} = 9.46$ Hz and the ^{1}H-NMR signal of H-12b is about 16 ppm upfield relative to **529**. Water quenching of **529** affords a mixture of **490c** and **520**.[355]

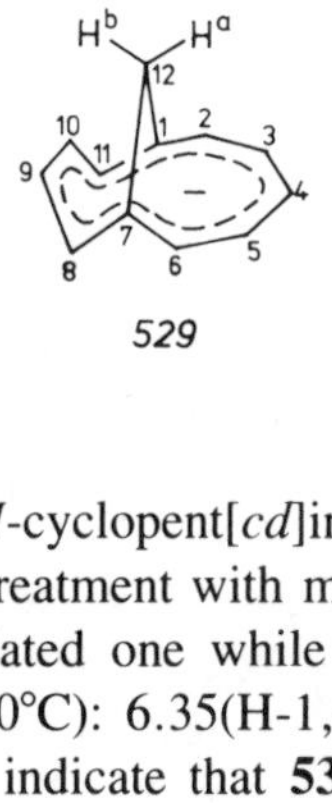

529

2. Tricyclic Systems

The synthesis of the anion of 1*H*-cyclopent[*cd*]indene (**25**) as lithium salt **530**, was performed by Hafner et al. through treatment with methyl-lithium.[21] The UV spectrum of **530** is in agreement with the calculated one while the ^{1}H-NMR spectrum presents the following signals (THF; δ ppm; $-50°$C): 6.35(H-1,4); 6.85(H-2,3); 7.30(m; H-5,6,7).[21] These olefinic-range chemical shifts indicate that **530** is not aromatic. Reactions of **530** (e.g., with benzophenone affording **531**) proceed at 1- (or 4-) position, in agreement with theoretical predictions.[356]

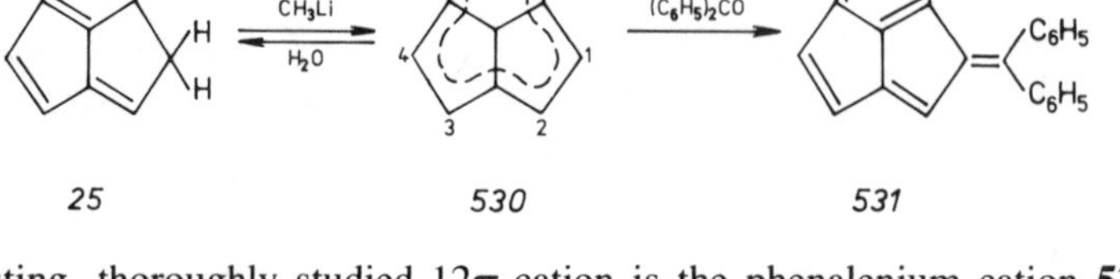

25   530   531

An interesting, thoroughly studied 12π-cation is the phenalenium cation **533**, first described in 1956 by Pettit.[357] Its synthesis started with diazoacetic ester addition to acenaphthene producing **532a**, which was further converted stepwise into isocyanate **532b**, amine **532c**, and chloroderivative **532d**. The chloroderivative **532d**, which showed no tendency to ionize spontaneously, could be converted into **533** treatment with silver perchlorate at 70°C.

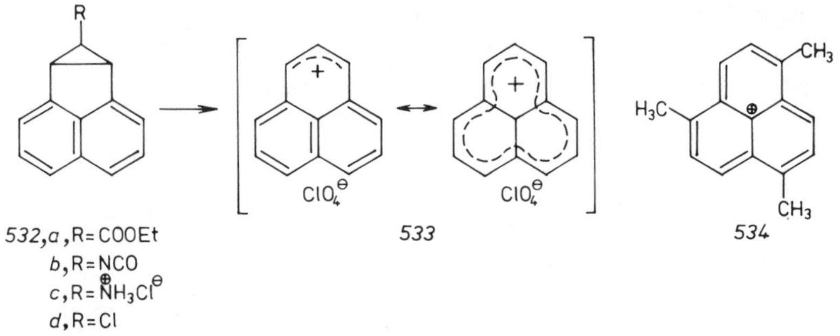

532,a, R = COOEt
b, R = NCO
c, R = $\overset{\oplus}{N}H_3Cl^{\ominus}$
d, R = Cl

533

534

Later, the same cation **533** was obtained[358] by reaction of phenalene with triphenylmethyl perchlorate. Similarly, a trimethylderivative **534** was obtained.[358] The phenalenium cation could also be generated on treatment of phenalene with FSO_3H at −78°C, with H_2SO_4 at −78°C, or with H_2SO_4 at room temperature.[359,360] Such an oxidation of hydrocarbons by FSO_3H occurs when the resulting cations are sufficiently stable. The corresponding formation of SO_2 suggested the redox reaction:[359]

$$RH + FSO_3H \rightarrow R^+ + SO_2 \uparrow + H_2O + F^-$$

The reactions with H_2SO_4 or $NO_2^+BF_4^-$ in acetonitrile proceed by analogous mechanisms.

Annelated phenalenium cations **536** and **538** were obtained on treatment of hydrocarbon **535** with oxygen and perchloric acid in acetic acid,[361] and of hydrocarbon **537** with H_2SO_4, respectively.[362]

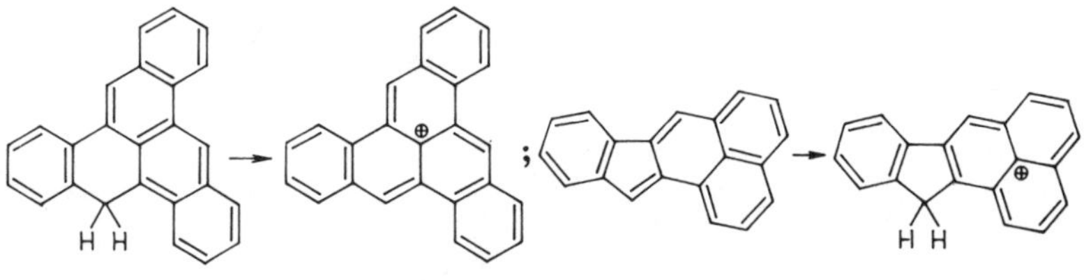

535

536

537

538

The phenalenium perchlorate **533** is a yellow solid which turns black and decomposes in moist air. The methylderivative **534** is an air-stable bronze solid.[358]

MO calculations for the phenalenium cation afforded a delocalization energy (DE) of 5.83 β[363] and a positive value of 0.073 for REPA.[364] It was calculated that the electron density in **533** is diminished with 1/6 of positive charge unit (0.1667) in positions 1,3,4,6,7,9 whereas at remaining positions no charge deficit results.[120]

The UV spectrum of **533** (in 60% H_2SO_4) indicates absorptions at λ_{max} 226; 378sh; 400 nm with long tailing.[357,365] The ^{1}H-NMR spectrum of **533**-hexachloroantimonate (in $AsCl_3$; δ ppm) presents signals at 8.48(H-2,5,8) and 9.30(H-1,3,4,6,7,9) with J_{ortho} = 7.2 Hz[366] whereas the perchlorate of **534** (in F_3C-COOH) at δ = 3.36 (CH_3 groups); 8.18 and 9.30 ppm with J_{ortho} = 8.3 Hz.[367]

Chemically, the cation **533** gives a mixture of phenalenone and phenalene on treatment with strong acids[357] or on hydrolysis.[358] The resistance to autoxidation in air depends strongly on the nature of the counter-anion.[366]

The unusual stability of the phenalenium cation **533** in salts such as the perchlorate or hexachloroantimonate, as well as of the corresponding anion and radical, suggest that there is substantial contribution of 10π-naphthalene-like limiting structures **533B** to **533D** with a *peri*-condensed allylic system. Irrespective whether zero, one, or two electrons occupy the nonbonding orbital of phenalene (an odd-alternant hydrocarbon), the DEs change relatively little and the stabilities are comparable.[368]

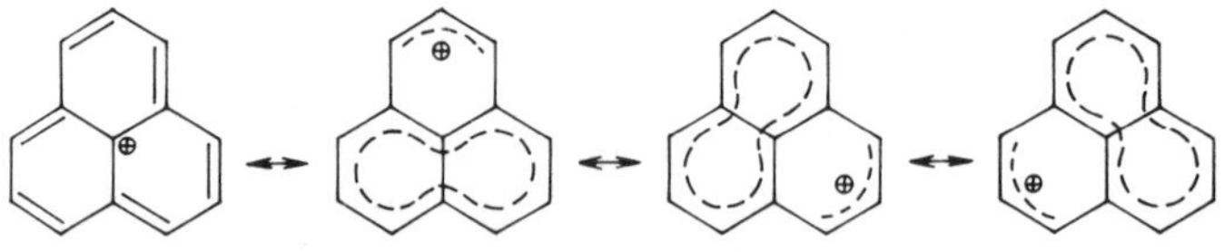

The generation of a methylene-bridged 12π-cation **540** from the 10,11-homophenalene (**539**) on treatment with FSO$_3$H at $-75°$C was described by Murata.[84]

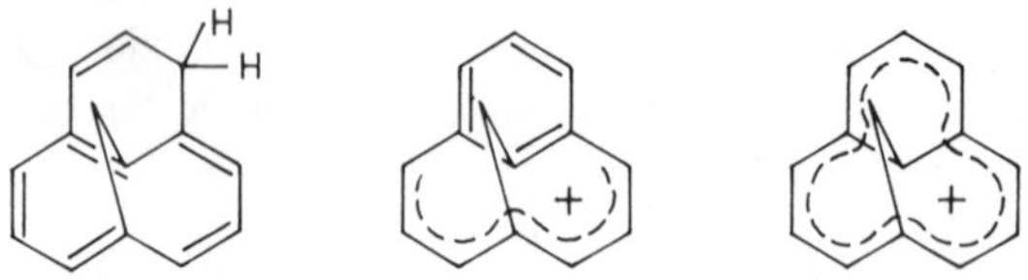

Interestingly, the ^{1}H-NMR spectrum is in agreement with the *meta*-cyclophanic structure **540A** and not with an *a priori* expected delocalized homophenalenium cation **540B**.[84] The cation **541** corresponding to the unknown hydrocarbon **155** could be obtained only in solution on treatment of the isomeric 2-H-benzo[*cd*]azulene (**158**) with triphenylmethyl fluoroborate[121] or with iodine when the presumable intermediate **542** loses hydrogen iodide.[121] Similarly, a tetramethyl cation **543** was obtained from the corresponding hydrocarbon.[122]

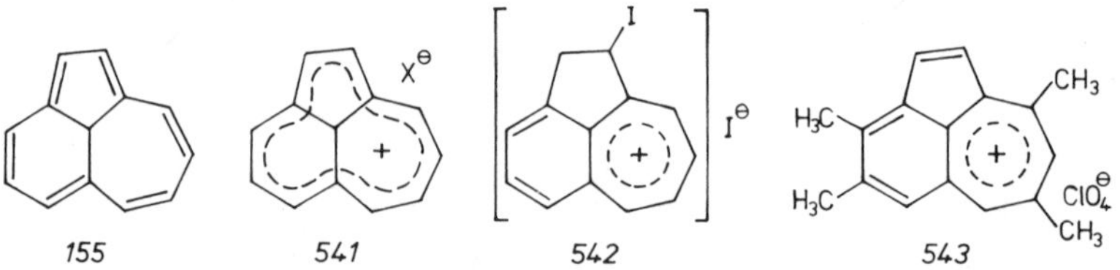

The UV spectrum of **541** (in CH$_3$CN) presents absorptions at 225, 245, 295, 350, and 680 nm.[121] The calculated electron densities[270] decrease from the five-membered ring towards the seven-membered one. One can conclude that, similar to phenalenium systems which include an aromatic naphthalene with a *peri*-condensed allylic system, the cation **541** consists of an aromatic azulene with a *peri*-condensed allylic system.

D. Bridged 14π-Annulene Ions

1. Bicyclic Systems

The dianions **544** and **545** of 1,7-methano[12]annulene (**146**) and 1,6-methano-[12]annulene (**153**) were obtained on polarographic or alkali metal reduction of the corresponding hydrocarbons.[369]

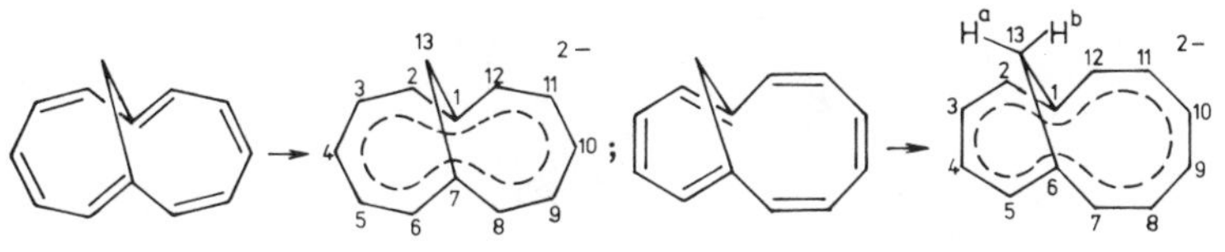

The ^{1}H-NMR spectra of **544** (d_8-THF; $-80°C$; δ ppm): -6.44 (H-13), 6.28 (H-3,5), 6.41 (H-4), and 7.16 (H-2,6) and of **545** (d_8-THF; $-80°C$; δ ppm), -6.08 (H-13b), -5.52 (H-13a), 6.51 (H-3,4), and 7.56 (H-2,5)[369] clearly indicate their aromatic character. Conversion of hydrocarbons **146** and **153**, compounds with localized π-bonds (see Section III.B.1), into their aromatic dianions is accompanied by an upfield shift of 10 to 12 ppm of bridge-CH$_2$ proton signals. The large difference between signals of H-13a and H-13b in **153** (about 4.7 ppm) is drastically reduced to about 0.5 ppm in the practically planar corresponding dianion **545**. The ^{13}C-NMR spectrum of **544** indicates a C$_{2v}$ symmetry, in agreement with its aromatic character.[369]

2. Tricyclic Systems

The phenalenium anion **546** was obtained initially as a red solution on treatment of phenalene (**157**) with phenyllithium[119] or by reduction of phenalene with potassium.[370]

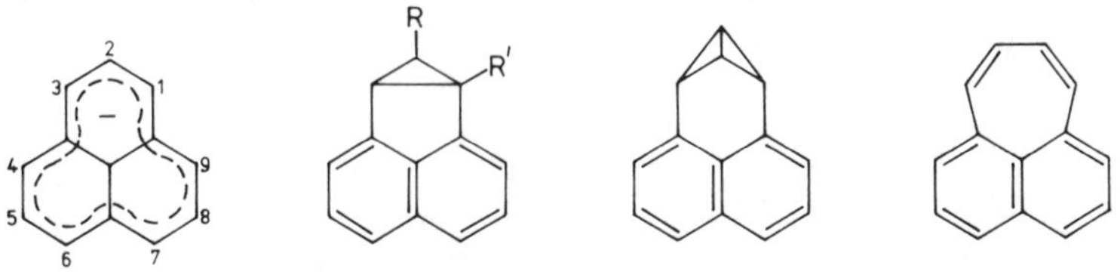

Wittig and co-workers obtained **546** on heating the isomeric lithium derivatives **547a** and **547b** or the bromide **547c** with sodium potassium alloy.[371] The lithium salt of **546** could be isolated as red crystals. The ^{1}H-NMR spectrum (lithium salt in ether) presents only two signals at δ = 5.17 (H-1,3,4,6,7,9) and 5.91 ppm (H-2,5,8) with J_{ortho} = 7.5 Hz;[366,371] this is a mirror image of the phenalenium cation spectrum and can be interpreted similarly, as resulting from a naphthalene moiety and an allyl anion.

The reaction of phenalenium anion with dichloromethane and butyllithium affords the 1,8-naphtho(CH)$_4$ hydrocarbon **548**, the first example of a "valece type" valence isomer of a nonalternant hydrocarbon, along with small amounts of pleiadiene **549**.[372,373]

An interesting reaction of phenalenium anion **546** with enamines affords enamines of methylene-phenalene **550** which cyclize under elimination of dimethylamine-producing pyrene (**201**) as described by Jutz et al.[305] The reaction of **546** with methyl iodide affords 1-methylphenalene (**554**) formed through an isomerization.[119]

550 551 201

546

552 553 554

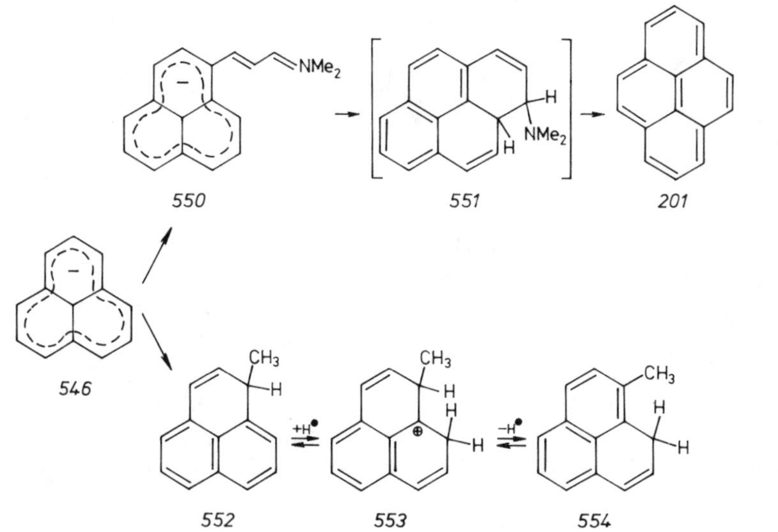

Another tricyclic 14π anion **555** was formed from 10,11-homophenalene (**539**) on treatment with butyllithium.[84] The ^{1}H-NMR spectrum, especially the low-field position of the CH_2 protons at $\delta = 2.64$ ppm, disproves a 14π delocalized homophenalenium anion structure **556** and is in agreement with the partially delocalized *m*-cyclophanic structure **555**.[84]

555 556 557A 557B

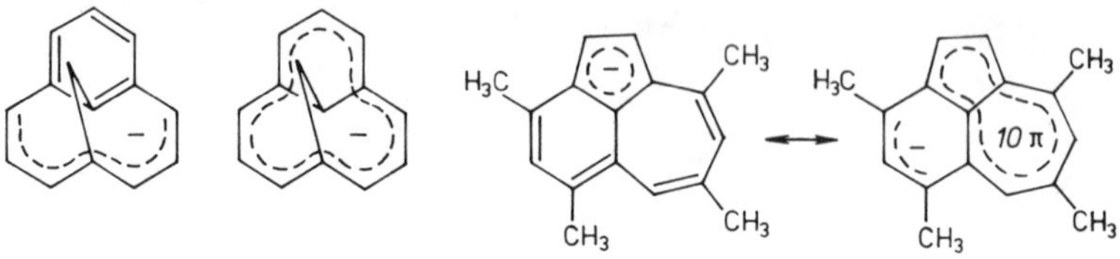

Treatment of 2*H*-benzo[*c,d*]azulene (**158**) with lithium triphenylmethide or with butyllithium gives rise to a deep blue-green solution containing most probably the nonisolable corresponding anion;[121] however, the tetramethyl anion **557** was obtained[122] as a green compound (sensitive to hydrolysis), containing a cyclic conjugated 14π-electron system which may be fragmented either into azulene + allyl anion, **557B** or into cyclopentadienide anion + octatetraene chain, **557A**.

The doubly bridged *syn*-tricyclo[9.4.1.1^{3,9}]heptadecaheptaenylium cation (**559**) was reported[11] to be obtained by hydride abstraction from the olefin **558** (produced via a double Wittig reaction). The aromatic character of **559** was proved by its diamagnetic ring current,[11] X-ray analysis (planarity, bond lengths of 1.389Å, and minimal overlap between C-1 to C-11 and C-3 to C-9,[374]) as well as by a high solvolytic stability.[11]

558 559

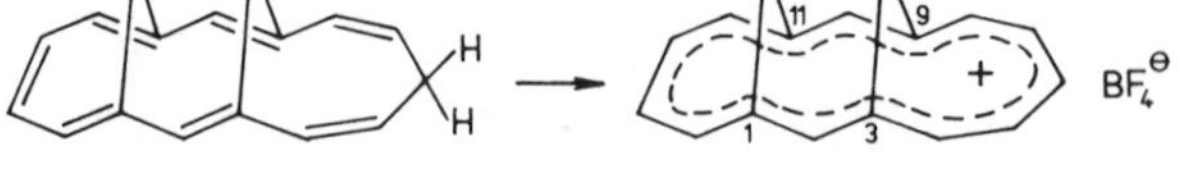

E. Bridged 16π-Annulene Ions

The dianion **560** of 1,6:8,13-propanediylidene[14]annulene (**196**) was obtained on reduction of the corresponding hydrocarbon with potassium.[375] The ^{1}H-NMR spectrum of **560**

Table 9
¹H-NMR DATA OF DIANIONS 562 IN ·
COMPARISON WITH NEUTRAL PRECURSORS
198 (δ ppm)[378]

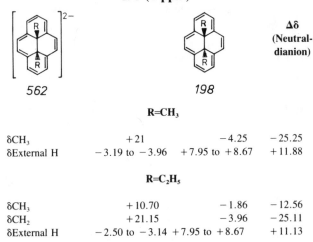

	562	198	Δδ (Neutral-dianion)

R=CH₃

	562	198	Δδ
δCH₃	+21	−4.25	−25.25
δExternal H	−3.19 to −3.96	+7.95 to +8.67	+11.88

R=C₂H₅

	562	198	Δδ
δCH₃	+10.70	−1.86	−12.56
δCH₂	+21.15	−3.96	−25.11
δExternal H	−2.50 to −3.14	+7.95 to +8.67	+11.13

(d_8-THF; δ ppm): 2.2 (bridge protons) and 5.5 (ring protons) indicates, in comparison with that of neutral hydrocarbon **196** (average value −0.9 ppm for the bridge protons and 7.6 ppm for ring protons) the change from a diamagnetic ring current to a paramagnetic one.[375]

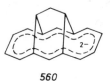

560

The important reduction of paramagnetic ring current due to deviations from coplanarity[376] can be seen on comparing the ¹H-NMR data of **560** (and the neutral hydrocarbon **196**) with the planar *trans*-15,16-dimethyl-dihydropyrene dianions (and the corresponding hydrocarbon)-*vide infra*.[375]

The dication **561** of *syn,syn*-1,6:8,17:10,15-trismethano[18]annulene (**394**) resulted[377] on oxidation of the neutral hydrocarbon with fluorosulfonic acid in ClSO₂F at −80°C.

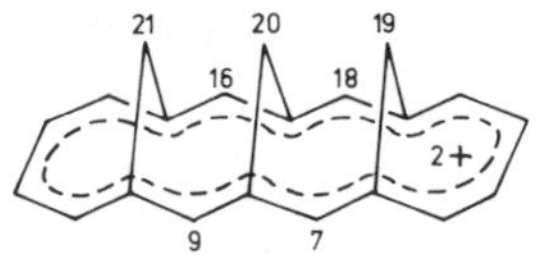

561

The ¹³C-NMR spectrum of **561** proved the C_{2v} symmetry as well as the buildup of charge at C-7,9,16,18 whereas the ¹H-NMR spectrum (ClSO₂F/FSO₃H/CD₂Cl₂; δ ppm): 4.35 and 7.58 (AX-system; H-19,21); 6.07(H-7,9,16,18); 7.01 to 7.35(H-2-5; 11-14); 8.25(H-20) proves the paramagnetic ring current, especially through the large downfield shift of bridge protons in comparison to **394**. The nearly equal chemical shifts for perimeter protons in **561** and **394** was attributed to a compensation of the expected upfield shift (due to paramagnetic ring) by the contrary effect of positive charges.[377]

Dianions **562** of *trans*-15,16-dialkyldihydropyrenes (**198**) were easily obtained on reduction of neutral hydrocarbons with potassium in tetrahydrofuran.[378] The conversion of a ($4n + 2$)π aromatic system into a $4n$π one has a dramatic effect on ¹H-NMR spectra, the diamagnetic ring current being replaced by a paramagnetic ring one (Table 9).

The huge values of above chemical shift differences between **198** and **562** reflect the important paramagnetic ring currents in planar anions **562**. The corresponding difference (e.g., $\Delta\delta$ for bridge H) for dianions **560** which deviate from coplanarity is much smaller (*vide supra*).[375] It should also be stressed that the chemical shift δ of about 21 ppm for the inner α-methylene protons in **562** is the greatest δ value yet observed for carbon protons.[378]

The dianions of the above type **562** show a normal reactivity; thus **562**, R=CH$_3$ can be protonated to afford *trans*-15,16-dimethyl-2,7,15,16-tetrahydropyrene **563a**[378] or can be converted to diester **563b** (with ethyl chloroformate) or to ditosylate **563c** (through reaction with formaldehyde followed by tosylation).[297]

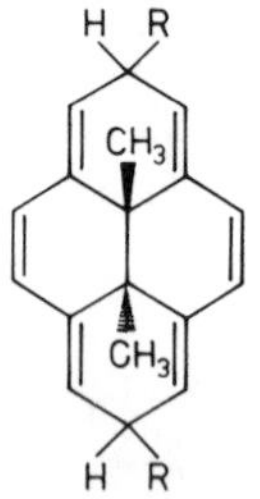

563,*a*,R=H
  b,R=CO$_2$Et
  c,R=CH$_2$OTs

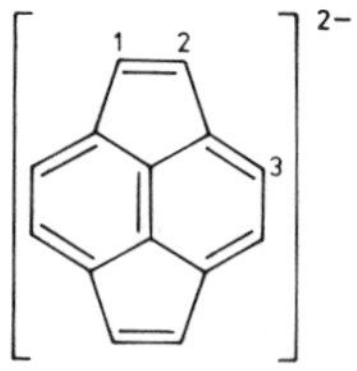

564

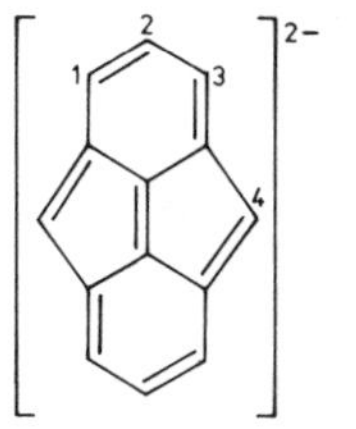

565

Other interesting 16π-species are the dianions of pyracylene (**161**) and of dibenzopentalene (**162**). These dianions **564** and **565**, respectively, though containing 16π-electrons, manifest 14π aromatic anion behavior because the central double bond (2π-electrons) remains quasi-separated from the peripheral aromatic 14π-electron system.

Actually, the first ion of pyracylene to be studied was the radical anion obtained on electrolysis.[379] Though nonalternant, it has a zero spin density at the inner C–C atoms since it possesses an alternant periphery.[379] Subsequent LCAO-MO calculations[146] afforded the following stabilization energies per C–C bond (SECC in electron volts): 0.3103 for neutral pyracylene, 0.3451 for the dication, and 0.4727 for dianion **564**, suggesting that pyracylene would be easily reduced and more difficultly oxidized. More recent information concerning the stabilities of pyracylene charged species resulted from calculations of topological resonance energies (TRE); thus the TRE; TRE-per π-electron and TRE-per ring bond in β units are -0.289; -0.024, and -0.017 for dications and 0.539, 0.034, and, respectively, 0.032 for dianions **564**.[380]

The dianion **564** was obtained in 1973 by Trost[381] on treatment of dihydropyracylene derivatives with butyllithium in THF at $-78°C$. From dihydroderivatives the anions **565** were obtained similarly.[139]

A dibenzo[*cd,gh*]pentalenide dianion **569**, an aromatic perturbed [12]annulenyl dianion, was obtained by Rabinovitz[382] through a new carbanionic rearrangement of hydrocarbon **566**:

Table 10
¹H-NMR DATA FOR DIANIONS 564, 565
AND 569 (δ ppm; JHz)

Compound	δ_1	δ_2	δ_3	δ_4	Ref.
564	6.00	6.00	6.75		139
565[a]	6.25	6.67	6.25	5.34	139
569[b]	6.41	6.68	6.41		382

[a] $J_{1,2} = J_{2,3} = 7.1$ Hz.
[b] $J_{1,2} = J_{2,3} = 7.3$ Hz.

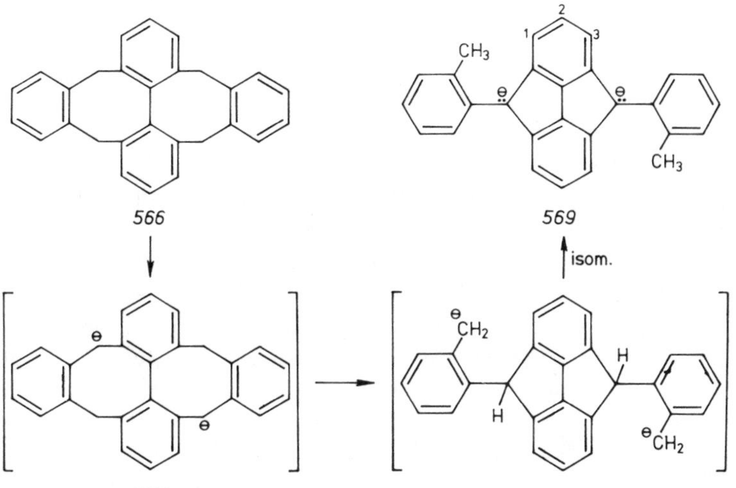

The ¹H-NMR data for dianions **564, 565,** and **569** are presented in Table 10.

The data from Table 10 corroborate well those of aromatic 14π-dianions, e.g., **545** (*vide supra*). It follows that in ions **564** and **565** the excess negative charge is located on the periphery, leaving the central double bond practically undisturbed. The calculated and experimental charge densities of **564** and **565** show an excellent agreement.[364] The diatropicity of **564** results also from Mallion's detailed calculations[148] performed by six different methods.

Two reactions of dianions **564** were reported, namely, deuteration and reaction with methyl iodide, both occurring at the inner ethylene bridge carbons.[381]

Another 16π-ion containing a 14π-electron periphery, the cation **570,** was isolated as an extremely stable tetrafluoroborate salt[383] in agreement with its large calculated REPA value of 0.067.[364]

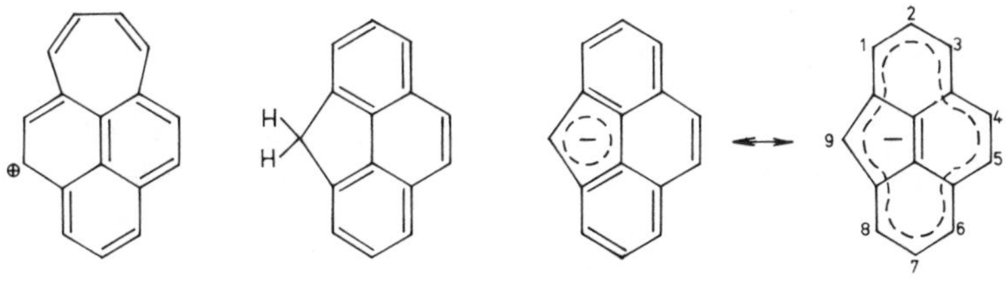

The anion **572** was obtained on treating hydrocarbon **571** with metallic potassium[384] or with butyllithium.[385-387] The ¹H-NMR characteristics of **572,** δ ppm: 6.06(s,H-9); 7.00(dd;

H-1,8); 7.41(m, H-2,3,6,7), and 7.73(s; H-4,5),[385] support its aromaticity, in agreement with the REPA value of 0.071.[364] The correct description of **572** is that of a perturbed 14π-system containing an inner double bond (**572B**).

Treatment of hydrocarbon **573** with butyllithium affords the anion **574**[388] whose ¹H-NMR spectrum: δ ppm: 6.88(H-1,2); 7.08 and 7.70(q; AB system); 7.20 and 7.33 (A₂B) for H-3,4,8,9 and H-5,6,7 is in agreement with a delocalized structure with C_{2v} symmetry possessing 14π-electrons. Calculations suggested[33] a significant resonance stabilization of anion **574** (DE/elect. = 0.3914β).

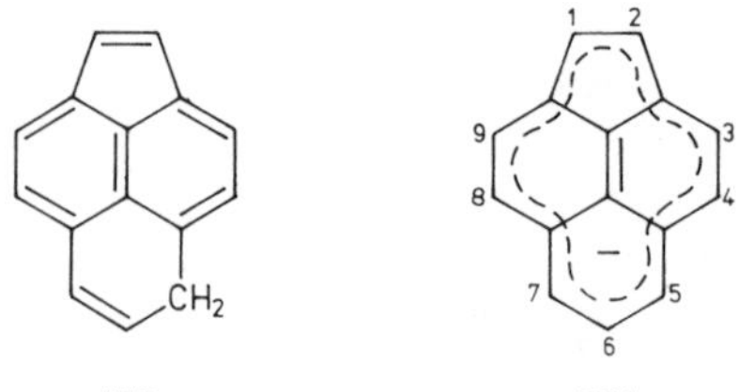

573   574

F. Bridged 18π-Annulene Ions

The tetracyclic cation **575** was reported[11] to have been synthesized analogously to **559** through hydride abstraction from the corresponding cycloheptatriene-like cyclopolyolefin (obtained in its turn through double Wittig reaction). The aromatic character of **575** was inferred from ¹H-NMR spectrum and the solvolytic stability.[11]

The dianion **576**, obtained on reduction of the corresponding hydrocarbon **378** with potassium[297] indicates the existence of a strong diamagnetic ring current through ¹H-NMR absorptions at δ, −5.91 and −5.99 for inner CH₃ groups; 2.20 and 3.17 for outer CH₃ groups, and 6.68(m); 8.20, 8.30, and 8.53 ppm (singlets) for peripheral protons.

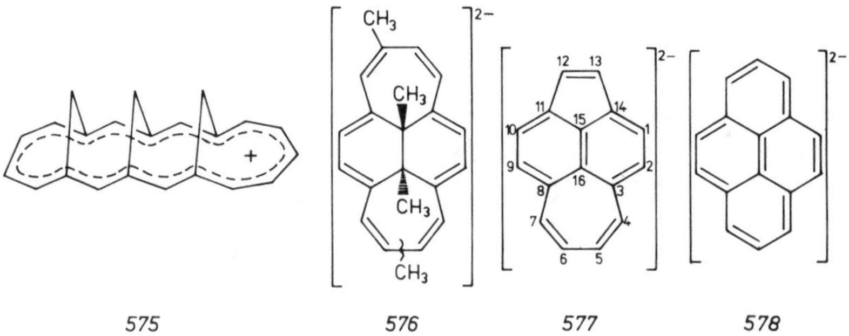

575   576   577   578

Another interesting dianion, **577**, was obtained on reduction of acepleiadylene (**203**) with lithium.[389] The ¹H-NMR data[389] clearly illustrate the difference between **203** and **577**; whereas acepleiadylene is a diamagnetic ethene-bridged 14π-annulene (δ ppm 6.89-H-5,6; 7.79-H-4,7; 7.83-H-12,13; 7.95-H-1,10; 8.33-H-2,9), the dianion **577** is a paramagnetic ethene-bridged 16π-system (δ ppm: −2.05-H-4,5,6,7; −0.33-H-2,9; 1.26-H-1,10; 1.53-H-12,13). Thus, an upfield shift (about 8 ppm) of ¹H-resonances is seen on passing from the neutral **203** to its dianion **577** (much more than expected for charge-induced shielding).

Lithium reduction of pyrene (**201**) conducted[390] to the *dianion* **578** with a ¹H-NMR multiplet around δ = 0.97 ppm, thus confirming a strongly paratropic character (16π peripheral system) similar to that of **577**.

The multistep synthesis of cycloocta[*def*]fluorene anion (**583**), an 18π-electron species related to **577** and **578** proceeds according to the following scheme in which the Grob oxidative decarboxylation of **580**, the valence isomerization of **581** into **582**, and the generation of anion **583** on treatment with butyllithium are the essential steps:[385-387,391,392]

571 579 580

581 582 583A 583B

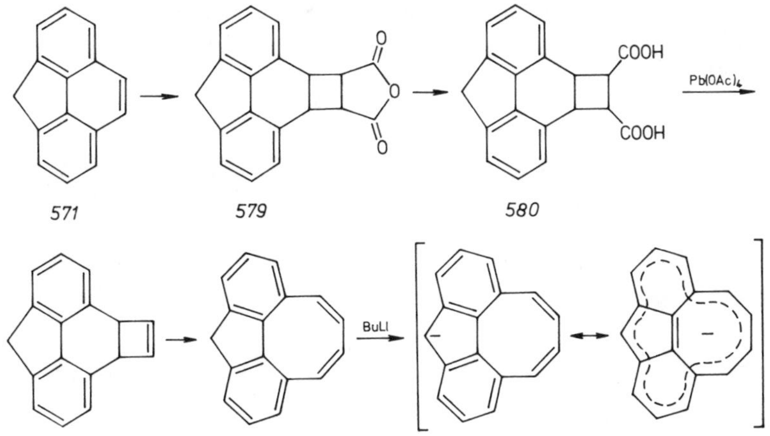

Conversion of **582** into **583** is accompanied by a paratropic upfield shift of proton signals ($\Delta\delta \sim 5.6$ ppm[391]). This shift should be attributed mainly to paratropicity and less to charge shielding, unlike the conversion of the related **571** into **572** where only charge shielding is relevant.[386,387,392] Thus the anion **583** can be considered to be an antiaromatic paratropic system with a 16π-electron periphery. In agreement with such a view is the very low value of $pK_a = 27$ for **582** as well as the results of calculations indicating a -0.011β value for REPE[35] and 0.049 for REPA.[364]

G. Bridged 20π-Annulene Ions

Reduction with lithium of acepleiadylene (**203**) proceeds via the dianion to the tetraanion **584**.[389] The ^{1}H-NMR spectrum of **584**[389] (δ ppm: 3.56-H-4,7; 4.28-H-5,6; 4.44-H-12,13; 4.90-H-2,9; 5.96-H-1,10) indicates a typical diatropic π-delocalized anion behavior. The passage from the dianion **577** (see above) to the tetraanion **584** leads to replacing the paramagnetic ring current by a diamagnetic one: the periphery has 18π-electrons.

584A 584B 585A *a*, R=R′=H 585B
b, R=H; R′=Ph
c, R=t-Bu; R′=H
d, R=Me; R′=H

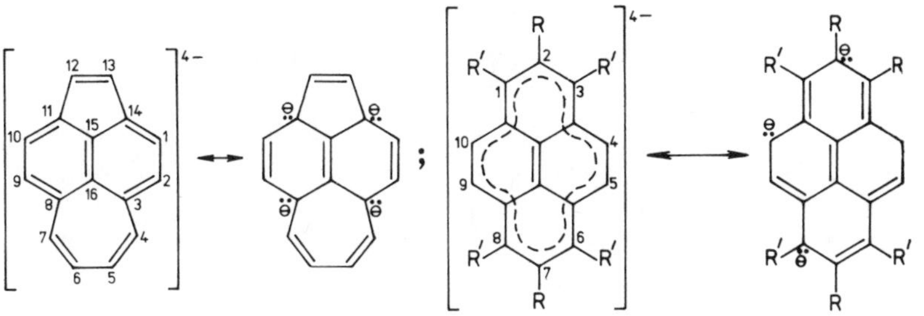

Sodium or potassium reduction of pyrene or of substituted derivatives affords the tetra-anions **585**.[387,392-394] The diatropicity of **585** results from the ^{1}H-NMR multiplet centered at about $\delta = 5.4$ ppm, in good agreement with a calculated value of $\delta = 4.9$ to 5.5 ppm.[393,394] Thus, a peripheral diatropic ring current was also established for anions **585** (14 carbon atoms and 18π-electron periphery) which are the first tetraionic species obtained from benzenoid hydrocarbons. The pattern of NMR spectra of **585** is in agreement with results of calculations[368,395] suggesting a symmetrical distribution of negative charge in **585b** (mainly in 1,3,6,8-positions) and an asymmetrical one in **585a,c,d** (mainly in 1,2,7,8- or 1,2,6,7-positions).[387] Newer data[394b] on the reduction of pyrene by alkali metals cast doubt on some of the earlier NMR spectral evidence of the tetraanion.

H. Miscellaneous

Treatment of hydrocarbon **586** with sodium conducted to the deep-blue dianion **587**[396] whose ^{1}H-NMR signals (δ = 7.75 and 8.04 ppm), practically in the same ranges as those of **586**, indicate the diamagnetic ring current existent in the anion **587** possessing a 30π-electron periphery.

586   587

I. Conclusions

The bridged annulene ions described above were grouped according to the total number of their π-electrons, but the physicochemical data indicate that this total number does not account for the physicochemical behavior; the ions may be classified better into diatropic (aromatic) and paratropic (antiaromatic) species in agreement with their conjugated π-electron periphery or conjugated subsystem. Thus, diatropic ions with 10π-electrons (e.g., cation **521** and anion **525**), 14π-electrons (e.g., dianions **544** and **545**, cation **559**), 18π-electrons (e.g., cation **575**, dianion **576**), as well as paratropic ions with 12π-electrons (e.g., anion **529**) and with 16π-electrons (e.g., dianions **560** and **562**, dication **561**) were synthesized and characterized.

An interesting observation is that in numerous cases ions with n π-electrons behave as $(n-2)\pi$ analogs due to an internal double bond insulated from the peripheral π-system. Thus, dianions of pyracylene **564** and dibenzopentalene **565**, as well as monocation **570** and monoanion **572** with 16π-electrons manifest 14π aromatic ion behavior; dianions of acepleiadylene **577** and of pyrene **578** as well as monoanion **583** of cycloocta[*def*]fluorene, though containing 18π-electrons, manifest characteristics of perturbed paratropic 16π-systems whereas 20π-tetraanions **584** and **585** indicate a 18π-diamagnetic behavior.

On the other hand, the aromaticity Hückel $(4n + 2)\pi$-electron rule is illustrated in the 10π-series by the related cation **521**, neutral hydrocarbon 1,6-methano[10]annulene (**3**), and anion **525**; this series constitutes a clear analogy to the well-known triad: tropylium cation, benzene, and cyclopentadiene anion from the 6π-series.

Interesting comparisons reflecting numerous similarities may be made between analogous ions (similarly charged) of isomeric hydrocarbons, e.g., dianions **544** and **545**, dianions **564** and **565** of pyracylene and dibenzopentalene, dianions **577** and **578** of acepleiadylene and pyrene, or between tetraanions **584** and **585** of the latter pair of hydrocarbons.

Other interesting ions are the strongly paratropic dianions **562** of dialkyldihydropyrenes showing outstanding δ values in their ^{1}H-NMR spectra.

In some cases charged species corresponding to very unstable (e.g., pyracylene **161** or even unsynthesized hydrocarbons (e.g., dibenzopentalene **162** or **155**) were obtained and characterized (the dianions **564**, **565** or the monocation **541**, respectively).

Finally, it should be mentioned that multicharged species (up to four negative charges) of polycyclic hydrocarbons such as pyrene or acepleiadylene are now under active investigation.

The study of bridged annulene ions has appreciably extended the area of aromatic and antiaromatic compounds constituting a notable advance into the fascinating field of annulene congeners.

Multiply charged carbocations of bridged annulenes (Vogel's type) were investigated by Pagni.[413] Doubly charged ions of bridged [4n]annulenes **375** (see above) and **153** were described by Müllen et al.[414] Alkali-metal reductions of pyrene to dianion salts were further investigated by Schnieders et al.[415]

REFERENCES

1. **Doering, W. v. E. and Goldstein, M. J.,** *Tetrahedron,* 5, 53, 1959.
2. **Vogel, E. and Roth, H. D.,** *Angew. Chem.,* 76, 145, 1964.
3. **Nelson, P. H. and Untch, K. G.,** *Tetrahed. Lett.,* p. 4475, 1969.
4. (a) **Vogel, E., Klug, W., and Brauer, A.,** *Org. Synth.,* 54, 11, 1974; (b) **Neidlein, R. and Gottfried, R.,** *Chem. Ztg.,* 107, 371, 1983.
5. **Banwell, M. G., and Papamihail, C.,** *Chem. Commun.,* p. 1182, 1981.
6. **Ledlie, D. B. and Bowers, L.,** *J. Org. Chem.,* 40, 792, 1975.
7. **Vogel, E., Feldman, R., and Düwell, H.,** *Tetrahed. Lett.,* p. 1941, 1970.
8. **Vogel, E., Deger, H. M., Sombroek, J., Palm, J., Wagner, A., and Lex, J.,** *Angew. Chem.,* 92, 43, 1980.
9. **Vogel, E.,** *Pure Appl. Chem.,* 54, 1015, 1982.
10. **Vogel, E.,** *Chimia,* 33, 57, 1979.
11. (a) **Vogel, E.,** *Isr. J. Chem.,* 20, 215, 1980; (b) **Jörgensen, F. S. and Paddon-Row, M. N.,** *Tetrahed. Lett.,* 24, 5415, 1983; (c) **Ermer, O.,** *Angew. Chem. Int. Ed. Engl.,* 16, 798, 1977; (d) **Ermer, O. and Mason, A. S.,** *Chem. Commun.,* p. 53, 1983.
12. **Klem, R.,** Ph.D. thesis, *Diss. Abstr.,* 32, 2072B, 1972.
13. **Chong, B.,** Ph.D. thesis, *Diss. Abstr.,* 32, 3845B, 1972.
14. **Vogel, E., Ippen, J., and Buch, V.,** *Angew. Chem. Int. Ed. Engl.,* 14, 566, 1975.
15. **Masamune, S., Brooks, D. W., Morio, K., and Sobczak, L.,** *J. Am. Chem. Soc.,* 98, 8277, 1976.
16. **Masamune, S. and Brooks, D. W.,** *Tetrahed. Lett.,* p. 3239, 1977.
17. (a) **Scott, L. T., Brunsvold, W. R., Kirms, M. A., and Erden, I.,** *J. Am. Chem. Soc.,* 103, 5216, 1981; (b) **Scott, L. T. and Brunsvold, W. R.,** *J. Am. Chem. Soc.,* 100, 4320, 1978.
18. **Rapoport, H. and Pasky, J. Z.,** *J. Am. Chem. Soc.,* 78, 3788, 1956.
19. **Rapoport, H. and Smolinsky, G.,** *J. Am. Chem. Soc.,* 82, 1171, 1960.
20. **McDowell, B. L., Smolinsky, G., and Rapoport, H.,** *J. Am. Chem. Soc.,* 84, 3531, 1962.
21. **Eilbracht, P. and Hafner, K.,** *Angew. Chem. Int. Ed. Engl.,* 10, 751, 1971.
22. **Hafner, K.,** *Pure Appl. Chem.,* Suppl. 2, 1, 1971.
23. **Gilchrist, T. L., Rees, C. W., Tuddenham, D., and Williams, D. J.,** *Chem. Commun.,* p. 691, 1980.
24. **Gilchrist, T. L., Rees, C. W., and Tuddenham, D.,** *J. Chem. Soc. Perkin I,* p. 3214, 1981.
25. (a) **Gilchrist, T. L., Tuddenham, D., McCague, R., Moody, C. J., and Rees, C. W.,** *Chem. Commun.,* 657, 1981; (b) Gilchrist, T. L., Rees, C. W., and Tuddenham, D., *J. Chem. Soc. Perkin I,* p. 83, 1983; (c) McCague, R., Moody, C. J., and Rees, C. W., *J. Chem. Soc. Perkin I,* p. 165, 1984.
26. (a) **McCague, R., Moody, C. J., and Rees, C. W.,** *Chem. Commun.,* p. 497, 1982; (b) *J. Chem. Soc. Perkin I,* p. 175, 1984.
27. **Lidert, Z. and Rees, C. W.,** *Chem. Commun.,* p. 499, 1982.
28. **McCague, R., Moody, C. J., and Rees, C. W.,** *Chem. Commun.,* p. 622, 1982.
29. **Lidert, Z. and Rees, C. W.,** *Chem. Commun.,* p. 317, 1983.
30. **Platt, J. R.,** *J. Chem. Phys.,* 22, 1448, 1954.
31. **Hanna, E. R., Finley, K. T., Saunders, W. H., and Boekelheide, V.,** *J. Am. Chem. Soc.,* 82, 6342, 1960.
32. **Jung, D. E.,** *Tetrahedron,* 25, 129, 1969.
33. **Zahradnik, R., Michl, J., and Pancir, J.,** *Tetrahedron,* 22, 1355, 1966.
34. **Hess, B. A. and Schaad, L.-J.,** *J. Org. Chem.,* 36, 3418, 1971.
35. **Gutman, I., Milun, M., and Trinajstic, N.,** *J. Am. Chem. Soc.,* 99, 1692, 1977.
36. **Randić, M.,** *J. Am. Chem. Soc.,* 99, 444, 1977.
37. **Garratt, P. J. and Sargent, M. V.,** *Adv. Org. Chem.,* 6, 1, 1969.
38. **Vogel, E.,** in *Proc. 12th R. A. Welch Found., Conf. Chem. Res., Organic Synthesis,* Houston, Tex., November, 1968, 215.
39. (a) **Dewar, M. J. S., Gleicher, G. J., and Thompson, C. C.,** *J. Am. Chem. Soc.,* 88, 1349, 1966; (b) **Roth, W. R., Böhm, M., Lennartz, H.-W., and Vogel, E.,** *Angew. Chem. Int. Ed. Engl.,* 22, 1007, 1983; (c) **Sabljic, A. and Trinajstic, N.,** *J. Org. Chem.,* 46, 3457, 1981.

40. **Allinger, N. L. and Sprague, J. T.,** *J. Am. Chem. Soc.,* 95, 3893, 1973.
41. **Lindner, H. J.,** *Tetrahedron,* 30, 1127, 1974.
42. **Espinosa-Müller, A. and Meezes, F. C.,** *J. Chem. Phys.,* 69, 367, 1978.
43. **Favini, G., Simonetta, M., Sottocornola, M., and Todeschini, R.,** *J. Chem. Phys.,* 74, 3953, 1981.
44. **Grunewald, G. L., Uwaydah, I. M., Christoffersen, R. E., and Spangler, D.,** *Tetrahed. Lett.,* p. 933, 1975.
45. **Gavezotti, A. and Simonetta, M.,** *Helv. Chim. Acta,* 59, 2984, 1976.
46. **Vogel, E. and Böll, W. A.,** *Angew. Chem.,* 76, 784, 1964.
47. **Blattmann, H.-R., Böll, W. A., Heilbronner, E., Hohlneicher, G., Vogel, E., and Weber, J.-P.,** *Helv. Chim. Acta,* 49, 2017, 1966.
48. **Dewey, H. J., Deger, H., Fröhlich, W., Dick, B., Klingensmith, K. A., Hohlneicher, G., Vogel, E., and Michl, J.,** *J. Am. Chem. Soc.,* 102, 6412, 1980.
49. (a) **Boschi, R., Schmidt, W., and Gfeller, J. C.,** *Tetrahed. Lett.,* p. 4107, 1972; (b) **Heilbronner, E.,** *Pure Appl. Chem.,* 44, 831, 1975.
50. **Briat, B., Schooley, D. A., Records, R., Bunnenberg, E., Djerassi, C., and Vogel, E.,** *J. Am. Chem. Soc.,* 90, 4691, 1968.
51. **Klingensmith, K. A., Puttman, W., Vogel, E., and Michl, J.,** *J. Am. Chem. Soc.,* 105, 3375, 1983.
52. (a) **Günther, H., Schmickler, H,. Bremser, W., Straube, F. A., and Vogel, E.,** *Angew. Chem.,* 85, 585, 1973; (b) **Günther, H. and Schmickler, H.,** *Pure Appl. Chem.,* 44, 823, 1975; (c) **Hunadi, R. J.,** *J. Am. Chem. Soc.,* 105, 6889, 1983.
53. **Dobler, M. and Dunitz, J. D.,** *Helv. Chim. Acta,* 48, 1429, 1965.
54. **Bianchi, R., Pilati, T., and Simonetta, M.,** *Acta Cryst.,* B36, 3146, 1980.
55. **Burgi, H. B., Scheffer, E., and Dunitz, J. D.,** *Tetrahedron,* 31, 3089, 1975.
56. **Coetzer, J.,** *Diss. Abstr.,* B29, 3671, 1969.
57. **Gerson, F., Heilbronner, E., Böll, W. A., and Vogel, E.,** *Helv. Chim. Acta,* 48, 1494, 1965.
58. **Gerson, F., Müllen, K., and Vogel, E.,** *Helv. Chim. Acta,* 54, 2731, 1971.
59. **Bremser, W., Grunder, H. T., Heilbronner, E., and Vogel, E.,** *Helv. Chim. Acta,* 50, 84, 1967.
60. **Bremser, W., Hagen, R., Heilbronner, E., and Vogel, E.,** *Helv. Chim. Acta,* 52, 418, 1969.
61. **Budzikiewicz, H., Roth, G., and Vogel, E.,** *Org. Mass Spectrom.,* 14, 140, 1979.
62. **Farnell, R. and Radom, L.,** *J. Am. Chem. Soc.,* 104, 7650, 1982.
63. **Cremer, D. and Dick, B.,** *Angew. Chem. Int. Ed. Engl.,* 21, 865, 1982.
64. **Bianchi, R., Morosi, G., Mugnoli, A., and Simonetta, M.,** *Acta Cryst.,* B29, 1196, 1973.
65. **Bianchi, R., Pilati, T., and Simonetta, M.,** *Acta Cryst.,* B34, 2157, 1978.
66. **Bianchi, R., Pilati, T., and Simonetta, M.,** *J. Am. Chem. Soc.,* 103, 6426, 1981.
67. **Vogel, E., Scholl, T., Lex, J., and Hohlneicher, G.,** *Angew. Chem. Int. Ed. Engl.,* 21, 869, 1982.
68. **Rzepa, H. S.,** *J. Chem. Res.,* 324, 1982 (S).
69. **Vogel, E.,** *Special Publ. No 21,* The Chemical Society, London, 1967, 113.
70. **Kaffory, M.,** *Acta Cryst.,* B34, 306, 1978.
71. (a) **Askenazi, P., Ginsburg, D., and Vogel, E.,** *Tetrahedron,* 33, 1169, 1977; (b) **Askenazi, P., Vogel, E., and Ginsburg, D.,** *Tetrahedron,* 34, 2167, 1978.
72. **Ginsburg, D.,** Propellanes. Sequel I, Department of Chemistry, Technion, Haifa, Israel, 1981, 55.
73. **Vogel, E., Böll, W. A., and Biskup, M.,** *Tetrahed. Lett.,* p. 1569, 1966.
74. **Scholl, T., Lex, J., and Vogel, E.,** *Angew. Chem. Int. Ed. Engl.,* 21, 920, 1982.
75. **Bornatisch, W. and Vogel, E.,** *Angew. Chem.,* 87, 412, 1975.
76. **Gleiter, R., Böhm, M. C., and Vogel, E.,** *Angew. Chem. Int. Ed. Engl.,* 21, 922, 1982.
77. **Effenberger, F. and Klenk, H.,** *Chem. Ber.,* 109, 769, 1976.
78. **Lammertsma, K. and Cerfontain, H.,** *J. Am. Chem. Soc.,* 100, 8244, 1978.
79. **Taylor, R.,** *J. Chem. Soc. Perkin II,* p. 1287, 1975.
80. **Klenk, H., Stohrer, W.-D., and Effenberger, F.,** *Chem. Ber.,* 109, 777, 1976.
81. **Vogel, E.,** *Chimia,* 22, 21, 1968.
82. **Paquette, L. A., Berk, H. C., and Ley, S. W.,** *J. Org. Chem.,* 40, 902, 1975.
83. **Neidlein, R. and Zeiner, H.,** *Chem. Ber.,* 115, 3353, 1982.
84. **Nakasuji, K., Katada, M., and Murata, I.,** *Tetrahed. Lett.,* p. 2515, 1978.
85. **Matsumoto, M., Otsubo, T., Sakata, Y., and Misumi, S.,** *Tetrahed. Lett.,* p. 4425, 1977.
86. **Neidlein, R. and Zeiner, H.,** *Chem. Ber.,* 115, 1409, 1982.
87. **Neidlein, R., and Zeiner, H.,** *Helv. Chim. Acta,* 65, 1285, 1982.
88. **Neidlein, R. and Zeiner, H.,** *Arch. Pharm. (Weinheim),* 315, 630, 1982.
89. **Neidlein, R. and Zeiner, H.,** *Chem. Ztg.,* 106, 233, 1982.
90. **Neidlein, R. and Zeiner, H.,** *Arch. Pharm. (Weinheim),* 315, 90, 1982.
91. (a) **Neidlein, R. and Zeiner, H.,** *Arch. Pharm. (Weinheim),* 313, 970, 1980; (b) **Marshall, J. A. and Conrow, R. E.,** *J. Am. Chem. Soc.,* 102, 4274, 1980.

92. **Neidlein, R. and Zeiner, H.**, *Arch. Pharm. (Weinheim)*, 315, 567, 1982.
93. **D'Arcy, B. R., Kitching. W., Olszowy, H. A., Wells, P. R., Adcock, W., and Kok, G. B.**, *J. Org. Chem.*, 47, 5232, 1982.
94. **Vogel, E., Schröck, W., and Böll, W. A.**, *Angew. Chem.*, 78, 753, 1966.
95. **Zeiner, H.**, *Ph.D. Thesis*, University of Heidelberg, 1981.
96. **Neidlein, R. and Zeiner, H.**, *Helv. Chim. Acta*, 65, 1333, 1982.
97. **Gebert, P. H., King, R. W., La Bar, R. A., and Jones, W. M.**, *J. Am. Chem. Soc.*, 95, 2357, 1973.
98. **(a) La Bar, R. A. and Jones, W. M.**, *J. Am. Chem. Soc.*, 96, 3645, 1974; (b) **Brinker, U. H., King, R. W., and Jones, W. M.**, *J. Am. Chem. Soc.*, 99, 3175, 1977.
99. **Fischer, E. O., Rühle, H., Vogel, E., and Grimme, W.**, *Angew. Chem.*, 78, 548, 1966.
100. **Mues, P., Benn, R., Krüger, C., Tsay, Y. H., Vogel, E., and Wilke, G.**, *Angew. Chem.*, 94, 879, 1982.
101. **Vogel, E., Grimme, W., and Korte, S.**, *Tetrahed. Lett.*, p. 3625, 1965.
102. **Vogel, E., Weyres, F., Lepper, H., and Rautenstrauch, V.**, *Angew. Chem.*, 78, 754, 1966.
103. **Itô, S., Ohtani, H., Narita, S., and Homa, H.**, *Tetrahed. Lett.*, p. 2223, 1972.
104. **Rautenstrauch, V., Scholl, H.-J., and Vogel, E.**, *Angew. Chem.*, 80, 278, 1968.
105. **Wesdemiotis, Ch., Schwartz, H., Budzikiewics, H., and Vogel, E.**, *Org. Mass Spectrom.*, 16, 89, 1981.
106. **Vogel, E., Korte, S., Grimme, W., and Günther, H.**, *Angew. Chem.*, 80, 279, 1968.
107. **Tanimoto, S., Schäfer, R., Ippen, J., and Vogel, E.**, *Angew Chem. Int. Ed. Engl.*, 15, 613, 1976.
108. **Vogel, E. and Sombroek, J.**, *Tetrahed. Lett.*, p. 1627, 1074.
109. **Vogel, E.**, in *Current Trends in Organic Synthesis*, Nokazaki, H., Ed., Pergamon Press, Oxford, 1983, 379.
110. **Ashkenazi, P., Peled, M., Vogel, E., and Ginsburg, D.**, *Tetrahedron*, 35, 1321, 1979.
111. **Scott, L. T. and Erden, I.**, *J. Am. Chem. Soc.*, 104, 1147, 1982.
112. **Scott, L. T. and Kirms, M. A.**, *J. Am. Chem. Soc.*, 104, 3530, 1982.
113. **Rzepa, H. A.**, *J. Chem. Res.*, 3301, 1982 (M).
114. **Vogel, E.**, *Pure Appl. Chem.*, 28, 355, 1971.
115. **Nelson, P. H., Untch, K. G., and Fried, J. H.**, U. S. Patent 3758583; *Chem. Abstr.*, 82, 31023, 1975.
116. **Vogel, E., Königshofen, H., Müllen, K., and Oth, J. F. M.**, *Angew. Chem. Int. Ed. Engl.*, 13, 281, 1974.
117. **Vogel, E., Mann, M., Sakata, Y., Müllen, K., and Oth, J. F. M.**, *Angew. Chem. Int. Ed. Engl.*, 13, 283, 1974.
118. **Lock, G. and Gergely, G.**, *Ber. Dtsch. Chem. Ges.*, 77B, 461, 1944.
119. **Boekelheide, V. and Larrabee, C. E.**, *J. Am. Chem. Soc.*, 72, 1245, 1950.
120. **Reid, D. H.**, *Q. Rev.*, 19, 274, 1965.
121. **Boekelheide, V. and Smith, C. D.**, *J. Am. Chem. Soc.*, 88, 3950, 1966.
122. **Hafner, K. and Schaum, H.**, *Angew. Chem. Int. Ed. Engl.*, 2, 95, 1963.
123. **Zahradnik, R. and Michl, J.**, *Coll. Czech. Chem. Commun.*, 30, 520, 3529, 1965.
124. **Dyatkina, M. E. and Shustorovich, E. M.**, *Dokl. Acad. Nauk S.S.S.R.*, 117, 1021, 1957.
125. **Clar, E.**, *Ber. Dtsch. Chem. Ges.*, 64, 2199, 1931.
126. **Federov, B. P.**, *Bull. Acad. Sci. U.S.S.R. Sci. Chem.*, p. 397, 1947.
127. **Dufraisse, C. and Girard, R.**, *Bull. Soc. Chim. France*, (5)1, 1359, 1934; (5)3, 1857 and 1894, 1936.
128. **Stubbs, H. W. D. and Tucker, S. H.**, *J. Chem. Soc.*, p.2936, 1951.
129. **Kloetzel, M. C. and Chubb, F. L.**, *J. Am. Chem. Soc.*, 72, 150, 1950.
130. **Anderson, A. G. and Anderson, R. G.**, *J. Org. Chem.*, 23, 517, 1958.
131. **Richter, H. F. and Feist, W. C.**, *J. Org. Chem.*, 25, 356, 1960.
132. **Trost, B. M.**, *J. Am. Chem. Soc.*, 88, 853, 1966.
133. **Trost, B. M. and Nelsen, S. F.**, *J. Am. Chem. Soc.*, 88, 2876, 1966.
134. **Nelsen, S. F., Trost, B. M., and Evans, D. H.**, *J. Am. Chem. Soc.*, 89, 3034, 1967.
135. **Trost, B. M. and Brittelli, D. R.**, *Tetrahed. Lett.*, p. 119, 1967.
136. **Trost, B. M., Bright, G. M.**, *J. Am. Chem. Soc.*, 89, 4244, 1967.
137. **Trost, B. M., Bright, G. M., Frihart, C., and Brittelli, D.**, *J. Am. Chem. Soc.*, 93, 737, 1971.
138. **Trost, B. M. and Kinson, P. L.**, *J. Am. Chem. Soc.*, 92, 2591, 1970.
139. **Trost, B. M. and Kinson, P. L.**, *J. Am. Chem. Soc.*, 97, 2438, 1975.
140. **Hafner, K., Diesel, H. D., and Richarz, W.**, *Angew. Chem. Int. Ed. Engl.*, 17, 763, 1978.
141. **Hafner, K., Meinhardt, K.-P., and Richarz, W.**, *Angew. Chem. Int. Ed. Engl.*, 13, 204, 1974.
142. **Mugnoli, A. and Simonetta, M.**, *J. Chem. Soc. Perkin II*, p. 822, 1976.
143. **Destro, R., Ortoleva, E., Simonetta, M., and Todeschini, R.**, *J. Chem. Soc. Perkin II.*, p. 1239, 1983.
144. **Brown, R. D.**, *J. Chem. Soc.*, 2391, 1951.
145. **Zahradnik, R. and Michl, J.**, *Coll. Czech. Chem. Commun.*, 30, 3550, 1965.
146. **Lo, D. H. and Whitehead, M. A.**, *Chem. Commun.*, p. 771, 1968.

147. **Coulson, C. A. and Mallion, R. B.,** *J. Am. Chem. Soc.,* 98, 592, 1976.
148. **Mallion, R. B.,** *Pure Appl. Chem.,* 52, 1541, 1980.
149. **Toyota, A. and Nakajima, T.,** *Bull. Chem. Soc. Jpn.,* 46, 2284, 1973.
150. **Das Gupta, A. and Dasgupta, N. K.,** *Can. J. Chem.,* 54, 3277, 1976.
151. **Aihara, J.,** *Bull. Chem. Soc. Jpn.,* 51, 3540, 1978.
152. **Das Gupta, N. K. and Birss, F. W.,** *Bull. Chem. Soc. Jpn.,* 51, 1211, 1978.
153. **Gupta, S. P. and Singh, P.,** *Bull. Chem. Soc. Jpn.,* 52, 2745, 1979.
154. **Tiwary, M. M., Upadhgay, R. K., and Srivastava, A. K.,** *Ind. J. Pure Appl. Phys.,* 18, 665, 1980.
155. **Vogel, E., Haberland, U., and Günther H.,** *Angew. Chem.* 82, 510, 1970.
156. **Vogel, E., Sombroek, J., and Wagemann, W.,** *Angew. Chem. Int. Ed. Engl.,* 14, 564, 1975.
157. **Vogel, E., Biskup, M., Vogel, A., Haberland, U., and Eimer, J.,** *Angew. Chem.,* 78, 642, 1966.
158. **Vogel, E., Deger, H. M., Helbel, P., and Lex, J.,** *Angew. Chem.,* 92, 943, 1980.
159. **Balci, M., Schalenbach, R., and Vogel, E.,** *Angew. Chem.,* 93, 816, 1981.
160. **Vogel, E., Nitsche, R., and Krieg, H.-U.,** *Angew. Chem.,* 93, 818, 1981.
161. **Okazaki, R., O-oka, M., Tokitoh, N., Shishido, Y., and Inamoto, N.,** *Angew. Chem.,* 93, 833, 1981.
162. **Vogel, E. and Reel, H.,** *J. Am. Chem. Soc.,* 94, 4388, 1972.
163. **Vogel, E., Vogel, A., Kubbeler, H.-K., and Sturm, W.,** *Angew. Chem.,* 82, 512, 1970.
164. **Vogel, E., Sturm, W., and Cremer, H.-D.,** *Angew. Chem.,* 82, 513, 1970.
165. **Lindsay, W. S., Stokes, P., Humber, L. G., and Boekelheide, V.,** *J. Am. Chem. Soc.,* 83, 943, 1961.
166. **Beokelheide, V. and Philips, J. B.,** *J. Am. Chem. Soc.,* 85, 1545, 1963.
167. **Boekelheide, V. and Philips, J. B.,** *Proc. Natl. Acad. Sci. U.S.A.,* 51, 550, 1964.
168. **Boekelheide, V. and Philips, J. B.,** *J. Am. Chem. Soc.,* 89, 1695, 1967.
169. **Boekelheide, V.,** in *Proc. 12th R. A. Welch Found. Conf. Chem. Res. Org. Synth.,* Houston, Tex., November 1968, 83.
170. **Renfroe, H. B., Gurney, J. A., and Hall, L. A. R.,** *J. Org. Chem.,* 37, 3045, 1972.
171. **Boekelheide, V. and Miyasaka, T.,** *J. Am. Chem. Soc.,* 89, 1709, 1967.
172. **Boekelheide, V. and Hylton, T. A.,** *J. Am. Chem. Soc.,* 92, 3669, 1970.
173. **Blaschke, H., Ramey, C. E., Calder, I., and Boekelheide, V.,** *J. Am. Chem. Soc.,* 92, 3675, 1970.
174. **Boekelheide, V. and Mitchell, R. H.,** *Jerusalem Symp. Quantum Chem. Biochem. III,* The Israel Academy of Science and Humanities, Jerusalem, 1971.
175. **Mitchell, R. H. and Boekelheide, V.,** *J. Am. Chem. Soc.,* 96, 1547, 1974.
176. **Blattmann, H. R. and Schmidt, W.,** *Tetrahedron,* 26, 5885, 1970.
177. **Mao, K. and Boekelheide, V.,** *J. Org. Chem.,* 45, 2746, 1980.
178. **Harris, T. D., Neuschwander, B., and Boekelheide, V.,** *J. Org. Chem.,* 43, 727, 1978.
179. **Boekelheide, V. and Tsai, C. H.,** *J. Org. Chem.,* 38, 3931, 1973.
180. **Mitchel, R. H. and Anker, W.,** *Tetrahed. Lett.,* 22, 5139, 1981.
181. **Tashiro, M. and Yamato, T.,** *J. Org. Chem.,* 46, 1543, 1981.
182. **Tashiro, M. and Yamato, T.,** *J. Am. Chem. Soc.,* 104, 3701, 1982.
183. **Huber, W., Lex, J., Meul, T., and Müllen, K.,** *Angew. Chem. Int. Ed. Engl.,* 20, 391, 1981.
184. **Huber, W., Irmen, W., Lex, J., and Müllen, K.,** *Tetrahed. Lett.,* 23, 3889, 1982.
185. **Dias, J. R.,** *Math. Chem.,* 14, 83, 1983.
186. **Clar, E.,** *Polycyclic Hydrocarbons,* Vol. 2, Academic Press, New York, 1964, 110.
187. **Coulson, E. A.,** *Chem. Ind.,* 60, 699, 1941.
188. **Weitzenböck, R.,** *Monatsh.,* 34, 193, 1913.
189. **Lock, G. and Walter, E.,** *Ber. Dtsch. Chem. Ges.,* 75, 1158, 1942; 77, 286, 1944.
190. **von Braun, J. and Rath, E.,** *Ber. Dtsch. Chem. Ges.,* 61, 956, 1928.
191. **Cook, J. W. and Hewett, C. L.,** *J. Chem. Soc.,* p. 365, 1934.
192. **Fleischer, K. and Retze, E.,** *Ber. Dtsch. Chem. Ges.,* 55, 3280, 1922.
193. **Baker, W., McOmie, J. F. W., and Norman, J. M.,** *J. Chem. Soc.,* p. 1114, 1951.
194. **Reid, D. H., Stafford, W. H., and Ward, J. P.,** *J. Chem. Soc.,* p. 1193, 1955.
195. **Boekelheide, V. and Vick, G. K.,** *J. Am. Chem. Soc.,* 78, 653, 1956.
196. **Boekelheide, V., Langeland, W. E., and Liu, Chu-Tsin,** *J. Am. Chem. Soc.,* 73, 2432, 1951.
197. **Gardner, P. D., Wulfman, C. E., and Osborn, C. L.,** *J. Am. Chem. Soc.,* 80, 143, 1958.
198. **Anderson, A. G., MacDonald, A. A., and Montana, A. F.,** *J. Am. Chem. Soc.,* 90, 2993, 1968.
199. **Anderson, A. G., Masada, G. M., and Montana, A. F.,** *J. Org. Chem.,* 38, 1439, 1973.
200. **Anderson, A. G., Montana, A. F., MacDonald, A. A., and Masada, G. M.,** *J. Org. Chem.,* 38, 1445, 1973.
201. **Jutz, C. and Schweiger, E.,** *Angew. Chem. Int. Ed. Engl.,* 10, 808, 1971.
202. **Jutz, C. and Schweiger, E.,** *Synthesis,* p. 193, 1974.
203. **Hafner, K., Diehl, H., and Richarz, W.,** *Angew. Chem. Int. Ed. Engl.,* 15, 108, 1976.
204. **Hafner, K.,** *Pure Appl. Chem.,* 28, 153, 1971.

205. **Reel, H. and Vogel, E.,** *Angew. Chem. Int. Ed. Engl.,* 11, 1013, 1972.
206. **Fujimori, M., Morita, N., Yasunami, M., Asao, T., and Takase, K.,** *Tetrahed. Lett.,* 24, 781, 1983; **Yasunami, M., Ameniya, T., and Takase, K.,** *Tetrahed. Lett.,* 24, 69, 1983.
207. **Hafner, K. and Schneider, J.,** *Liebigs Ann. Chem.,* 624, 37, 1959.
208. **Hafner, K., Fleischer, R., and Fritz, C.,** *Angew. Chem. Int. Ed. Engl.,* 4, 69, 1965.
209. **Diesel, H.-D.,** Ph.D. thesis, Darmstadt, 1978.
210. **Hafner, K. and Bangert, K. F.,** *Liebigs Ann. Chem.,* 650, 98, 1961.
211. **Munday, R. and Sutherland, I. O.,** *Chem. Commun.,* p. 569, 1967.
212. **Munday, R. and Sutherland, I. O.,** *J. Chem. Soc. (C),* 1427, 1969.
213. (a) **Hafner, K. and Flach, J.,** *Erdöl, Kohle, Erdgas, Petrochem.,* 31, 89, 1978; (b) **Flach, J.,** Ph.D. thesis., Darmstadt, 1976.
214. **Kramer, W.,** Ph.D. thesis, Heidelberg, 1981.
215. **Destro, R., Pilati, T., and Simonetta, M.,** *Acta Cryst.,* B33, 940, 1977.
216. **Gramaccioli, C. M., Mimun, A. S., Mugnoli, A., and Simonetta, M.,** *J. Am. Chem. Soc.,* 95, 3149, 1973.
217. **Günther, H. V., Puttkammer, H., Deger, H. M., Hebel, P., and Vogel, E.,** *Angew. Chem.,* 92, 944, 1980.
218. **Destro, R. and Simonetta, M.,** *Acta Cryst.,* B33, 3219, 1977.
219. **Destro, R., Pilati, T., and Simonetta, M.,** *Tetrahedron,* 36, 3301, 1980.
220. **Müllen, K. and Reel, H.,** *Helv. Chim. Acta,* 56, 363, 1973.
221. (a) **Murata, I.,** in press, quoted in ref. 11; (b) for a substituted derivative see also **Renfroe, H. B.,** *J. Am. Chem. Soc.,* 90, 2194, 1968.
222. **Mugnoli, A. and Simonetta, M.,** *Acta Cryst.,* 30B, 2896, 1974.
223. **Bremser, W., Roberts, J. D., and Vogel, E.,** *Tetrahed. Lett.,* p. 4307, 1969.
224. **Casalone, G., Gavezotti, A., Mugnoli, A., and Simonetta, M.,** *Angew. Chem. Int. Ed. Engl.,* 9, 519, 1970.
225. **Gavezotti, A., Mugnoli, A., Raimondi, M., and Simonetta, M.,** *J. Chem. Soc. Perkin II,* p. 425, 1972.
226. **Günther, H., Schmickler, H., Königshofen, H., Recker, K., and Vogel, E.,** *Angew. Chem.,* 85, 261, 1973.
227. **Gramaccioli, C. M., Mugnoli, A., Pilati, T., Raimondi, M., and Simonetta, M.,** *Acta Cryst.,* B28, 2365, 1972.
228. **du Vernet, R. and Boekelheide, V.,** *Proc. Natl. Acad. Sci. U.S.A.,* 71, 2961, 1974.
229. **Johnson, C. E. and Bovey, F. A.,** *J. Chem. Phys.,* 29, 1012, 1958.
230. (a) **Haigh, C. W. and Mallion, R. B.,** *Org. Magn. Reson.,* 4, 203, 1972; (b) **Mallion, R. B.,** *Pure Appl. Chem.,* 52, 1541, 1980.
231. **Blattman, H.-R., Boekelheide, V., Heilbronner, E., and Weber, J.-P.,** *Helv. Chim. Acta,* 50, 68, 1967.
232. **Spanget-Larsen, J. and Gleiter, R.,** *Helv. Chim. Acta,* 61, 2999, 1978.
233. **Boekelheide, V., Murrel, J. N., and Schmidt, W.,** *Tetrahed. Lett.,* p. 575, 1972.
234. **Gerson, F., Heilbronner, E., and Boekelheide, V.,** *Helv. Chim. Acta,* 47, 1123, 1964.
235. **Hanson, A. W.,** *Acta Cryst.,* 18, 599, 1965.
236. **Choutecky, J., Hochman, P., and Michl, J.,** *J. Chem. Phys.,* 40, 2439, 1964.
237. **Nakajima, T.,** *Pure Appl. Chem.,* 28, 219, 1971.
238. **Birss, F. W. and Dasgupta, N. K.,** *Can. J. Chem.,* 49, 2840, 1971.
239. **Dasgupta, A. and Dasgupta, N. K.,** *Tetrahedron,* 28, 3587, 1972.
240. **Gastmans, J. P., Fromenteau-Gastmans, D., De Groote, R. A. M.,** *Tetrahed. Lett.,* p. 3339, 1974.
241. **Dasgupta, A. and Dasgupta, N. K.,** *Theor. Chim. Acta,* 33, 177, 1974.
242. **Dasgupta, A. and Dasgupta, N. K.,** *Can. J. Chem.,* 52, 155, 1974.
243. **Chatterjee, S. and Dasgupta, N. K.,** *Bull. Chem. Soc. Jpn.,* 49, 1832, 1976.
244. **Graovac, A., Gutman, I., Randic, M., and Trinajstić, N.,** *J. Am. Chem. Soc.,* 95, 6267, 1973.
245. **Gastmans, J. P., Fromenteau-Gastmans, D., and Slade, S. F.,** *Can. J. Chem.,* 57, 2864, 1979.
246. **Dewar, M. J. S. and de Llano, C.,** *J. Am. Chem. Soc.,* 91, 789, 1969.
247. **Ray, P. R., Muchopadyay, A. K., and Mukherjee, N. G.,** *J. Ind. Chem. Soc.,* 57, 608, 1980.
248. **Yamaguchi, H. and Nakajima, T.,** *Bull. Chem. Soc. Jpn.,* 47, 1898, 1974.
249. **Robertson, J. M. and White, J. G.,** *J. Chem. Soc.,* p. 358, 1947.
250. **Alger, T. D., Grant, D. M., and Paul, E. G.,** *J. Am. Chem. Soc.,* 85, 5397, 1966.
251. **Breslow, R., Grubbs, R., and Murahashi, S. I.,** *J. Am. Chem. Soc.,* 92, 4139, 1970.
252. **Pullman, B., Pullman, A., Berthier, G., and Pontis, J.,** *J. Chem. Phys.,* 49, 20, 1952.
253. **Sidman, W.,** *J. Am. Chem. Soc.,* 78, 1261, 1956.
254. **Pitt, D. A., Petro, A. J., and Smith, C. P.,** *J. Am. Chem. Soc.,* 79, 5633, 1957.
255. **Herndon, W. C.,** *J. Am. Chem. Soc.,* 98, 887, 1976.
256. **Radvilavicius, C. and Bolotin, A. B.,** *Liet. Fiz. Rinkinys,* 8, 159, 1968.

257. **Yamaguchi, H. and Nakajima, T.,** *Bull. Chem. Soc. Jpn.,* 44, 682, 1971.
258. **Jones, A. J., Gardner, P. D., Grant, D. M., Litchman, W. M., and Boekelheide, V.,** *J. Am. Chem. Soc.,* 92, 2395, 1970.
259. **Dauben, H. J., Wilson, J. D., and Laity, J. L.,** *J. Am. Chem. Soc.,* 91, 1991, 1969.
260. **Hanson, A. W.,** *Acta Cryst.,* 13, 215, 1960.
261. **Hanson, A. W.,** *Acta Cryst.,* 21, 97, 1968.
262. **Schneider, W. G., Bernstein, H. J., and Pople, J. A.,** *J. Am. Chem. Soc.,* 80, 3497, 1958.
263. **Schaefer, T. and Schneider, W. G.,** *Can. J. Chem.,* 41, 966, 1963.
264. **Trost, B. M. and Herdle, W. B.,** *J. Am. Chem. Soc.,* 98, 4080, 1976.
265. **Gerson, F. and Heinzer, J.,** *Helv. Chim. Acta,* 49, 7, 1966.
266. **Craig, D. P. and Macoll, A.,** *J. Chem. Soc.,* p. 964, 1949.
267. **Craig, D. P.,** *J. Chem. Soc.,* p. 3175, 1951.
268. **Baumgartner, P., Weltin, E., Wagniere, G., and Heilbronner, E.,** *Helv. Chim. Acta,* 48, 751, 1965.
269. **Randić, M.,** *Tetrahedron,* 30, 1905, 1977.
270. **Zahradnik, R.,** *Angew. Chem. Int. Ed. Engl.,* 4, 1039, 1965.
271. **Lindner, H. J.,** *Chem. Ber.,* 102, 2456, 1969.
272. **Fleischer, R., Hafner, K., Wildgruber, J., Hochmann, P., and Zahradnik, R.,** *Tetrahedron,* 24, 5943, 1968.
273. **Vogel, E.,** *Festschrift für Leo Brandt,* Westdeutscher-Verlag, Köln, 1968, 117.
274. **Phillips, J. B., Molyneux, R. J., Sturm, E., and Boekelheide, V.,** *J. Am. Chem. Soc.,* 89, 1704, 1967.
275. **Mitchell, R. H., Calder, I., Huisman, H., and Boekelheide, V.,** *Tetrahedron,* 31, 1109, 1975.
276. **Boekelheide, V. and Sturm, E.,** *J. Am. Chem. Soc.,* 91, 902, 1969.
277. **Kamp, D. and Boekelheide, V.,** *J. Org. Chem.,* 43, 3475, 1978.
278. **Blattmann, H.-R., Meuche, D., Heilbronner, E., Molyneux, R. J., and Boekelheide, V.,** *J. Am. Chem. Soc.,* 87, 130, 1965.
279. **Naef, R. and Fischer, E.,** *Helv. Chim. Acta,* 57, 2224, 1974.
280. **Schmidt, W.,** *Helv. Chim. Acta,* 54, 862, 1971.
281. **Muszkat, K. A. and Fischer, E.,** *J. Chem. Soc. B.,* p. 662, 1967.
282. **Rodd, E. H., Ed.,** *Chemistry of Carbon Compounds,* Vol. 3 (Part B), New York, 1956, 1502.
283. **Cook, J. W. and Schoental, R.,** *J. Chem. Soc.,* p. 170, 1948.
284. **(a)Dehmlow, E. V. and Lissel, M.,** *Liebigs Ann. Chem.,* p. 182, 1979; (b) **Nakasuji, K., Katada, M., and Murata, I.,** *Angew. Chem. Int. Ed. Engl.,* 18, 946, 1979.
285. **Kuthan, J., Donnerova, Z., and Skala, V.,** *Collect. Czech. Chem. Commun.,* 34, 2398, 1969.
286. **Turner, R. B., Lindsay, W. S., and Boekelheide, V.,** *Tetrahedron,* 27, 3341, 1971.
287. **Tarbell, D. S. and Huang, T.,** *J. Org. Chem.,* 24, 887, 1959.
288. **(a) Anderson, A. G., Masada, G. W., and Kao, G. L.,** *J. Org. Chem.,* 45, 1312, 1980; (b) *J. Org. Chem.,* 47, 5426, 1982.
289. **Anderson, A. G. and Kao, L. G.,** *J. Org. Chem.,* 47, 3589, 1982.
290. **Ginsburg, D.,** *Propellanes,* Verlag-Chemie, Weinheim, 1975.
291. **Alscher, A., Bremser, W., Cremer, D., Günther, H., Schmickler, H., Storm, W., and Vogel, E.,** *Chem. Ber.,* 108, 640, 1975.
292. **Bianchi, R., Mugnoli, A., and Simonetta, M.,** *Acta Cryst.,* B31, 1283, 1975.
293. **Batich, C., Heilbronner, E., and Vogel, E.,** *Helv. Chim. Acta,* 57, 2288, 1974.
294. **Kolc, J., Michl, J., and Vogel, E.,** *J. Am. Chem. Soc.,* 98, 3935, 1976.
295. **Gerson, F., Müllen, K., and Vogel, E.,** *J. Am. Chem. Soc.,* 94, 2924, 1972.
296. **Tanner, D., Wennerström, O., and Vogel, E.,** *Tetrahed. Lett.,* 23, 1221, 1982.
297. **Mitchell, R. H. and Boekelheide, V.,** *Chem. Commun.,* 1557, 1970.
298. **Hafner, K., Hafner-Schneider, G., and Bauer, F.,** *Angew. Chem. Int. Ed. Engl.,* 7, 808, 1968.
299. **Friebe, W.-G.,** Ph.D. thesis, Darmstadt, 1973.
300. **Das Gupta, A. and Das Gupta, N. K.,** *Can. J. Chem.,* 53, 3777, 1975.
301. **Wagemann, W., Iyoda, M., Deger, H.-M., Sombroeck, J., and Vogel, E.,** *Angew. Chem.,* 90, 988, 1978.
302. **Pilati, T. and Simonetta, M.,** *Acta Cryst.,* B33, 851, 1977.
303. **Gayoso, J. and Boncekkine, A.,** *Tetrahed. Lett.,* p. 2447, 1971.
304. **Jutz, C. and Kirchlechner, R.,** *Angew. Chem.,* 78, 493, 1966.
305. **Jutz, C., Kirchlechner, R., and Seidel, H.-J.,** *Chem. Ber.,* 102, 2301, 1969.
306. **Murata, I., Nakasuji, K., Yamamoto, K., Nakazawa, T., Kayane, Y., Kimura, A., and Hara, O.,** *Angew. Chem. Int. Ed. Engl.,* 14, 170, 1975.
307. **Cava, M. P. and Schlessinger, R. H.,** *Tetrahedron,* 21, 3051, 1965.
308. **Kolc, J. and Michl, J.,** *J. Am. Chem. Soc.,* 95, 7391, 1973.
309. **Lawson, J., du Vernet, R., and Boekelheide, V.,** *J. Am. Chem. Soc.,* 95, 956, 1973.

310. du Vernet, R. B., Otsubo, T., Lawson, J. A., and Boekelheide, V., *J. Am. Chem. Soc.*, 97, 1629, 1975.
311. du Vernet, R. B., Wennerström, O., Lawson, J., Otsubo, T., and Boekelheide, V., *J. Am. Chem. Soc.*, 100, 2457, 1978.
312. Cram, D. J. and Dewhirst, K. C., *J. Am. Chem. Soc.*, 81, 5963, 1959.
313. Otsubo, T., Gray, R., and Boekelheide, V., *J. Am. Chem. Soc.*, 100, 2449, 1978.
314. Otsubo, T., Stusche, D., and Boekelheide, V., *J. Org. Chem.*, 43, 3466, 1978.
315. Kamp, D. and Boekelheide, V., *J. Org. Chem.*, 43, 3470, 1978.
316. Gleiter, R., Spanget-Larsen, J., Thulstrup, E. W., Murata, I., Nakasuji, K., and Jutz, C., *Helv. Chim. Acta*, 59, 1459, 1976.
317. Haddon, R. C., *Tetrahedron*, 28, 3613 and 3655, 1972.
318. Sondheimer, F. and Shani, A., *J. Am. Chem. Soc.*, 86, 3168, 1964.
319. Shani, A. and Sondheimer, F., *J. Am. Chem. Soc.*, 89, 6310, 1967.
320. Vogel, E., Biskup, M., Pretzer, W., and Böll, W. A., *Angew. Chem.*, 76, 785, 1964.
321. Vogel, E., Pretzer, W., and Böll, W. A., *Tetrahed. Lett.*, p. 3613, 1965.
322. Gerson, F., Heinzer, J., and Vogel, E., *Helv. Chim. Acta*, 53, 95, 1970.
323. Zwaard, A. W. and Klosterziel, H., *Tetrahed. Lett.*, 23, 4151, 1982.
324. Vogel, E., Biskup, M., Vogel, A., and Günther, H., *Angew. Chem.*, 78, 755, 1966.
325. Vogel, E., Haberland, U., and Ick, J., *Angew. Chem.*, 82, 514, 1970.
326. Vogel, E. Brocker, U., and Junglas, H., *Angew. Chem.*, 92, 1051, 1980.
327. Vogel, E., Kuebart, F., Marco, J. A., Andree, R., Günther, H., and Aydin, R., *J. Am. Chem. Soc.*, 105, 6982, 1983.
328. Ganis, P. and Dunitz, J. D., *Helv. Chim. Acta*, 50, 2369, 1969.
329. Destro, R., Gavezotti, A., and Simonetta, M., *Acta Cryst.*, B38, 1352, 1982.
330. Blackburne, I. D. and Katritzky, A. R., *Acc. Chem. Res.*, 8, 300, 1975.
331. Vogel, E., Beerman, D., Balci, E., and Altenbach, H.-J., *Tetrahed. Lett.*, p. 1167, 1976.
332. Schäfer-Ridder, M., Wagner, A., Schwamborn, M., Schreiner, H., Devrout, E., and Vogel, E., *Angew. Chem.*, 90, 894, 1978.
333. (a) Lippa, W. J., Crawford, H. T., Radlick, P. C., and Helmkamp, G. K., *J. Org. Chem.*, 43, 3813, 1978; (b) Lippa, W. J., *Diss. Abstr.*, 38B, 4243, 1978.
334. Gölz, H. J., Muchowsky, J. M., and Maddox, M. L., *Angew. Chem. Int. Ed. Engl.*, 17, 855, 1978.
335. Maddox, M. L., Martin, J. C., and Muchowsky, J. M., *Tetrahed. Lett.*, 21, 7, 1980.
336. Vogel, E., Feldmann, R., Düwell, H., Cremer, H.-D., and Günther, H., *Angew. Chem. Int. Ed. Engl.*, 11, 217, 1972.
337. Grimme, W., Reisdorf, J., Jünemann, W., and Vogel, E., *J. Am. Chem. Soc.*, 92, 6335, 1970.
338. Boekelheide, V. and Pepperdine, W., *J. Am. Chem. Soc.*, 92, 3684, 1970.
339. Destro, R., Simonetta, M., and Vogel, E., *J. Am. Chem. Soc.*, 103, 2863, 1981.
340. Schlägl, R. and Widhalm, M., *Chem. Ber.*, 115, 3042, 1982.
341. Hilken, G., Kinkel, T., Schwamborn, M., Lex, J., Schmickler, H., and Vogel, E., *Angew. Chem. Int. Ed. Engl.*, 21, 784, 1982.
342. Warner, P. and Winstein, S., *J. Am. Chem. Soc.*, 91, 7785, 1969.
343. Lammertsma, K. and Cerfontain, H., *J. Am. Chem. Soc.*, 102, 4528, 1980.
344. Lammertsma, K. and Cerfontain, H., *J. Am. Chem. Soc.*, 102, 3257, 1980.
345. Smith, R. J., Miller, T. M., and Pagni, R. M., *J. Org. Chem.*, 47, 4181, 1982.
346. Lammertsma, K., *J. Am. Chem. Soc.*, 104, 2070, 1982.
347. Grimme, W., Hoffmann, H., and Vogel, E., *Angew. Chem.*, 77, 348, 1965.
348. Grimme, W., Heilbronner, E., Hohlneicher, G., Vogel, E., and Weber, J.-P., *Helv. Chim. Acta*, 51, 225, 1968.
349. Destro, R., Pilati, T., and Simonetta, M., *J. Am. Chem. Soc.*, 98, 1999, 1976.
350. Simonetta, M., *Pure Appl. Chem.*, 52, 1597, 1980.
351. Kemp-Jones, A. V., Jones, A. J., Sakai, M., Beeman, C. P., and Masamune, S., *Can. J. Chem.*, 51, 767, 1973.
352. Grimme, W., Kaufhold, M., Dettmeier, U., and Vogel, E., *Angew. Chem.*, 78, 643, 1966.
353. Radlick, P. and Rosen, W., *J. Am. Chem. Soc.*, 88, 3461, 1966.
354. (a) Raddlick, P. and Rosen, W., *J. Am. Chem. Soc.*, 89, 5308, 1967; (b) Takahashi, K., Kagawa, T., and Takase, K., *Chem. Lett.*, p. 701, 1979; (c) Takahashi, K., Kagawa, T., and Takase, K., *Chem. Commun.*, p.862, 1979; (d) Takahashi, K., Takase, K., and Kagawa, T., *J. Am. Chem. Soc.*, 103, 1186, 1981.
355. Staley, S. W. and Orvedal, A. W., *J. Am. Chem. Soc.*, 95, 3348, 1973.
356. Zahradnik, T., Michl, J., and Koutecky, J., *Collect. Czech. Chem. Commun.*, 29, 1932, 1964.
357. Pettit, R., *Chem. Ind.*, p. 1306, 1956.

234   *Annulenes, Benzo, Hetero-, Homo-Derivatives and Their Valence Isomers*

358. Bonthrone, W. and Reid, D. H., *J. Chem. Soc.*, p. 2773, 1959.
359. Larsen, J. W., Bouis, P. A., Watson, C. R., and Pagni, R. M., *J. Am. Chem. Soc.*, 96, 2284, 1974.
360. Pagni, R. M., Bouis, P. A., and Easley, P., *Tetrahed. Lett.*, p. 2671, 1975.
361. Clar, E. and Stewart, D. G., *J. Chem. Soc.*, p. 23, 1958.
362. Reid, D. H., *Angew. Chem.*, 67, 761, 1955.
363. Windgassen, R. J., Jr., Saunders, W. W., and Boekelheide, V., *J. Am. Chem. Soc.*, 81, 1459, 1959.
364. Hess, B. A. and Schaad, L. J., *Pure Appl. Chem.*, 52, 1471, 1980.
365. Pettit, R., *J. Am. Chem. Soc.*, 82, 1972, 1960.
366. Prinzbach, H., Freudenberg, V., Scheidegger, U., *Helv. Chim. Acta*, 50, 1087, 1967.
367. Bonthrone, W. and Reid, D. H., *J. Chem. Soc. (B)*, p. 91, 1966.
368. Streitwieser, A., Jr., *Molecular Orbital Theory for Organic Chemists*, John Wiley & Sons, New York, 1961, 46.
369. Oth, J. F. M., Müllen, K., Königshofen, H., Mann, M., Sakata, Y., and Vogel, E., *Angew. Chem. Int. Ed. Engl.*, 13, 284, 1974.
370. Shannon, R. L. and Cox, R. H., *Tetrahed. Lett.*, p. 1603, 1973.
371. Wittig, G., Rautenstrauch, V., and Wingler, F., *Tetrahed. Suppl.*, 7, 189, 1966.
372. Pagni, R. M. and Watson, C. R., *Tetrahed. Lett.*, p. 59, 1973.
373. Murata, I. and Nakasuji, K., *Tetrahed. Lett.*, p. 47, 1973.
374. Bianchi, R., Destro, R., and Simonetta, M., *Acta Cryst.*, B35, 1002, 1979.
375. Gerson, F., Müllen, K., and Vogel, E., *Angew. Chem.*, 83, 1014, 1971.
376. Untsch, K. G. and Pople, J. A., *J. Am. Chem. Soc.*, 88, 4811, 1966.
377. Deger, H. M., Müllen, K., and Vogel, E., *Angew. Chem.*, 90, 990, 1978.
378. Mitchell, R. H., Klopfenstein, C. E., and Boekelheide, V., *J. Am. Chem. Soc.*, 91, 4931, 1969.
379. Trost, B. M., Nelsen, S. F., and Brittelli, D. R.,, *Tetrahed. Lett.*, p. 3959, 1967.
380. Predrag, I. and Trinajstić, N., *J. Org. Chem.*, 45, 1738, 1980.
381. Trost, B. M., Buchner, D., and Bright, G. M., *Tetrahed. Lett.*, p. 2787, 1973.
382. Dagan, A. and Rabinovitz, M., *J. Am. Chem. Soc.*, 98, 8268, 1976.
383. Yamamoto, K., Kayane, Y., and Murata, I., *Bull. Chem. Soc. Jpn.*, 50, 1964, 1977.
384. Cox, R. H., Janzen, E. G., and Gerlock, J. L., *J. Am. Chem. Soc.*, 90, 5906, 1968.
385. Willner, I. and Rabinovitz, M., *Tetrahed. Lett.*, p. 1223, 1976.
386. Rabinovitz, M. and Wilner, I., *Pure Appl. Chem.*, 52, 1575, 1980.
387. Ravinovitz, M., Willner, I., and Minsky, A., *Acc. Chem. Res.*, 16, 298, 1983.
388. Murata, I., Yamamoto, K., Morioka, M., Tamura, M., and Hirotsu, T., *Tetrahed. Lett.*, p. 2287, 1975.
389. Beker, B. C., Huber, W., and Müllen, K., *J. Am. Chem. Soc.*, 102, 7803, 1980.
390. Müllen, K., *Helv. Chim. Acta*, 61, 2307, 1978.
391. Willner, I. and Rabinovitz, M., *J. Org. Chem.*, 45, 1628, 1980.
392. Rabinovitz, M. and Minsky, A., *Pure Appl. Chem.*, 54, 1005, 1982.
393. Minsky, A., Klein, J., and Rabinovitz, M., *J. Am. Chem. Soc.*, 103, 4586, 1981.
394. (a) Minsky, A., Meyer, A. Y., and Rabinovitz, M., *J. Am. Chem. Soc.*, 104, 2475, 1982; (b) Eliasson, B., Lejon, T., and Edlund, U., *Chem. Commun.*, p. 591, 1984.
395. Meyer, A. Y., *Theor. Chim. Acta*, 9, 401, 1968.
396. Minsky, A. and Rabinovitz, M., *Tetrahed. Lett.*, p. 5341, 1981.
397. Gibbard, H. C., Moody, C. I., and Rees, C. W., *J. Chem. Soc. Perkin I*, p. 731, 1985.
398. Gibbard, H. C., Moody, C. J., and Rees, C. W., *J. Chem. Soc. Perkin I*, p.735, 1985.
399. Bischof, P., Gleiter, R., Haider, R., and Rees, C. W., *J. Chem. Soc. Perkin II*, p. 1001, 1985.
400. Lidert, Z., McCague, R., Moody, C. J., and Rees, C. W., *J. Chem. Soc. Perkin I*, p. 383, 1985.
401. Gatti, C., Barzaghi, M., and Simonetta, M., *J. Am. Chem. Soc.*, 107, 878, 1985.
402. Andréa, R. R., Cerfontain, H., Lambrechts, H. J. A., Louwen, J. N., and Oskam, A., *J. Am. Chem. Soc.*, 106, 2531, 1984.
403. Marshall, J. A. and Conrow, R. E., *J. Am. Chem. Soc.* 105, 5679, 1983.
404. Brown, R. D., *J. Chem. Soc.*, p. 2391, 1951.
405. Fieser, L. and Fieser, M., *J. Am. Chem. Soc.*, 55, 3010, 1933.
406. Vogel, E. et al., *Nachr. Chem. Techn. Lab.*, 32, 576, 1984; *Angew. Chem.*, in press.
407. Mitchell, R. H., Chaudhary, M., Dingle, T. W., and Williams, R. V., *J. Am. Chem. Soc.*, 106, 7776, 1984.
408. Gerson, F., Huber, W., and Lopez, J., *J. Am. Chem. Soc.*, 106, 5808, 1984.
409. Anderson, A. G., Jr., Davidson, E. R., Daugs, E. D., Kao, L. G., Lindquist, R. L., and Quenemon, K. A., *J. Am. Chem. Soc.*, 107, 1896, 1985.
410. Vogel, E., Kürschner, U., Schmickler, H., Lex, J., Wennerstrom, O., Tanner, D., Norinder, U., and Krüger, C., *Tetrahed. Lett.*, 26, 3087, 1985.

411. **Flitsch, W. and Peeters, H.,** *Tetrahed. Lett.,* p. 1461, 1975.
412. **Okazaki, R., Hasegawa, T., and Shishido, Y.,** *J. Am. Chem. Soc.,* 106, 5271, 1984.
413. **Pagni, R. M.,** *Tetrahedron,* 40, 4161, 1984.
414. **Müllen, K., Meul, T., Vogel, E., Kürschner, U., Schmickler, H., and Wennerstrom, O.,** *Tetrahed. Lett.,* 26, 3091, 1985.
415. **Schneiders, C., Müllen, K., and Huber, W.,** *Tetrahedron,* 40, 1701, 1984.
416. **Lo, D. H. and Whitehead, M. A.,** *Chem. Commun.,* p. 771, 1968.
417. **Lo, D. H. and Whitehead, M. A.,** *Can. J. Chem.,* 46, 2027, 1968.
418. **Chung, A. L. H. and Dewar, M. J. S.,** *J. Chem. Phys.,* 42, 756, 1965.
419. **Destro, R., Pilati, T., Simonetta, M., and Vogel, E.,** *J. Am. Chem. Soc.,* 107, 3185, 1985; 107, 3192, 1985.

ADDENDA

ADDENDUM TO CHAPTER 2

From its heat of hydrogenation, bicyclo[6.2.0]decapentaene (**35**) was found to be anti-aromatic with a resonance energy of 3.7 kcal/mol; the formally single bond of the 1,4-bridge has antibonding character.[1] Aromatic stabilities of bridged polyenes and of the corresponding cyclopolyenes were computed, obtaining evidence for substantial homoaromatic delocalization.[2]

In addition to examples presented in Chapter 2, Section F.2, Hosoya[3] described a method for designing non-Kekulé pericondensed benzenoid hydrocarbons, i.e., diradicals which have the same numbers of starred and nonstarred carbon atoms ("concealed non-Kekuléans" in the terminology of Cyvin and co-workers[4]).

References
1. **Roth, W. R., Lennartz, H. W., Vogel, E., Leiendecker, M., and Oda, M.,** *Chem. Ber.,* 119, 837, 1986.
2. **Jurić, A. and Trinajstić, N.,** *Croat. Chem. Acta,* 59, 617, 1986.
3. **Hosoya, H.,** *Croat. Chem. Acta,* 59, 583, 1986.
4. **Brunvoll, J., Cyvin, B. N., and Cyvin, S. J.,** *J. Chem. Inf. Comput. Sci.,* in press.

ADDENDUM TO CHAPTER 3

A brief exposition on graph-theoretical applications, including valence isomerism, has appeared.[1] The numbers of connected cubic multigraphs with 4 to 12 vertices were calculated by means of a computer program and of a recursive formula for counting disconnected cubic multigraphs.[2] By improving the constructive method, the numbers and the structures of constitutional isomers of $(CH)_{14}$ valence isomers were found.[3] An exposition on symmetry in chemical structures and reactions[4] included a discussion of polyhedral molecules (tetrahedrane, cubane, dodecahedrane) and of reaction graphs. By using several different graph-theoretical methods, the numbers of substituted isomers of dodecahedrane,[5-7] of other polyhedral $(CH)_{2k}$ molecules,[7] and of several unsaturated valence isomers of annulenes were calculated.[8]

Two new books[9,10] have been published in the area bordering chemistry and graph theory.

References
1. **Balaban, A. T.,** *J. Chem. Inf. Comput. Sci.,* 25, 334, 1985.
2. **Balaban, A. T., Vancea, R., Motoc, I., and Holban, S.,** *J. Chem. Inf. Comput. Sci.,* 26, 72, 1986.
3. **Balaban, A. T. and Deleanu, C.,** *Rev. Roum. Chim.,* in press.
4. **Balaban, A. T.,** in *Symmetry Unifying Human Knowledge,* Hargittai, I., Ed., Pergamon Press, New York, 1986, p. 999.
5. **Paquette, L. A. et al.,** *J. Am. Chem. Soc.,* 105, 5441, 1983.
6. **Brocas, J.,** *J. Am. Chem. Soc.,* 108, 1135, 1986.
7. **Balaban, A. T.,** *Rev. Roum. Chem.,* 31, 679, 1986.
8. **Balaban, A. T.,** *Rev. Roum. Chem.,* 31, 695, 1986.
9. **Gutman, I. and Polansky, O. E.,** *Mathematical Concepts in Organic Chemistry,* Springer, Berlin, 1986.
10. **Trinajstić, N., Ed.,** *Mathematics and Computational Concepts in Chemistry,* Harwood Ltd., Chichester, and Halsted Press (Wiley), New York, 1986.

ADDENDUM TO CHAPTER 4

The ring current effects reviewed earlier by Vogler[1] were shown by theoretical calculations (Pariser-Parr-Pople's method)[2] to be larger for charged than for neutral Hückel systems, in agreement with Müllen's experimental data.[3]

Interesting developments for annulenes with zero-atom bridges were reported by Hafner and co-workers: the first nonbenzenoid linearly condensed tetracyclic system with Hückel perimeter (14 π-electrons) and partial to 10 π-electronic structure was obtained on protonation of the tetra-t-butyl derivative of dicyclopenta[a,e]-pentalene (cf. **42** in Chapter 2); only three of the four five-membered rings participate in the cyclic conjugation of the cation.[4] The 1,3,5-7-tetra-t-butyl derivative of *sym*-indacene (**303** in Chapter 5) is a crystalline compound with a weak antiaromatic character, as shown by its dienic reactivity in cycloadditions, its X-ray diffraction data, and its ^{1}H-NMR peaks which appear at higher field than in the dihydroderivative.[5]

Pleiaheptalene, a pericondensed tricyclic 16 π-electron system with three seven-membered rings, has alternating bond lengths and presents temperature-dependent properties.[6] Starting from the lower vinylog, aceheptylene (**1**), Müllen and co-workers[7] prepared the anions **2** (R = H or Me) which are the first 14 π-homologs of the cyclopentadienide anion, and (via R = CH$_2$OH, CH$_2$OTs, and solvolytic ring expansion), an orange-colored oily compound, **3**, which has a diatropic [14]annulenic periphery: the methyl resonance in the ^{1}H-NMR spectrum occurs at δ = −0.68 ppm, while the remaining protons resonate at 6.53 to 7.26 ppm. Double methylation of **1** followed by cycloadditions led to the tetracyclic compounds **4** and **5** with [14]annulenic peripheries.

Ab initio heats of formation of the truncated tetrahedron and of other (CH)$_{12}$ valence isomers were calculated,[13] and it was suggested that thermal rearrangements may lead to such compounds by comparison with the formation of (homo)triquinacene from (homo)diademane.[14]

References

1. **Christl, M., Mattauch, B., Irngartinger, H., and Goldmann, A.,** *Chem. Ber.,* 119, 950, 1986.
2. **Christl, M., Kemmer, P., and Mattauch, B.,** *Chem. Ber.,* 119, 960, 1986.
3. **Wingert, H. and Regitz, M.,** *Chem. Ber.,* 119, 244, 1986.
4. **Maier, G., Bauer, I., Huber-Patz, U., Jahn, R., Kallfass, D., Rodewald, H., and Irngartinger, H.,** *Chem. Ber.,* 119, 1111, 1986.
5. **Erden, I.,** *Synth. Commun.,* 14, 989, 1984.
6. **Waldron, R. F., Barefoot, A. C., and Lemal, D. M.,** *J. Am. Chem. Soc.,* 106, 8301, 1984.
7. **Maier, G., Euler, K., Irngartinger, H., and Nixdorf, M.,** *Chem. Ber.,* 118, 409, 1985.
8. **Bertsch, A., Grimme, W., and Reinhardt, G.,** *Angew. Chem. Int. Ed. Engl.,* 25, 377, 1986.
9. **Escher, A. and Neuenschwandler, M.,** *Angew. Chem.,* 96, 983, 1984.
10. **Fessner, W-D. and Prinzbach, H.,** personal communciation.
11. **Paquette, L. A., Miyahara, Y., and Doecke, C. W.,** *J. Am. Chem. Soc.,* 108, 1716, 1986.
12. **Galucci, J. C., Doecke, C. W., and Paquette, L. A.,** *J. Am. Chem. Soc.,* 108, 1343, 1986.
13. **Schulman, I. M., Disch, R. L., and Sabio, M. L.,** *J. Am. Chem. Soc.,* 108, 3258, 1986.
14. **Meijere, A. de,** *Tetrahed. Lett.,* p. 4057, 1976.

ADDENDUM TO CHAPTER 5
(BRIDGED ANNULENES)

The electronic structure of 1,5-methano[10]annulene was investigated by He[1]-photoelectron spectroscopy.[1]

A comparison between the molecular orbitals of bridged [10]annulene and the orbitals of [6], [10], and [18]annulene was reported.[2]

The dinorcaradienic isomers as well as their equilibria with methano[10]annulenes were investigated.[3-5]

A methyl derivative of the hydrocarbon **155**; Chapter 5, Section C.1.b; namely 9b-methyl-9b-*H*-benzo[*cb*]azulene (**1**) was synthesized recently by Hafner and Kühn;[6] the ^{1}H-NMR spectrum (temperature independent up to the decomposition temperature of 80°, with δ_{CH_3} = 4.75; $\delta_{\text{perimeter H}}$ = 3.88 − 4.69 ppm) is in agreement with the π-bond structure **1**. Compound **1** is "the first tricyclic [12]annulene which proves to be an antiaromatic [12]annulene *per excellence* among the 12π systems".[6]

The structures of bridged [14]annulenes (Vogel's type) were thoroughly investigated by Gerson and co-workers.[7,8]

New syntheses of dicyclohepta[*cd,gh*]pentalene (**206**; Chapter 5, Section D.1) were described by Vogel and co-workers.[9,10] The tricyclic [14]annulene **2** was discussed recently by Müllen.[11]

The tetraanions of acepleiadylene and pyrene (**584** and **585**, respectively; Chapter 5, Section H.7) were further investigated.[12]

A tetrakismethano[24]annulene **3** was obtained recently by Wilke and Neidlein in Ni^{II}-induced cyclotetramerization of cyclopropabenzene.[13]

References

1. **Roth, W. R. and Sustmann, R.,** *Chem. Ber.,* 119, 751, 1986.
2. **Haddon, R. C. and Raghavachari, K.,** *J. Am. Chem. Soc.,* 107, 289, 1985.
3. **Bianchi, R., Pilati, T., and Simonetta, M.,** *Helv. Chim. Acta,* 67, 1707, 1984.
4. **Gatti, C., Barzaghi, M., and Simonetta, M.,** *J. Am. Chem. Soc.,* 107, 878, 1985.
5. **Budzelaar, P. H. M., Kraka, E., Cremer, D., and Schleyer, P. v. R.,** *J. Am. Chem. Soc.,* 108, 561, 1986.
6. **Hafner, K. and Kühn, V.,** *Angew. Chem. Int. Ed. Engl.,* 25, 632, 1986.
7. **Gerson, F., Huber, W., and Lopez, J.,** *J. Am. Chem. Soc.,* 106, 5803, 1984.
8. **Gerson, F., Knöbel, J., Lopez, J., and Vogel, E.,** *Helv. Chim. Acta,* 68, 371, 1985.
9. **Vogel, E., Markowitz, G., Schmalstieg, L., Itô, S., Breuckmann, R., and Roth, W. R.,** *Angew. Chem.,* 96, 719, 1984.
10. **Vogel, E., Wieland, H., Schmalstieg, L., and Lex, J.,** *Angew. Chem.,* 96, 717, 1984.
11. **Müllen, K.,** *Pure Appl. Chem.,* 58, 177, 1986.
12. **Mortensen, J. and Heinze, J.,** *Tetrahed. Lett.,* 26, 415, 1985.
13. **Mynott, R., Neidlein, R., Schwager, H., and Wilke, G.,** *Angew. Chem. Int. Ed. Engl.,* 25, 367, 1986.

INDEX

D

T

Z